# The New Accounting Manual
## A Guide to the Documentation Process

# Subscription Notice

This Wiley product is updated on a periodic basis with supplements to reflect important changes in the subject matter. If you purchased this product directly from John Wiley & Sons, Inc., we have already recorded your subscription for this update service.

If, however, you purchased this product from a bookstore and wish to receive (1) the current update at no additional charge, and (2) future updates and revised or related volumes billed separately with a 30-day examination review, please send your name, company name (if applicable), address, and the title of the product to:

Supplement Department
John Wiley & Sons, Inc.
One Wiley Drive
Somerset, NJ 08875
1-800-225-5945

For customers outside the United States, please contact the Wiley office nearest you:

Professional & Reference Division
John Wiley & Sons Canada, Ltd.
22 Worcester Road
Rexdale, Ontario M9W 1L1
CANADA
(416) 675-3580
1-800-567-4797
FAX (416) 675-6599

John Wiley & Sons, Ltd.
Baffins Lane
Chichester
West Sussex, PO19 1UD
UNITED KINGDOM
(44) (243) 779777

Jacaranda Wiley Ltd.
PRT Division
P.O. Box 174
North Ryde, NSW 2113
AUSTRALIA
(02) 805-1100
FAX (02) 805-1597

John Wiley & Sons (SEA) Pte. Ltd.
37 Jalan Pemimpin
Block B #05-04
Union Industrial Building
SINGAPORE 2057
(65) 258-1157

# The New Accounting Manual
## A Guide to the Documentation Process

Athar Murtuza

**JOHN WILEY & SONS, INC.**

New York ● Chichester ● Brisbane ● Toronto ● Singapore

*Library of Congress Cataloging in Publication Data:*

Murtuza, Athar.
    The new accounting manual: a guide to the documentation process /
    by Athar Murtuza.
        p.    cm.
    Includes index.
    ISBN 0-471-30370-4 (alk. paper)
      1.  Accounting—Handbooks, manuals, etc.—Authorship.  I.  Title.
  HF5657.M88  1995
808\.066657—dc20                                                    94-34136

Dedicated
to my parents
and
their parents

# ABOUT THE AUTHOR

Athar Murtuza, PhD, CMA is currently associate professor of accounting at the W. Paul Stillman School of Business at Seton Hall University. In addition to accounting documentation, his research and consulting interests include effective communication of accounting information, linguistic approaches to accounting, and the use of accounting as a decision-making tool. An expert in cost and management accounting, Dr. Murtuza has written chapters for several professional books, including John Wiley & Sons' *Handbook of Budgeting, Third Edition* by Robert Rachlin and H.W. Allen Sweeny, and has been published widely in both professional and academic journals, including *Management Communication Quarterly*. He also has been an invited speaker at some of the accounting industry's largest associations, such as the American Accounting Association, Institute of Management Accounting, and Information Resource Management Association.

# PREFACE

*The New Accounting Manual: A Guide to the Documentation Process* seeks to change the popular perception of the documentation process as well as the process itself by asking that those involved with it pay attention to the process as well as to the end product, the procedure manual. Within the documentation process is the potential for illuminating the functions, processes, and procedures of the organization. This potential turns the documentation process into an opportunity for organizational learning, since it is well-recognized that questioning the status quo facilitates innovation.

In the past, those assigned the chore of documenting procedures have merely recorded the status quo without asking any questions about it. This led to the process being accorded scant respect. One could ignore the state of affairs in the area of procedure documentation were it not for the outpouring of reports describing organizations busily redesigning their work. Given such changes in the way various processes are being performed throughout organizations, why not improve the process used for procedure documentation as well? It seems well suited to spark questions concerning the status quo within the accounting departments. Through such questions, one could turn the documentation process into an occasion for learning.

The basic idea expressed here was put forth in my paper, "Procedures Documentation Ought to be Illuminative Not Just Archival" published in a Sage Publications, Inc. journal *Management Communication Quarterly*, Vol. 8, Issue 2, Nov. 1994, pp. 225–243.

A bookish germination such as this comes about with the support and encouragement of many. Among them Virginia, whose support cannot be sufficiently acknowledged with mere words, their efficacy is, after all, limited—using words also spoken on a somewhat similar endeavor in an earlier, more uncertain time; certainly, Miriam; as well as the rest of my family, here, there, and in so many places.

The ideas about accounting and its role within an organization were planted when I took my first course in accounting from Professor H. Thomas Johnson when he was at Washington State University. I learned a lot from the work he has done since he first taught me.

Professor Joseph Wilkinson's books on accounting information systems were of greatest assistance in helping me understand the working of the various components of the accounting system. His books helped give form to this book.

Deeply felt gratitude to Alfred King, a fellow CMA, for serving as the catalyst in getting the book published; to David Rosenthal at Seton Hall University for his contribution to Chapter 5; to Bill Brunsen for helping to keep the ideas developing between the bouts of production grading; and last but not the least, to Sheck Cho, and Victoria Hofstad at John Wiley & Sons for their patience and advice.

Athar Murtuza

South Orange, New Jersey

# CONTENTS

# INTRODUCTION

# DOCUMENTATION AS AN ILLUMINATIVE ACT

It would be a truism to say that organizational change is becoming globally endemic. Corporations are busily reshaping the processes and procedures they rely on to get things done. Even though organizations have begun to reinvent what they do and how they do it, these emerging management concepts and organizational practices have yet to reach procedures documentation, an organizational communication genre. The winds of change have yet to reach the procedures documentation process.

Academics do not even teach procedures documentation, let alone teach it as an activity whose uses and value reside in the entire process and transcend the end product, the documentation itself. Like academics, managers continue to see documentation as an afterthought rather than the learning organization paradigm and extension of process redesign that it is. Professors and practitioners both mistakenly see procedures documentation as merely archival.

*The New Accounting Manual*: *A Guide to the Documentation Process* argues that seeing procedures documentation as merely archival has isolated this organizational communication genre from the developments taking place within contemporary organizations. Such isolation has prevented the procedures documentation from becoming a useful tool, an occasion to review and prune outdated, inefficient, or redundant procedures, and to think of newer ways to do things in view of changing environments and technologies.

By not limiting procedures documentation to merely the final act of writing the procedures manual, but seeing it broadly as an illuminative tool, an organization can learn how to streamline its own processes. The value of procedures documentation resides in the process of creation as well as in the document created.

Current practice only mirrors the existing process, thereby perpetuating the status quo with all its flaws. Rather than simply reflecting the status quo like a mirror, procedures documentation should also illuminate like a lamp the adequacy of current practices and processes. The axiom that an unexamined life is not worth living can apply not only to individuals but to organizations as well.

Attempts aimed at continuous improvements or even periodic review and updating of procedures can go a long way in preventing the accumulation of deadwood by replacing them with productive and efficient procedures and policies. Such ongoing reviews and incremental steps have had a major role in the emergence of Japan as a manufacturing giant. Notwithstanding the demonstrated efficacy of periodic procedures review not only in Japan but also here in the United States, documentation experts have not added such reviewing to the process governing procedures documentation. They continue to limit the documentation process to the actual writing of the manuals rather than as an occasion for

1

periodic organizational self-examination. Such failure wastes potential opportunities to examine, reflect, consult, improve, learn, and prosper.

## WHAT IS THIS BOOK ABOUT?

This book explains the process involved in documenting the accounting manual without limiting the documentation process to merely describing the various procedures and policies in place within the accounting departments. The current perception of documentation as essentially archival permits irrelevant, redundant, non-value-adding procedures to continue within an organization. All too often, archival documentation becomes obsolete even as it is written because it makes no provision for the changes that are constantly taking shape within the accounting department, throughout the organization, and in the environment in which the organization operates. The world is constantly changing and what does not change with it soon comes to regret its immobility.

The documentation process can be an occasion to examine the very nature of accounting functions within the organization, to review how they are being carried out, and to reengineer them if necessary. Such a perception of what the documentation process ought to be is quite practical. The practicality of activity management, process redesign, and learning paradigm has been well-documented in recent years. But to date no one has shown how these emerging ways of doing business can be incorporated as part of the process dealing with the documentation of the accounting procedures manual.

Incorporating the knowledge of the new ways of doing work as an integral part of procedures documentation process can change the documentation from being an afterthought, a necessary nuisance, into a chance for improvement not only for the accounting department but for the entire organization.

So even though we will focus on the process that governs the documentation of accounting policies and procedures, in effect we will be seeking to affect the role accountants perform for the organization. If the accounting department itself is not operating efficiently and does not seek ongoing improvements, then the efforts aimed at organization-wide improvement through such means as activity-based costing and process reengineering cannot have optimum impact. To increase the potential of success for organization-wide improvements, do not ignore the work of the accounting department. The process concerned with procedures documentation provides an ideal occasion to institutionalize continuous improvement and learning within the controller's department as well as the rest of the organization. Procedures documentation can become a safeguard against obsolescence.

**Outline.**    In keeping with its objective, this book starts by devoting a chapter to discussing the past, present, and the potential future of the documentation process. This chapter traces the historical origin as well as the current perception of organizational procedures documentation. It then looks toward a new perception of the documentation process.

A major tenet of the redefinition of the procedures documentation process is that it cannot be limited to studying the procedures isolated from the department where they originate. Any discussion of the process governing the documentation of accounting procedures must start first by defining accounting and the role performed by accountants. At the present time, accounting is in the midst of great changes and we should study not only the present status and current functions, but also the functions that accountants may be expected to perform in the days to come. The accounting procedures are no more than the means to carry out the roles assigned by a given organization to its accounting department. Hence, one cannot study the accounting procedures, let alone document them, without first understanding the functions performed by the accounting department for the organization.

The discussion of the role accountants perform is followed by an examination of the new ways of doing work that have emerged in recent years. These new ways of working may be thought of as new competencies that professionals and workers must learn in order to survive and prosper in today's environment. Among the competencies reviewed are:

Continuous improvement
Activity-based management/Activity-based costing
Process reengineering
Learning organization paradigm
Changing information technology
Process consulting

Following this discussion of the new competencies, we will look at how existing accounting procedures ought to be reviewed to see whether they are adequate or whether they need changing with the help of new technologies available in Chapter 10 "Deconstructing Accounting Procedures: A Tool for Organizational Learning."

Deconstruction is a technique used in literary and social science contexts to analyze, or take apart, written works in order to demonstrate underlying assumptions and to evaluate their adequacies. Deconstructing written texts can shed light on potential contradictions and incorrect assumptions that may be present in the works. Deconstructing accounting procedures can reveal outdated assumptions and underlying inadequacies that may have become a part of the way things are done. Without a periodic questioning and reevaluation of accounting procedures, professional evolution is stymied and extinction becomes a real possibility. The techniques of deconstruction lend themselves well to questioning the various procedures in place and whether they are adequate and effective.

Such deconstruction of selected accounting procedures is followed by a look at the impact internal control and auditing needs can have on various accounting procedures. A great many procedures carried out by and for accountants seek to maintain proper control of the organizational assets and to help auditors, both internal and external, perform proper audits. As such, those conducting a review of accounting procedures must be made aware of internal control and the factors auditors look for in their audits. Through such awareness, the accounting procedures can be designed to meet internal control and audit needs.

Following an overview of internal control and auditors' expectations, we examine the various processes that comprise an accounting system. An accounting system is not monolithic. It is a patchwork quilt: its efficacy and efficiency depends on how well the patches, the subsystems, are arranged. They must be well integrated to be of optimum use.

Given the great organizational diversity that exists in the workplace, no attempt is made to provide sample documentation for various accounting functions. One size cannot fit everyone! Instead of outlining the specific way various functions ought to be documented, the book describes the nature of various accounting functions, such as those dealing with sales and the collection of accounts receivable. While all organizations are concerned with collecting accounts receivable, payroll, and accounts payable, they do not perform them in an identical manner. Each organization needs to have its own unique set of procedures to help fulfill its particular needs. All organizations must pay their bills, accordingly, the existence of the accounts payable function is fairly universal. Each organization, however, is likely to structure the functions in keeping with its unique constraints. If the goals to be attained with the help of various accounting functions are clear, then procedures can be developed to help attain these goals in keeping with organizational constraints. Accordingly, an overview of an accounting system, rather than an accounting manual containing a set of sample procedures, is provided. When armed with the knowledge that an overview of accounting functions can provide, managers in charge of reviewing procedures or developing new ones can do a more effective job of ascertaining whether the current procedures are adequate and in line with the needs of their particular organization before they document the procedures.

Finally, the techniques that can help make the actual documentation more effective are discussed. The physical aspects of manuals, such as the page layout, graphics, and fonts can go a long way in making the manual easy to read. This last chapter will review logistical aspects of preparing accounting manuals. Equally important for procedures documentation may be something called hypertext. Using information technology to document procedures in hypertext can improve their accessibility, which in turn could improve their usefulness as well as the frequency of their use.

In keeping with the focus of teaching expertise rather than simply providing a set of hypothetical documentations, an extensive bibliography is provided so that readers can get additional information on the topics discussed in the book. In addition to the bibliography, there is also an appendix that lists an extensive body of the published literature dealing with activity-based costing following Chapter 8.

# CHAPTER 1

# PROCEDURE MANUALS: PAST, PRESENT, AND FUTURE

## 1.1 HISTORICAL ORIGIN

In recent years JoAnne Yates, a management communication expert who has written extensively on procedures documentation and other forms of organizational communication, has shown how developments in office technology, such as the invention of the telegraph, the typewriter, the mimeograph, and the vertical filing system, combined with the growth of various communication genres, the executive summary, the interoffice memo, and the procedures manual helped bring about the advent of modern control systems between 1870 and 1920.

The growth of office technology made documentation more accessible. The growing accessibility of documentation allowed managers to record procedures and to communicate their expectations to employees. This downward flow of internal communication, in turn, permitted performance measurement and evaluation, thus leading to better organizational control and coordination. The upward flow of information in quantitative terms enabled upper management to monitor and evaluate processes and individuals at lower levels. It made possible improved managerial control of organizational processes and assets. In addition to upward and downward communication flows, due to the new office technology, it also became possible to communicate laterally, thus promoting the staff functions within organizations.

Yates has convincingly shown that the developments in office technology and the newer organizational communication genres such as the procedures manual permitted the emergence of a formal communication system, which became an important control mechanism. Indeed, it allowed formal internal communication to become the dominant tool for managerial control. The documentation of procedures, policies, and rules, and the reporting of financial and operational information on a regular basis led to the emergence of an organizational "memory," which facilitated organizational learning. She notes, however, that the advent of technology alone was not enough, the vision to use the new tools in creative ways was also necessary. Nor would gains have been realized by simply extending the communication patterns current during the period from 1870 to 1920. According to Yates's research, attention to potential human problems can facilitate organizational and technological changes. Moreover, the documentation of the past for its own sake is not very useful for making decisions in the present, but systematic recording and analysis of data allows informed decisions to be made.

Notwithstanding the negative aspects of bureaucracy, the antidote is not to dispense with the bureaucracy itself, but to create conditions that will allow an organization to adjust to changes in the environment and to take advantage of the evolution of technology. In the last twenty years, American management has relied on a number of techniques, such as quality circles, participative management, activity management, continuous learning, and process reengineering to ensure that organizations can adapt to meet the demands of external changes. But these management changes have not yet found their way to the process involved in the documentation of procedures, thereby making manuals, the end product of the process, an ignored resource for organizational learning.

## 1.2 PROCEDURES MANUALS GET NO RESPECT!

Today, accounting manuals that set forth organizational policies and procedures are not perceived as a learning tool; instead they are often the neglected afterthoughts of management.

Most firms have procedures and policy manuals, however, they are rarely used in an optimum manner. The procedural and policy manuals on hand are all too often inadequate because of poor writing, cumbersome organization, and out-of-date information. The presence of such unusable manuals represent a two-fold waste: on the one hand, the preparation of the inadequate manuals was a waste of resources; on the other hand, by not using them, the firm is denying itself optimum use of a potentially useful learning tool.

## 1.3 REASONS FOR THE INADEQUATE DOCUMENTATION

Among the reasons for the inadequacy and the resulting sub-optimum use of manuals is that documentation is given low priority in the world of controllers and managers. Because of the lack of, and at times absence of, respect given to the preparation of the documentation, manuals remain inadequate and underutilized. Within the universe of controllers and managers, other tasks, other crises, seem to get a faster track. As a result, the writing and maintenance of manuals remains an afterthought.

Another reason for the poor quality of documentation is its absence in the educational training of future accountants, as well as engineers, programmers, and other professionals. Their professional education does not prepare them for the writing of manuals. College accounting courses focus on the preparation of financial reports for disclosure but ignore the effective communication of those reports. Many of the labels hurled at accountants (bean-counters and number-crunchers, for example), can be attributed to accounting professionals' inattention to the disparities between the information they provide and the actual needs of the users. It is no surprise, therefore, that accountants are unprepared and untrained for writing manuals. Too often, they are not motivated even to undertake documentation, let alone to do it well. It naturally follows that the documentation that results is not good and does not get used.

Accounting documentation does not have to be inadequate. But if it is to change into a productive tool, then both the perception of procedures manuals and the documentation process itself must change.

## 1.4 THE STAGNANT PERCEPTION OF MANUALS

In addition to the poor writing in manuals, their cumbersome organization and the failure to keep them current results in them being accorded low respect. An examination of the literature written in the last twenty-five years to instruct readers on how to write operating and systems manuals shows that there has been very little change in the way experts view them. The literature calls for manuals to be understandable, accurate, and professional looking. The experts think of manuals as archival instruments that merely describe current procedures. The same perception permeates advertisements for workshops on the preparation of organizational manuals.

Given the stagnant perception of procedures documentation and the low priority accorded it, it is no surprise that accounting manuals remain an ignored, wasted resource. As a result, their potential remains unrecognized and unused. The manuals have many traditional uses but their poor quality keeps them from being optimally utilized.

## 1.5 THE MANY USES OF MANUALS

There is no doubt that accounting manuals have many potential uses. Some are:

- Organizational Memory. Manuals can serve as the archive of a firm by providing knowledge about the development of a firm's policies and procedures. Such knowledge can prevent the firm from repeating its procedural mistakes.
- Reference Source. The manuals can also serve as a reference source, keeping employees from wasting time devising policies that already exist every time an infrequently used procedure has to be applied.
- Training Instrument. Manuals can help in the training of new employees. Instead of wasting the scarce, expensive time of experienced workers, newly hired employees can get their initial orientation by reading the operations and procedures manuals.
- Prevention of Work Interruption. Proper, well-written, current documentation can also allow work to continue if a key employee is absent or resigns. The key employee's departure does not lead to a crisis since others may refer to the manual for information and guidance.
- Regulatory Compliance. By documenting various regulatory requirements, manuals can also permit firms to comply with them. Workers have less chance of complying with requirements if they do not know what they are.

Accounting periodicals have documented many instances where manuals have been used successfully in various ways. In one instance, the productivity of a firm improved through the use of a procedures manual that provided those involved in the budgetary process with needed instruction about various steps in the process. Clearly, organizations, regardless of their size, should have a budget manual as part of their financial planning systems and procedures, so that employees do not have to start from scratch every year. The manual provides for an organized, centralized approach to the budget process by letting the workers know what needs to be done when, in what order, and by whom. It can be argued that manuals are needed not just for budgetary processes, but for almost all the tasks undertaken by administrative units, in general, and the controller's department, in particular.

There are many other ways that procedures manuals can be useful. The Coca-Cola Company, for instance, found a comprehensive, easy-to-use manual to be an extremely important tool in maintaining financial controls in a multinational environment. The recently reported experience at Dow Chemical confirms the usefulness of accounting documentation to ensure consistent financial reporting. The giant chemical firm found a policy and procedures manual especially useful in policy formulation, as well as the implementation and administration of corporate strategy.

Even though manuals are potentially useful, instances of their successful use are the exceptions rather than the rule. They are, for the most part, an underutilized resource, but the organizational documentation does not have to be neglected. However, if the manuals are to change into a productive tool and regain their original use as a form of organizational memory and a learning tool in the 1990s, then the perception of documentation must not be limited to an archival role, nor should the process be limited to merely recording the procedures.

## 1.6 MAKING DOCUMENTATION AN ILLUMINATIVE ACT

It is common to find procedures and policies lingering on longer after exhausting their usefulness. A periodic review and updating of manuals can go a long way in preventing the accumulation of deadwood by replacing it with productive and efficient procedures and policies in keeping with state-of-the-art technology. Such periodic reviews will allow the documentation process to become an occasion to learn.

The preparation and maintenance of procedures and operating manuals ought to be seen as the occasion for analyzing how the accounting department is carrying out its designated tasks. The writing should be an occasion for organizational self-examination and for learning. In a changing world, assigned departmental roles are subject to change. And even if the basic roles remain unchanged, certainly the implementation of them can evolve over time.

Before the procedures are documented they must be reviewed. Such a review can proceed by asking questions such as:

What is the given procedure seeking to accomplish?

Whose needs is the particular procedure seeking to meet? Can those needs be met in other ways?

Does the procedure enable the organization to better serve the customers? Does it add any value?

Are all the steps involved necessary?

Can the same goals be accomplished in another way?

How can technology be used cost-effectively to expedite the process? Can new technology provide an alternative approach that accomplishes the same goals?

The periodic review and analysis of the procedures used to carry out the functions assigned to accounting departments must be conducted with a view toward eliminating procedural gridlock, incorporating new technologies, and enhancing organizational productivity. Such analysis can ensure that the deadwood, non-value-adding procedures are dropped or improved. A periodic review of what accounting procedures are meant to do, how they do it, and the extent to which they are meeting users' current and evolving needs can keep the procedures from becoming obsolete. By seeing to it that obsolescence does not infect procedures, accounting departments can prevent the spread of obsolescence to the information system of an organization and to the organization itself.

The obsolescence that plagued American accounting systems in the 1980s, and continues to do so, could have been avoided by periodic review of the users' needs, the information provided, and the procedures used to generate the reports of the accounting information systems.

While writing procedures manuals, don't simply describe the way things are but think also about how they could be. The procedures analysis should become an occasion for cultivating employees' experience and knowledge in order to innovate. The centerpieces of the Japanese response to a constantly changing world are "knowledge-creating" companies who innovate continuously. They do so not by "processing" facts and figures of objective knowledge or by delegating to various departments such as research and development, marketing, or strategic planning, the task of innovation. Japanese firms like Honda and Canon create knowledge by taking a holistic approach that taps into the knowledge, experience, insights, and intuitions of all its employees and which emphasizes organizational self-knowledge and a collective sense of identity and purpose.

The task of organizational reexamination can benefit from the emerging tools such as activity management, process redesign, and the paradigm of learning organizations.

To summarize, accounting procedure documentation should not be an afterthought but a deliberate, illuminative act, an occasion for periodic self-examination on the part of the organization.

The preparation of manuals ought to be the result of teamwork, which uses collective knowledge and experience of the involved employees to periodically reexamine what the firm does and to explore ways of improving and evolving.

The documentation must not perpetuate the status quo. In a constantly changing world, maintaining the status quo can be the kiss of death. The fate of the dinosaurs testifies to that effect. Likewise, there is a lesson to be learned from Sears as it painfully pays in the 1990s for having played Rip Van Winkle in the American retailing scene in the 1970s and the 1980s.

## 1.7 THE TIME IS RIGHT FOR THE NEW PERSPECTIVE

This book brings a new perspective to the writing of accounting manuals by arguing that the documentation process can and should be an occasion to upgrade or even to redesign procedures in keeping with the changing environment and developing technology. That such a perspective is needed is borne out by the prevailing environment and the health of business organizations.

In an article called "Make Your Office More Productive," published in the February 25, 1991 issue of *Fortune* magazine, author Ronald Henkoff argues that "the imperative" to improve the productivity of white-collar workers is becoming screamingly obvious in the face of the economic hardships plaguing banks, thrifts, insurance companies, accounting firms, airlines, and major retailers. The improvement of white-collar productivity cannot be induced by firing low-paid workers or by opening the coffers to pay for high-tech automation. According to Henkoff, productivity can be induced by removing bottlenecks, avoiding mistakes, and by focusing on meeting the needs of the customers. Resorting to new technology should follow, not precede, removal of non-value-adding activities. According to an official of Baldrige Award–winning Motorola who is quoted by Henkoff, mistakes are made and procedural gridlocks are created not by incompetent or lazy employees but by the way they have been told to do their work, due to procedures that are obsolete, too complex, or redundant. By examining and streamlining procedures used for the monthly balancing of its accounts, Motorola was able to close its books in four days instead of the eight it used to require. The halving of the time meant an annual saving of $20 million.

## 1.8 ISOLATION OF THE DOCUMENTATION PROCESS

Seeing manuals as merely archival has also isolated the documentation process from the developments taking place in the world at large. Among the progressive organizations these days, the vogue is for activity management, process redesign, continuous improvement, and organizational learning. But so far no one has connected the new competencies developing in the organizational world with the process involved in the documentation of procedures. By studying the developments taking place in various managerial disciplines, those charged with writing accounting manuals can use the new techniques to broaden the perception of the documentation, making it a valuable contributor to the company's health rather than merely a record of company history.

## 1.9 CONCLUSION

Organizations, both public and private, must realize that the process dealing with the documentation of operating manuals can itself be useful for organizational learning and evolution. At present the focus of documentation is merely on the end product, rather than the process that leads up to it. A broadening of the perception of the documentation process and an investment of resources to develop better manuals should more than pay for itself by making operations smoother and more efficient and by preventing organizational obsolescence. To live up to their potential usefulness, manuals must be current, available, readable, and easy to understand. Even if the actual writing is done by one individual, the preparation of manuals should be the result of teamwork, using the collective knowledge and experience of the involved employees to periodically reexamine what the firm does and to explore ways of improving and evolving. The documentation must not perpetuate the status quo, and it does not have to!

# CHAPTER 2

# THE NATURE OF ACCOUNTING

## 2.1 DEFINITION OF ACCOUNTING

According to the media's portrayal of the accountant there is little more to accounting than number crunching, bean counting, pencil pushing, and the wearing of green eyeshades, but there is much more to accounting.

The definition of accounting quoted most often is the one issued by a committee of the American Accounting Association, an organization of accounting academics, in 1966, in *A Statement of Basic Accounting Theory* (ASOBAT). The committee defined accounting as "the process of identifying, measuring, and communicating economic information to permit informed judgements and decisions by users of the information." This popular definition sees accounting essentially as financial information that reports on and measures the various activities performed by an organization. But it ignores the impact accounting has in shaping the activities that presumably are measured and reported by means of accounting statements. Furthermore, the definition ignores the role accountants perform as members of the management team. Even though the definition mentions communication of information alongside its identification and measurement, in practice, accounting remained synonymous with the preparation of the financial reports and not with its communication and effective uses.

The National Association of Accountants (NAA), now renamed Institute of Management Accountants (IMA), through its *Statements on Management Accounting* (SMAs), issued a definition of accounting that encompasses much more than preparing financial statements. According to SMA 1B, "Objectives of Management Accounting," accountants, especially those who work in controllers' departments, not only provide information to managers, but actively participate in the management process. Given such participation in the management process, the IMA came forth with a different definition, which reads: "Management accounting is the process of identification, measurement, accumulation, analysis, preparation, interpretation, and communication of financial information used by management to plan, evaluate, and control within an organization and to assure appropriate use of and accountability for its resources." Management accounting is taken to mean information used by managers internally, in contrast to financial statements that are intended for external users such as stockholders, investors, lenders, and regulatory bodies, such as the Securities Exchange Commission. In contrast to the definition provided by ASOBAT, the IMA's includes interpretation of information as a part of the accounting itself.

The SMA 1B agreed with ASOBAT in saying accounting is information, but identifies additional uses of accounting. One objective of accounting, according to SMA 1B, is to provide information to internal users for use in planning, evaluating, and controlling operations. In addition, accounting information is meant to track the status of organizational resources and obligations, assets and liabilities. Finally, accounting is a form of communication with the stakeholders outside the organization, including stockholders and regulatory agencies. The SMA 1B also made explicit that accountants are a part of the management team. Not only do controllers help coordinate the organizational affairs by providing information that lets the managers know the state of the affairs, they are also actively involved in organizational decision making, operating, strategic, and tactical.

## 2.2 ACCOUNTING AS A SYMBOLIC REPRESENTATION

Neither definition addresses the question of how accounting relates to the organization and its activities. Consequently, it is necessary to go beyond the definitions put forth by ASOBAT and the IMA and argue that accounting is a symbolic construct of organizational reality. In other words, accounting reflects and illuminates how an organization is doing and it does so by recording and reporting how resources are deployed within an organization. These resources include assets listed in the balance sheet, as well as intangible assets not directly listed in accounting statements. Some assets are more tangible than others: steel used to make cars and eggs used to bake cakes are more tangible than the effort expended by the designer to design a new car model. Resources can also refer to intellectual and creative elements that are and must be a part of the organizational resource repertoire.

A major portion of a manager's job is making decisions that best use the organization's resources to help realize the organization's goals. Effective use of resources depends on tracking their use. Given the wide diversity of resources, this is not an easy task. An organization may find it difficult to directly track the efforts being expended to design a car, to better motivate its labor crews, and to win over customers. Not only the deployment of its resources but also the specific benefits derived from their deployment are difficult to track. Since physically following the resources being deployed within an organization is not very practical for managers and decision makers, they must rely on abstractions, sometimes called surrogates or analogues, to represent those resources, their deployment, and the results. Accounting, acting as a surrogate or an analogue, translates the deployment of resources into financial terms: instead of following millions of pounds of steel down an assembly line, decision makers can translate into financial terms the conversion of the steel into automobiles by representing the steel and the work done to it in terms of money spent to acquire and convert the steel into cars.

Symbolic representations can become a trap when human beings take them for the whole of the real thing. The symbolic representations have their limitations in that they represent only a part, albeit a representative part, of the total phenomenon. There is a lot more involved in converting steel into cars than just the dollars paid to acquire steel and to convert it. The challenge for accountants is not to limit information to dollars and cents but to convey all that is involved in converting steel into cars and selling them profitably.

Even with the limitations of surrogates, it is much easier to track the money associated with the deployment of the various organizational resources than to track the resource use directly. Such translation of resource deployment into monetary terms can *approximate* the extent to which those resources were expended and the benefits that resulted from their deployment. There are times when the benefits resulting from the uses of organizational resources may not be easily translated into monetary terms: how does one place a value on what is accomplished by the state's spending on higher education? Despite such difficulties, tracking the monetary resources is a useful tool for managers since it can help them make more informed decisions by symbolically describing the organizational activity.

Accounting departments within organizations are assigned the job of tracking the resources in monetary, that is symbolic, terms. They do so by means of accounting records and financial statements that translate organizational activities in terms of their costs. In so doing they reveal what is going on in an organization, where the company is headed, how it is handling its resources, and what it achieved by deploying its resources.

The accounting reports not only analyze the organizational activity, they also help shape it. Since the perception of managers' performance is based on the activities that are measured, the measures that get reported determine the actions of the managers. Managers, in other words, tailor their performance so that it will be reported favorably. Accounting, therefore, must not be seen only as a neutral reporting of the managerial and organizational performance, but as helping to determine and shape organizational behavior.

## 2.3 INADEQUACIES OF FINANCIAL ACCOUNTING

The branch of accounting responsible for accounting record keeping and for preparing financial statements is called financial accounting. Since accounting records contain a large amount of data describing the countless transactions that are constantly occurring within an organization, the raw data cannot be used effectively by users. It is summarized for use by means of financial reports such as balance sheets and income statements.

Some users of accounting information are content with these summary reports. Such users, mostly investors, are external to the organization's day-to-day operations; they do not want to be bothered by the petty operational details. These users invest money by buying stocks and bonds issued by a corporation and trust the corporate managers with using the investment well and judiciously. Government, through a network of laws and regulatory agencies, seeks to ensure that managers are reporting the corporate performances accurately. The interests of such external investors are served by the financial statements, but corporate managers need more detailed information. To supply the information needs of internal managers, managerial accounting came forth. It is different from financial accounting, even though the same data may be used for producing managerial accounting reports and statements.

Managerial accounting provides managers with detailed information that tracks the use of specific resources in particular segments of the organization, and can be used by managers to make more informed decisions.

Financial statements are historical records. Like the scores at the end of games, they are, at best, of limited help to managers in day-to-day decision making or in planning for the future. They record and summarize what occurred in a given fiscal period, but they do so after the fact. As a result, they are not enough to satisfy the information need of the managers, who must make decisions that deal with the present and the future. The managers' interest in the game is not confined to the score at the end of the game. An income statement gives the total revenue of the firm, but managers want to know how the sales were in each store, in each department, and by product. The total sales revenue for the previous period as recorded in the income statement does not reveal such necessary details. The statements for a period, taken by themselves, do not help much in performance evaluations; they do not offer help in product pricing, nor do they reveal whether the organizational goals were met. To overcome the inadequacies of financial statements is the goal of managerial accounting.

## 2.4 DIFFERENCES BETWEEN MANAGERIAL AND FINANCIAL ACCOUNTING

There are a number of differences between managerial and financial accounting. Managerial accounting:

- Focuses on the needs of internal users
- Places more emphasis on the future
- Is not bound by generally accepted accounting principles (GAAP)
- Is not mandatory through government regulations
- Emphasizes relevance and flexibility in its reports
- Can and does use non-monetary information
- Emphasizes the parts of the organization rather than the whole

Despite these differences between managerial and financial accounting, they generally rely on the same data base and the same department is charged with both kinds of accounting.

To summarize, accounting is a symbolic construct of the organizational activity; it tracks monetarily the organization's performance and the deployment of its resources. But accounting is information, and its value directly corresponds to the uses it serves for those receiving it. It must get communicated and be understood in order to be optimally used.

## 2.5 ACCOUNTING AND COMMUNICATION

The importance of effective communication skills for accountants is a popular topic. Popular belief holds that communication skills of accountants need only be in the area of writing and speaking, but effective communication of accounting information requires more than the ability to write or to speak. Those whose skills are limited to good writing and speaking may still be unable to explain to movers and shakers in the organization why their newest factory is plagued by exorbitant cost and production variances. The accountant must be able to observe, to investigate, to understand, to report, to explain, and to help eradicate—in other words, to effectively communicate—the impact of recurring variances on the organizational performance.

The impediments to effective communication of accounting information can arise throughout the communication loop: they can be caused by organizational, system-generated, behavioral, and cognitive factors, and they are not limited to weak writing and poor speaking skills. Accountants must be aware of the various other impediments that can occur anywhere in the communication loop. The impediments may have to do with receivers' perceptions, they may be due to the inappropriate selection of channels used to transmit messages, or they may be caused by distorting noises anywhere in the environment.

## 2.6 COMMUNICATION VERSUS INFORMATION

General communication is not the same as the communication of information. The two, communication and information, are different and yet they are interdependent. Without information, communication may occur but will be not meaningful. Therefore, the effectiveness of an information system is related to knowing what information is needed by whom and for what purpose. Then, an information system can be created that processes specific inputs and outputs that satisfy the needs of the users. The effectiveness of the information system, in other words, depends on first knowing the users' needs and then collecting, processing, and subsequently communicating information to them.

An accounting information system must reckon with the requirements for effective communication so that the data it accumulates and transmits will amount to meaningful and usable information. Otherwise, the numbers crunched out are just the sound and fury of number crunching, signifying nothing.

## 2.7 IMPEDIMENTS TO COMMUNICATION OF INFORMATION

In a brief but very useful book, the management expert, Henry Mintzberg described impediments that can disrupt and undermine the effective use of information by managers. He notes that effective communication can be impeded by factors in the particular information system or organization as well as by behavioral and cognitive factors specific to senders and receivers. In other words, the system may be providing information that is obsolete, late, or unreliable. It may be aggregating the data, making it too general to satisfy a manager's specific needs. Furthermore, the system may be too limited to fully answer the needs of everyone in the organization. Like the system, the organization can impede effective communication through its culture or by making its organizational objectives too rigid, and thereby making information communication dysfunctional. Besides the system and the organization, the manager's own cognitive abilities can restrict his receptivity and ability to use the information being communicated to him. His prior experiences may cause his cognition to filter out certain information and make him less open to information delivered. We will first discuss the impediments that pertain to accounting information itself; subsequently, the impediments attributable to the organization and behavioral factors are discussed.

**(a) *The accounting system is too limited*.** In recent years a number of experts have pointed out that too many corporate accounting systems are driven by the needs of external financial reporting, and this makes the output created by the controller of little use to internal corporate decision makers. For product pricing, performance measurement, and optimum utilization of corporate resources, externally aimed financial reporting is not of much use. It is no surprise that the potential users are turning to other information sources to satisfy their needs. The problem of the irrelevance of accounting reports to many of their potential users is severe enough that the American Institute of CPAs, through its future issues committee and strategic planning committee, is looking into the problem in the hope of redefining the nature of accounting statements.

**(b) *The accounting reports aggregate far too much*.** As anybody familiar with financial reports knows, the income statements may say something about the parent corporation but whatever they report about the specific segments gets buried in the aggregate. The financial statements report the final score of the game, but little else; coaches and managers, on the other hand, are interested in everything, including the final score. Though managerial accounting texts talk about reports like segmented income statements and about responsibility accounting, the efficacy of these managerial accounting tools remains limited since the system is driven by the needs of external users. Activity-based costing has emerged in recent years to satisfy information needs that the myopic, none-too-communicative, much-too-aggregated forms of cost allocations, which used to be and still are the norm for cost systems, were unable to satisfy.

**(c) *There is a time lag in reporting*.**    There is a considerable time lag between the time accounting reports get prepared and the occurrence of events and activities that led to those reports. The time lag affects not only financial statements, but also hinders the optimum use of variances reported. Standard cost variances translate production into accounting terms. More specifically, they are aimed at explaining what led the budgeted income to be at variance with the actual income. But that translation of production into accounting takes time. Time lags impede communication. The increasing call for non-accounting performance measures in recent years is, in part, meant to solve the problem of information received too late and little understood by production managers.

**(d) *Some accounting information may be unreliable*.**    It is well known that historical costing is the rule, per GAAP. This keeps the allocation of fixed costs from reflecting the costs accurately, and keeps them from being current. Those familiar with joint costs allocation will know how inaccurate and subjective they are. Their inaccuracy does not impinge on external, aggregated reports, even though the fictions of joint cost allocations are of no use to product pricing. Then, there are problems resulting from the practice of income management. Financial information can be, and often is, manipulated in order to make performance measures, such as return on investment, appear more favorable.

The next group of impediments are those that result from the nature of the organization. In terms of the communication model, organizations constitute the environment within which the senders and receivers seek to communicate. The environment can induce distortions and create noise that will distort or impede the communication flow. An organization's structure and its politics impact on the information communicated within the organization. As a result, the information obtained and used can be affected. Often, managers are all too willing to side step the formal information if organizational objectives are rigid and ill-considered. Organizational politics can also lead to managers distorting or covering up information instead of letting it flow. The fast-paced nature of managerial work favors quicker, verbal channels, but the formal information flow relies on written documentation, which can keep managers from using information effectively.

Consider the organizational role budgets play to appreciate how organizational structure and politics affect the communication of information. Budgets are, among other things, vehicles that communicate information, which in turn allows for improved planning and control. However, games played with budgets, well known to managers and extensively discussed, lead to information distortion and dysfunctional communication. Such game-playing impacts roles budgets can play, nullifying them as instruments for communicating organizational reality translated into financial terms.

The message and the environment by themselves do not complete a communication loop: the sender and the receiver must also be present in order for communication to occur. Behavioral and cognitive factors associated with the sender and the receiver of the information must also be considered. The participants involved in information communication suffer from cognitive limitations. Usually they are unreceptive to information that may change the organizational status quo and their place in the organization. Their perceptions are shaped by their prior experience, education, and training. Their self-interests may lead them to ignore, distort, or tune out information. Accountants involved in auditing are well aware of how misconceptions regarding the auditors' role leads managers to defensive behavior and sometimes to outright hostility. Such hostilities certainly impede communication.

A large amount of the information communicated by accountants is directed to non-accountants, whose receptivity to accountants and accounting messages is suspect. Adding to the problem of effective communication is the burden stemming from the fact that a very large number of accounting terms do not mean what they imply. There are gaps between the denotation and connotation of the accounting concepts and vocabulary. It is not uncommon to find individuals who see net income as something real, the proverbial bottom line. Unfortunately, it is something made up to describe the performance of an organization; it is an approximation of the increase in the value of a firm within a given time period. Most people do not know that the income statement's bottom line, the net income, can never be deposited in the firm's bank account. Similarly, terms like "retained earning," "assets," and "expensing the costs incurred" are abstractions that do not mean what they imply. When terms like net income are used in an otherwise grammatical, well-written sentence, they can be, and usually are, misunderstood.

In recent years, the subject of how individuals make decisions and how information affects the decision-making process has been researched and discussed often. It is well known to researchers that an organization's culture can influence how decisions are made and how information flows. Only when information supplied is congruent with the decision-making behavior can it be effectively communicated and acted upon. Very little of such research has found its way into accounting curriculums and, consequently, accountants are unlikely to be familiar with such research.

This chapter started by discussing the common stereotypes used to describe accountants. The negative perceptions can be countered if accountants stop seeing their role as effectively limited to developing financial information. Instead they must become responsive to the various users of the information supplied and must develop information to meet their specific needs. Currently, accountants seem aware only of those users' needs that can be met by the aggregated reports issued after the fact. By becoming more communication-conscious, accountants will be able to respond to users who need more than the proverbial bottom line. In the next chapter, we will discuss how an exclusive focus on the bottom line is driving users to go elsewhere to satisfy their needs and what may be done about it.

# CHAPTER 3

# ROLES PERFORMED
# BY ACCOUNTANTS

## 3.1 SCOREKEEPER, ATTENTION DIRECTOR, AND PROBLEM SOLVER

A classic definition, dating back to the 1950s, of the roles performed by the controller's department identifies them as scorekeeping, attention directing, and problem solving. The scorekeeping role deals with preparing financial statements; to do so the controller's staff has to ensure that the journals and ledgers are up to date. Accurate ledgers are used to ensure that customers get billed, accounts receivable are collected, bills are paid, payroll is met, and long-term uses of fixed assets recorded. In addition, accountants see to it that inventory gets properly valued.

The attention-directing role is performed by way of annual budgets, cost standards, and variance analysis, which serve to provide employees with the goals that must be met and also provide standards against which employees' performance can be judged. In addition, budgets serve as a road map for the organization. The problem-solving role of accountants is performed through tasks such as capital budgeting and special reports dealing with investment in the plant and equipment, as well as strategic choices.

Nearly four decades have gone by since the definition cited was put forth. The three roles defined for accountants in the 1950s have remained largely intact, however, a changing environment has added an additional role to those ascribed to accountants, namely, the regulatory compliance expected of business firms by governmental agencies. Additionally, the changes in technology have necessitated that accountants keep up with information technology. Indeed, technological changes are so revolutionary and so frequent that the efforts to keep up with the emergent technology can be seen as a separate, distinct role performed by accountants.

## 3.2 REGULATORY COMPLIANCE

While the regulatory compliance expected of business firms is nothing new, it has steadily increased since the 1950s. Taxes are facet of governmental control exerted on business, and they must be complied with by all companies. But now there are regulations from a host of agencies demanding compliance. Businesses must consider GAAP, GATT, GAAS, and, of course, FASBs. They must also know about the changing expectations of the IRS, INS, EEOC, EPA, FTC, and SEC, to name just a few regulatory bodies. These compliance and regulatory issues create work for accountants, among others.

Businesses must not only comply with local, regional, state, national, and international regulations, they must follow their own organizational rules, procedures, and policies, hence the advent of internal auditors.

Given the regulatory compliances expected of a business these days, much of it impacting them financially, accounting departments are deeply involved with compliance, be it internal or external. Indeed, a significant number of resources expended by the controller's department are devoted to compliance. Given this, it is time to recognize regulatory compliance as a distinct role expected of organizational controllers.

## 3.3 ACCOUNTING ROLES ARE NOT EQUAL

The roles assigned to the controller's department do not carry equal weight. Over the last half-century, the scorekeeping function has overwhelmed the controller's staff to the extent that it defines virtually everything they do.

Much of the time available to accounting departments is taken up with routines pertaining to the scorekeeping responsibilities, and little time is devoted to improving departmental operations, efficiency, and effectiveness, or to performing managerial/cost accounting functions.

During the 1960s, 1970s, and even into the 1980s, the financial accounting needs became the driving engine for the production of accounting information in American businesses, and this led to a situation in which the other two roles, attention directing and problem solving, were performed using the scorekeeping imperatives. The results were not good; the inadequacies of financial accounting, discussed in chapter 2, make it unsuitable for managerial needs. Hence, the usurpation of accounting systems by the needs of external reporting led to a management/cost accounting obsolescence. The impact of such accounting obsolescence was widespread, and it adversely influenced the ability of American businesses to compete in the global village.

## 3.4 THE ADVENT OF ACCOUNTING OBSOLESCENCE

Accounting as it is currently practiced has been attacked as obsolete. This accounting obsolescence has occurred because the information being prepared by accountants fails to meet the needs of many of its users. The recent accounting obsolescence came about because the financial accounting needs became the driving force for the production of accounting information, while the needs of other users were ignored. Accounting obsolescence could have been prevented if those responsible for management accounting systems had taken corrective steps and not allowed the problem to fester.

The problem's severity moved the American Institute of Certified Public Accountants (AICPA) to act, according to Thomas Rimerman. Because the users of financial statements have moved to other information sources, and because of the potential threat posed by such moves to the accountants' role, the AICPA asked its future issues committee, chaired by Rimerman, to study and suggest solutions to the problem.

The charges of obsolescence have also been brought up by the accounting academics. Gary Sundem, the current president of the American Accounting Association, an organization representing the accounting academics, recently described accounting as having become outdated, like the Ptolemaic system in astronomy. Sharon McKinnon and William Bruns, Jr., in their book, *The Information Mosaic*, have demonstrated that throughout North America managers and decision makers are forced to rely upon personally gathered information because formal accounting information systems do not provide them with the needed information. These charges are supported by countless articles published in managerial accounting periodicals, such as *Management Accounting*, in the last decade.

## 3.5 OBSOLESCENCE AS A COMMUNICATIONAL PROBLEM

If one does not learn from history one is forced to repeat it. Accordingly, it is in the best interest of accountants to study the causes of the current obsolescence of accounting systems for managerial users. Such knowledge can make the accounting function more relevant and prevent future obsolescence.

The perception promoted in the professional literature is that the recent accounting obsolescence came about because needs of external users became the driving engine for the production of accounting information in American businesses. However, the usurpation by external financial reporting needs was not the only cause of the disparity between the needs of the managers and the accounting information supplied them. The insistence of the accounting profession that accounting's role be limited to the preparation of information for disclosure also helped bring about the obsolescence.

Given the accounting discipline's mindset, it was inevitable that the discipline's performance came to be seen as lacking in relevance. The unflattering reputation of accountants has its origin in the insistence of the profession to see itself as being concerned exclusively with the preparation of financial information for disclosure.

Accounting obsolescence is in reality a communicational problem. It results from the failure of accountants to communicate effectively with the users of accounting information and to keep abreast of their changing needs. According to communication models, communication is achieved when the message sent by the sender is understood by the receiver. An effective information system depends on understanding what information is needed, who needs it, and for what purposes, and then making sure that the various parties understand the meaning of each input and output. An accounting information system must reckon with the requirements for effective communication so that the data it accumulates and transmits amounts to meaningful and usable information.

Since accounting information is meant to be used by those receiving it, the lack of its use by the receivers is a failure of the information communication process. Again, if the accounting information supplied is incorrect and is responsible for wrong policy decisions, then, too, effective information communication is lacking. Both types of failures commonly occur.

## 3.6 THE ACCOUNTING PROFESSION IS USER-UNFRIENDLY

To find evidence to support the charge that the accounting profession perceives its professional role as being limited to preparing information for disclosure, one needs only to read the Financial Accounting Standards Board's second *FASB Statement of Concept*. Its twenty-second paragraph repeats a statement originally made in the thirty-fourth paragraph of the first *FASB Statement of Concept*. It reads: "The information should be comprehensible to those who have a reasonable understanding of business and economic activities and are willing to study the information with reasonable diligence."

The paragraph underscores that accountants have a user-unfriendly attitude. The excerpt connotes that if the disclosed accounting information is incomprehensible to users, then the fault lies with the users' lack of reason or their unwillingness to be diligent.

But such an assumption on the part of the information suppliers is simply a user-unfriendly attitude.

A similar user-unfriendly orientation is echoed in FASB's second concepts statement in paragraph 40: "The financial statement is a tool and, like most tools, cannot be of much direct help to those who are unable or unwilling to use it or who misuse it. Its use can be learned...[and the tools] can be used by all—nonprofessionals as well as professionals—who are willing to learn to use it properly." Again, the user, or rather the receiver of accounting information, is being asked to take on the entire burden of learning the use of the tool. The receiver of the information is told, in effect, that if he is unable to use a tool, then it must be due to his poor learning habits. Accountants shirk responsibility for ensuring the communicability or the effective use of the information supplied by them.

The FASB statement fails to acknowledge that the accounting tools may be badly designed and may not be the right tools for the task at hand. Tools can also be improved upon—were it not so then we would still be living in the Stone Age. Furthermore, the reader is led to believe that all tools are created equal. However, a reading of the automotive or hardware sections of the Sears catalog will show us that some tools are easier to use than others!

Clearly, the accounting obsolescence that has plagued American businesses came about in considerable measure because accountants, the tool makers, kept supplying the wrong tools for the tasks performed by American managers in the previous decades. Like much of the information supplied by accounting systems in the last decade to corporate managers throughout America, the FASB statements of concepts show an implicit disregard for the needs of the users of accounting information. Furthermore, paragraph 34 of the first statement lets it be known that investors and creditors are the only users of accounting information that matter for accountants; managers, workers, policy makers, and the public at large, all potential users of accounting information, are summarily ignored. Ignoring the receivers of information is not apt to result in an effective communication process.

Even though FASB concepts are not on the bestsellers list, they do reflect accountants' attitudes, since they are written by those who represent accountants. They offer tangible evidence of what accountants do and how they think.

## 3.7 THE NARROW PERCEPTION OF COMMUNICATION

It is a measure of accountants' prevailing disregard for the requirements of effective communication that even when they talk about communication, they limit it to writing and speaking skills. In an AICPA publication more than two decades ago, it was argued that: "To the CPA the ability to express himself well is more than the hallmark of an educated man, it is a professional necessity. The inability to express his feelings in understandable, explicit, unambiguous, intelligible English can be self-defeating, potentially misleading, and possibly disastrous to clients, creditors, and investors. . . . Candidates who cannot write the English language at least as well as a minimum threshold should be denied admission to the profession, if need be on this account alone[1]". The authors, while recognizing the importance of communication skills for professional accountants, unfortunately are thinking primarily in terms of good writing skills. They neglect other elements of the process pertaining to organizational communication of information and are typical of those who have discussed the communication skills needed by accountants in various academic and professional journals over the last ten years.

Complaints of accountants' inadequate communication skills will persist until we realize that effective communication of accounting information can require a lot more than reading, writing, and arithmetic and until such realization becomes part of accounting communication. At some point in their education and training accounting students must be made familiar with all the elements involved in the development, communication, and effective use of accounting information. Such an approach would ideally explore the complex nature of information communication and would teach that accounting is not merely a set of numbers, but rather helps to define organizational reality. Accountants must be sufficiently aware of those needs that lead to the generation of accounting reports. Beyond an appreciation of such needs, the accountant needs to appreciate the roles played by all the component elements of the accounting communication loop: the controller, who sends the information; the users, who receive accounting information; the way the information is used; the impacts, both intended and unintended, of the reports that the controller's staff sends out; and an awareness of organizational structures and their impact on information communication. Through such know-how, the utilization of the accounting information generated could be enhanced and made more effective.

Teaching the methods of controlling or removing the impediments to effective use of information to accountants will better prepare them to appreciate and understand the specific needs of their clients. Such needs are not uniform and are constantly changing. Not everyone's needs are met by standard financial accounting statements. One size does not fit all, but those who prepare accounting statements do not seem aware of this exigency.

## 3.8 A DIFFERENT PERCEPTION OF COMMUNICATION

Recently, the perception of the role accountants are expected to play has come under review by accounting educators analyzing the accounting curriculums. Such analysis led to recommendations made by the American Accounting Association's Committee on the Future Structure, Content, and Scope of Accounting Education, also known as the Bedford Committee Report. It called for a redefinition of the role accountants play, arguing that they must not just prepare accounting information for disclosure, but also be concerned with its effective distribution and use. The realization of the new gestalt envisioned for the profession will entail major changes in the way communication has hitherto been incorporated into the accounting curriculum.

The focus of the Bedford Committee's report is best explained by Norton Bedford himself. He suggests that future accountants be made aware that the "quality of economic decision making depends not only on the quality of information developed, but also on the ability of accounting information producers, inventors, creators, or developers to distribute that information effectively and efficiently. . . . Accounting information must be distributed at the right time in the right form to the right decision maker who is told by the accountant, as part of the information distribution function, how to use the information."[2] Accounting academics do not teach about managing the communication of information. They need to start teaching the management of information distribution just as they now teach students how to develop financial statements.

## 3.9 WHAT WILL IT TAKE TO MAKE ACCOUNTANTS USER-FRIENDLY?

In the long term, the implementation of the new accounting as envisioned by the Bedford Report can ensure that accountants are user-friendly, that they take time out to understand the needs of those who use the information being supplied, and that they will work with the users to find out what they need. The typical users may not know precisely the information they need to help them understand the organization's status and plan for its growth. A collaborative approach taken by accountants with the information users can help overcome that problem. Hopefully, the accountants of the next century will work with the information users to ascertain what information is needed, in what form, and in what frequency, and how can it be best transmitted to the users.

While the future will be brighter if the Bedford Report is implemented, the present remains a problem. To change the situation, the following steps ought to be undertaken.

**(a) Use the Emerging Techniques to Redesign Accounting Procedures.** If accountants are swamped with day-to-day chores, the reason may well be from a lack of adequate resources. But an equally likely reason is that the chores are designed to be time consuming. Even today, most accounting and auditing texts talk about vouchers that need five and more copies to take care of the purchases and the related payments. When you create five or more copies of a document, you are creating activities that will need five or more employees to handle the tasks. In other words, you are drowning yourself in paperwork. If accounting departments want to free their resources, they must redesign the departmental procedures and cut down on the non-value-adding activities. Ford Motor Company has done just that, and reduced the number of employees needed in its accounts payable department by two-thirds. The key to Ford's performance was process redesign, effective use of information technology, and a relationship with vendors that was cooperative rather than adversarial.

**(b) Adopt an Information Resource Management Perspective.** Accounting departments and management information systems (MIS) have remained separate and are often at odds with each other. Their turf battles rage, even though both work for the same firm. What is needed is the adoption of the concept known as information resource management, which is based on the realization that information cuts across organizational boundaries and departmental functions. Under such an approach, accounting and MIS are not seen as self-contained departments, but as elements of a system bonded together and devoted to providing appropriate information to users. Indeed, the organization's information system is too important to be left entirely to the care of either the accounting or the system personnel, it should involve the entire organization to help ensure the relevance of its output. Managers function better when they are well informed, but that will not happen if what they are being supplied is not what they need.

**(c) Know the Users and Their Needs.** One of the major topics in the literature today is the need for firms to supply the customers' needs. "Customerize" has been coined by a major corporation in its advertisements to emphasize its regard for its customers' needs. The more adept a firm is in anticipating and meeting the needs of its customers, the more it is likely to succeed. What is true for firms and the products they supply holds true for the accounting department and the users of its products. Doctors, while treating their patients, do not expect all of them to be suffering from the same illness, nor do they expect patients to know what medications and treatments are needed to cure their various ailments. Each patient is handled uniquely and doctors work with their patients in order to ascertain the illnesses and then prescribe the treatments. It is time for accountants to work like doctors. It is also time that they practice within the accounting departments what is being preached for the firms at large, namely, a customer orientation. Doing so may mean being willing to use non-monetary performance measures and to abandon a preoccupation with financial reporting, if this is not the type of information needed by the users.

**(d) Appreciate the Importance of Communication.**   Closely related to understanding the needs of the users of accounting information is an awareness of what may impede the communication of accounting information. The prevention of future occurrences of accounting obsolescence demands that accountants become familiar with all the elements of the process involved in the development, communication, and effective use of accounting information. Such an orientation will ensure that accounting is not seen merely as number crunching but as a tool that helps to define organizational reality, translates its performance in communicable terms, and allows managers to make informed decisions.

### 3.10 USER-FRIENDLY ACCOUNTING MANUAL

Given the accounting department's insistence that its role is essentially limited to providing financial scores, as well as its propensity to ignore what the users actually need, it should be no surprise that accounting procedures, the building blocks of the functions accountants perform, reflect that same attitude. Their underlying emphasis may well be a major reason for the lack of respect accorded the accounting manuals. If the documentation process is to be an illuminative occasion, then it must be relevant and user-friendly. Fortunately, the competencies currently available make it easier to change the underlying focus of the accounting procedures manual, and they are the subject of the next section of the book.

### ENDNOTES

1. Roy, R.H. and J.H. MacNeill. *Horizons for a Profession*. New York: AIC PA, 1987.
2. Bedford, Norton M. "The Education of Future Accounting Professionals," *Georgia Journal of Accounting*, Spring 1988, pp. 109–124.

# CHAPTER 4

# ATTAINING EVOLUTIONARY COMPETENCE

## 4.1 WHAT IS EVOLUTIONARY COMPETENCE?

*Evolutionary competence* is the ability of an organism to survive the changing environment by adapting to the changes. In a biological context, we know of species that are no longer around because they were unable to change with the environment. Such failure to adapt, cope, and evolve rendered them extinct. The evolutionary competence that allows species to continue has its counterpart in the organizational context as well. A number of organizations are able to evolve, while many more fail to do so and are found no more. Indeed, there is evidence that proves that institutional life span tends to be short: a survey taken by Shell Oil showed that about one-third of the firms listed among the Fortune 500 in 1970 no longer existed by 1983.

A considerable body of literature exists that argues that institutional survival need not be a matter of random chance. Such literature highlights organizations that are able to confront changing, and indeed, hostile environments, and still manage to prevail, not just survive. Such documented cases pertaining to firms that display evolutionary competence demonstrate that skills to organizationally evolve can be learned, indeed institutionalized.

This chapter introduces tools and techniques that are actually being practiced and that are helping firms to evolve and prosper. In addition to an overview of these cutting-edge tools is a discussion of the elements that are common to all of these various techniques being used to help institutions grow and become more productive. A brief discussion is also provided on why acquiring such competence may be imperative for firms at this point in time. Following this introductory chapter, in subsequent chapters, we take a more detailed look at the more important of the several evolutionary competencies mentioned in this chapter.

## 4.2 A SHORT LIST OF EVOLUTIONARY COMPETENCIES

Over the years a number of approaches aimed at improving organizational performance have evolved. Some examples are:

*Quality Circles and Suggestion Systems.* The two are attempts dating back to the 1960s to involve employees in improving the efficiency of the institutional processes and thus enhance productivity. Quality circles are groups, usually small, of employees meeting regularly to identify, analyze, and solve problems involving quality, costs, performance measurement, and other problems related to their work and its environment. Suggestion systems are designed to capture, evaluate, implement, and reward ideas generated by individuals or groups seeking to improve work processes. Even though one does not hear much about quality circles and suggestion systems these days, they have been incorporated within more recent approaches aimed at improved institutional efficiencies.

*Continuous Improvement.*    Continuous improvement, variously called total quality management (TQM), total quality environment (TQE), and *kaizen* (in Japan), is the philosophy of work management that owes its fame to Edward Deming. It encompasses much more than simple work techniques, rather it involves attitudes and is properly seen as the philosophy of work itself. It stresses that managers and workers ought to focus on continuous improvement and ongoing elimination of activities that are non-value adding. As has often been noted, just such a focus is behind the relentless advance of Japanese firms like Sony, Toyota, and Honda. The search for continuous improvement starts with a permanent dissatisfaction with the status quo. Such identification with change leads to relentless, continuous improvement in all areas and all levels of the organizational hierarchy. It is a collective effort and has to be coordinated and supported by all levels of the institutional hierarchy.

*Activity-based Management and Activity-based Costing.*    As their names imply, the two focus on activities undertaken by firms to produce and deliver products and service to their customers. Activity-based management focuses on activities such as the measure of efficiency, while activity-based costing assigns costs to products and services by evaluating activities needed to make them. The focus on activities is often tied to improving the time needed to complete the production process. Despite their common focus, the two are technically different: one is concerned primarily with learning the costs incurred to produce and deliver a product, while the other does not involve the allocation of costs. However, the proponents of activity-based costing try to portray it not just as a costing tool, but as a comprehensive way to improve management by illuminating and eliminating activities that are wasteful and do not add value to the product or service.

*Just-in-time Inventory Systems (JIT).*    JIT refers to an inventory system that has material needed for making products arrive from the vendors as it is needed. Indeed, the supplies are delivered not to a receiving warehouse but to the production centers within the plant. The implementation of such a system can reduce a firm's investment in inventory, often the largest component of the working capital. It will no longer purchase and store inventories needed to produce products, and so cash is no longer tied up in inventories or in the facilities needed to store and look after them. Instead the burden is out-sourced to the vendors who supply the firm with the needed raw material.

The process of implementing JIT requires high levels of coordination and cooperation between the company and its vendors. Firms have to devote considerable resources to planning and scheduling work so that raw material needed is available when needed. For JIT to work well, a firm and its suppliers have to be able to communicate about the exact specifications of the material required and about reliable delivery. JIT cannot work if a supplier ships bucket seats when the cars on the assembly line require bench seats. Nor will it work if the truck carrying the steering columns is stranded in traffic.

*Process Reengineering.*    Another approach to institutional evolution is the emerging discipline of process redesign, or reengineering. In contrast to continuous improvement, it seeks more immediate changes. It incorporates and brings forward techniques associated with task analysis, operational auditing, and work simplification. It focuses on organizational processes instead of specific functions, and is based on the premise that by reexamining why and how things are being done and by imaginatively using emergent technology, an organization can achieve major productivity gains and systemically change the way its work is performed. The reengineering process starts by asking why something is done in a given way. Following such an analysis, the reengineering process focuses on achieving the same goals by more efficient means. Fundamental to such redesign is the desire to eliminate non-value-adding steps from the process and to approach the process from a cross-functional, rather than a departmental, perspective. Use of emergent technologies, particularly information technologies, is likewise crucial to process redesign.

A large number of current work practices exist because of the tradition that they break work into its component tasks and then let different specialists take care of their respective assignments, usually narrow segments of the overall process. Such thinking focuses on individual tasks much more than on the underlying goal, and the result is costly delays, as well as an excess of stakeholders in different departments all fighting for their own turf. A major rationale for implementing reengineering within an organization is to get past such a functionally based, narrowly focused orientation and think in terms of broader processes.

*Learning Organization Paradigm.*    Paralleling the development of activity-based cost systems, process redesign, and continuous improvement is the recent emergence of what is being called a learning organization paradigm. It focuses much more on the future, rather than on simply improving the status quo. However, all the approaches listed emphasize individual and institutional learning. Indeed, the term "learning paradigm" can well be seen as the label for the common denominator underlying the various approaches used by institutions to attain evolutionary competence. The term "learning" must not be equated with simply digesting information and taking in details. Learning organizations allow their workers to expand their ability to deal with current problems and visualize future opportunities. A learning organization is concerned with things as they could be and not just as they are; it seeks to anticipate opportunities, not just deal with crises as they occur.

While each of these techniques has its distinctive features, they have much in common: they all seek to help an institution evolve and improve. Furthermore, implementation of one approach does not exclude others from being used at the same firm. An organization may well devise a strategy that combines elements from several of these approaches. The choice of the specific approach to implement depends on the needs of the firm, in the context of its particular resources and strategic orientation. They all have some basic common elements, which are surveyed in the next section.

## 4.3 BUILDING BLOCKS OF INSTITUTIONAL EVOLUTIONARY COMPETENCE

The most important elements needed to fashion and implement an institution's evolutionary strategy include: customer orientation, an institution-wide sense of shared vision, facilitative leadership, empowerment of the employees, an organizational culture supportive of ongoing participation by all levels of employees, a training program that enables employees to be involved and keep developing their skills, and, finally, an effective reward structure that encourages employee creativity. All of these elements must be part of any efforts to make institutions more efficient. Each of these are necessary building blocks.

**(a) Customer Focus.**   One usually thinks of a customer as one who purchases products. It has long been known that in order to prosper a business must practice strategies aimed at pleasing its customers and providing them with value. But the attitude that the customer is always right need not be limited to externally based sales, because ensuring that external customers receive a good value cannot be accomplished if it is limited only to the final transaction. It is also important to satisfy internal customers, the users of information within the organization. Making external customers happy can only be the result of a chain of customer-supplier relationships that stretches throughout the organization and culminates finally in giving the purchasers of goods or services what they need. Every employee performs work that results in creating or providing services or products that in turn are used by others to repeat the process. Each employee acts to add an incremental value, a building block, in the process that finally ends in giving external customers what they want.

This orientation implies that employees must view the next step in the process, next link in the chain, as their respective customers and work towards keeping these internal clients happy by providing them with products and services that they actually need. The customer for an area accountant is the department in the factory he or she is assigned to cover. It is incumbent on the accountant to keep that department supplied with information it actually needs in a timely fashion and in a form that is understandable to the production area manager. The bottom line for this manager is not the same as the bottom line for an investor in the company's stock.

**(b) Fostering a Shared Vision.**   It is the responsibility of the organizational leader to develop a clear vision of what the company is, what it should be doing, and where it should go. However, simply getting senior managers to have a clear vision of institutional mission is not enough. The vision must be in tune with the institution's unique nature and makeup, and it must be shared throughout the organization. Only when the vision is communicated to all the employees does it become conducive for focused growth. Understanding clearly what a strategy for change is meant to achieve is the responsibility of the top manager, as is selecting the optimum change strategy. But the senior management must share the strategy with others, foster a consensus about what it is meant to accomplish, and ensure that skills and resources are available to implement the changes. Symbols and rituals can effectively communicate the vision of change and growth to all levels of the institution—indeed Japanese companies have done just that, making symbols and rituals an important component for continuous improvement.

**(c) Facilitative Leadership.**   It is well known that a large ship requires a good deal of time in order to change direction. Institutional leaders, like captains of large ships, cannot just will the change. A firm, like an ocean liner, takes time to change direction. Senior executives invariably realize the limited ability they have to mandate change, however, they can leverage their ability to change the institution's direction by fostering an organizational culture and environment that supports change. In so doing, top managers are exercising a facilitative leadership role. A very important ingredient of facilitative leadership is the communicational climate within the firm and communication of the results of activities undertaken. A good communicational climate will encourage individuals and groups to cooperate with each other, while communicating results openly will ensure the motivation to be proactive and work towards change. Leaders who practice facilitative leadership know key stakeholders and include them in decision making, demonstrate constancy of purpose, and intervene when necessary. Intervention may be required if the employees' dedication to change is uneven, if they do not fully understand the necessity and the nature of change, or if they are lacking in the skills to implement changes.

**(d) Employee Empowerment.**   Employee empowerment refers to the delegation of authority to levels where activity is taking place. It contrasts with wisdom dating back to nineteenth-century gurus such as Adam Smith and Frederick Taylor. Through empowerment, frontline employees are allowed to make adjustments without seeking approval from higher-level supervisors. They can deviate from established procedure if a given situation calls for such deviation without waiting for a clearance from the supervisors. When empowered, employees do not feel a need to cover up problems; instead, they seek to solve the problems in an expeditious manner. Empowerment comes about when an institution is willing to trust its employees to do the right thing. For such trust to develop, the corporate culture must promote shared vision and risk taking. It also requires properly trained employees, who know more than a narrowly defined function and are capable of understanding the whole process. Empowerment is also dependent on properly rewarding employees for taking on a higher level of responsibility and for continuously upgrading their skills.

**(e) Employee Involvement.**   As mentioned earlier, senior management cannot mandate change on their own, they must have help from their employees. Efforts to get employees involved must rely on voluntary, collaborative efforts within a non-threatening, relaxed environment. All too often employees will resist changes. Such resistance is caused by uncertainty about what lies ahead or what the changes imply for their jobs. In the face of ambiguity, employees understandably become anxious. Related to uncertainty is a perceived threat to employees' self-interest. Both managers and employees may feel that the changes will lead to loss of their jobs, or at least a reduction in their power and perks. Another reason may be the fear that existing relationships and social networks will be disturbed.

Differences in perceptions about the need for changes can also lead to uneasiness. Lower-level employees may not know why the changes have to be made or they may see it as just another fad. Senior managers may be assessing the situation differently based on information provided only to upper-level managers. Such perceptual differences can lead employees to react negatively to changes being instituted. A reward system that rewards individual instead of team efforts can also be a barrier to participation.

**(f) Training the Employees.**   Institutional evolution requires skill building and employee training. New competencies are needed on the part of the employees to keep up with the changes taking place within an organization. If the work is redesigned but the employees do not learn the new skills, the redesign will be counterproductive. As part of the total quality management (TQM) implementation, employees need to learn about problem solving, statistics, and data collection techniques. But, in addition to such specific skills needed to implement TQM, employees also need to cultivate a learning attitude, since such a willingness makes optimum learning possible. A learning attitude will entice workers to expand their ability to deal with current problems, visualize future opportunities, and anticipate opportunities, rather than just dealing with crises as they occur.

**(g) Reward Structure.** Employee learning and participation can be facilitated with the help of a proper reward and recognition structure. Traditional forms of reward are meant to reward individuals, however this can be a disincentive for employees to become team players. Many of the approaches being used to promote organizational development at present are meant to be team based. Given this trend, it is better to create a reward system that recognizes and promotes team efforts and participation.

## 4.4 IMPLEMENTING A STRATEGY FOR CHANGE

If proposed changes are meeting resistance among employees, it could be because they are drastic in nature and were imposed without any warning. Given such imposition from above, resistance on the part of employees is to be expected. Resistance can be minimized through participation, education, communication, and facilitation. Sharing with employees the rationale behind the changes is a good first step in fostering cooperation among all levels of employees. Dissatisfaction within the organization can be a great motivator for implementing evolutionary changes.

Changes become more acceptable to employees if they are involved in the planning and implementation of change. Their participation allows them to better understand the change itself and the reasons for making the change.

Open communication between various administrative layers should also help reduce employees' fears and uncertainty. Communication is also essential for employees' participation in the planning and implementation of the changes. Furthermore, changes ought to be made gradually. Abrupt imposition of major changes can cause resistance among the employees. Management should also take care to ensure it is not promoting change as a magic, instantaneous cure without any discomfort or side effects. The changes must not be seen by employees as change for the sake of change or, even worse, as being faddish.

At the same time, managers should not wait indefinitely for the resistance to subside. There is always going to be fear, anger, and uncertainty in the wake of change and if the choice is entirely up to employees, they will opt for the path of least uncertainty. While educating the employees and establishing open communication, managers may sometimes find it necessary to mandate the change, even if employees are unenthusiastic. Acceptance of the changes will grow as employees recognize the benefits.

Implementation must be well planned and evaluated strategically prior to being put into effect. Changes can be less than optimal if plans are the result of oversimplification or if the difficulties involved are underestimated. They will also be counterproductive if they rely entirely on technology while ignoring human factors.

Care should be taken to ensure that implementation takes place as planned. Once the change has been implemented, efforts should be made to measure its impact.

## 4.5 WHY THE IMPERATIVE TO EVOLVE?

Why bother with change? Look at the news headlines for confirmation that in the 1990s a company must change for the better, or its very survival is threatened. The sense of prosperity of the 1980s has given way to a downturn in the 1990s that shows every intention of being long-lived. The downturn in economic activity has affected both the manufacturing and the service sectors. The extent of the stagnation is evident by the wave of business failures, bankruptcies, downsizing, and mergers that dominate the headlines and the magazine covers. The bluest of the blue chips, IBM, Sears, and GM, are fighting for survival. Sears started the catalog business, but it is no longer king of the hill; it is not even an important player. But Sears' loss of leadership in the catalog business is not from a lack of market. Americans have only to look at their mail to realize that catalog shopping is flourishing, and if they watch television, they learn of new channels for stay-at-home shopping. Sears' shrinking business came about from its unwillingness to cultivate an evolutionary competence in the face of a changing environment, and not from the disappearance of the market.

The operating environment has changed in the 1990s. International competition, the end of the Cold War, new technologies, and worldwide demographic changes are all making the world a different place than it was in the quarter century following World War II. Back then, the increases in demands allowed American firms to ignore quality issues and even the customers' concerns. They could usually rely on marketing to pull the sales through for them. But to their chagrin, foreign competition began delivering products and services that far exceed customer expectations.

There is a tendency that would have us retreat into a buy-American isolationism and to erect artificial barriers to counter the shrinking of the globe. But neither such ostrich-like behavior nor the short-term remedies of "trimming the fat" and emphasizing "lean and mean" management is likely to develop viable institutional evolutionary competence.

Management experts, such as Peter Drucker, seem to be calling for institutional organization that incorporates built-in structures to respond to change, indeed make the ability to change faster than the competition a cardinal virtue and a strategic advantage. According to current theory, knowledge is power and knowledge is always evolving. The experts are calling for organizations to force change, to periodically examine what they do, and to justify to themselves and their stakeholders the adequacy or viability of what they do and what they know. In making their arguments, the experts cite cutting-edge organizations that are able to bail out before the markets reach maturity or saturation, and can do so because of their willingness to cultivate organizational learning, the essence of any evolutionary strategy.

Throughout the ages and all over the world, educational institutions have fostered learning and continue to do so globally. Their nearly universal presence argues that the structure educational institutions provide and the environment they create makes for improved learning. Giving individuals books and letting them read if and when they so desire is not seen as the optimum way to promote learning. Similarly, organizations must also provide structures, occasions, and incentives to foster institutional learning. The role being argued in this book for the procedures documentation process does just that, it structures organizational learning, it institutionalizes evolution.

## 4.6 CONCLUSION

Next we will discuss the emerging information technology and how best to employ it. Then, in the following chapters, we take up the various approaches to promote institutional evolution that were introduced in this chapter. We will describe at greater length, continuous improvement, activity-based management, activity-based costing, process reengineering, and the learning organization paradigm. These competencies are not mutually exclusive. Firms can use elements from all. No matter which approach an institution may elect to follow in order to evolve in the contemporary environment, it has to rely on technology in general, and information technology in particular, to do so.

Before we discuss the above topics, we take up the role of informational technology and the impact it can have on accounting functions.

This section of the book concludes with a discussion of the requirements for effectively managing group processes. Procedures documentation is a group process that benefits if those involved are informed about ways that permit groups to work effectively. Using the collective experience and expertise of employees at all levels will not only produce better, streamlined procedures, it could lead to greater employee involvement, improved employee motivation, and, most importantly, to an innovative environment. The writing of manuals suggests itself as a natural way to channel employees' participation and involvement and to promote ongoing improvements. But the employees' participation cannot be willed into existence. It must be nurtured. Often, worker participation is against a firm's own tradition, and these traditions must be broken before employees can contribute effectively. The acquired habits—bad ones in particular—of the supervisors must be overcome if subordinates are to feel free enough to participate. To help encourage participation, management must provide occasions that lend themselves to participation and learning. The writing and reviewing of procedures and policies can be seen as such an occasion, but getting the members of an organization to examine what they do and to suggest change is not an easy task, especially for accountants. A survey of the available research on facilitating group processes will help accountants and others within an organization work more effectively.

# CHAPTER 5

# INFORMATION TECHNOLOGY (IT) AS AN AGENT OF CHANGE

## 5.1 INTRODUCTION

This chapter is meant to shed light on a perplexing problem today's corporate controllers confront: how does one determine whether to apply information technology (IT) to help solve departmental problems and to improve operating procedures. If IT is to be used, then which technology? Such a choice would not be difficult were it not for the mixed record of IT. It is this mixed record that can lead a controller to see the promise of IT as either marketing hype or an enabler.

## 5.2 THE PROMISE AND THE PERFORMANCE

Since computers were first introduced in the 1950s, their potential for changing procedures and processes has been loudly proclaimed. Unfortunately, this prediction has not always been fulfilled. Incidents of complete failure of IT systems or significant variances between promised and attained benefits are not uncommon. The much-touted advantages of IT failed to appear in manufacturing as well as in white-collar jobs, such as those in a controller's department. Indeed, instances have been recorded when large investments in IT actually lowered productivity, and thus profitability.

## 5.3 TECHNOLOGY IS NOT AT FAULT

Instances of the disappointing performance of IT notwithstanding, the emergent technology has indeed compiled an impressive record. In the manufacturing sector, it has helped evolve newer tools such as robots, sensors, and automatic, self-correcting testing. It has helped enhance material handling with automatic storage and retrieval systems in warehouses around the world. In addition, the impact of IT is visible through computer-aided design and manufacture, material requirement planning, production scheduling, and a variety of decision-support systems. In the world of white-collar work, IT is also changing how things get done through word processing, spreadsheets, databases, automated filing, transaction systems, computer conferencing, electronic mail, computer bulletin boards, and expert systems.

There should be no doubt that in the future, the price-to-performance ratio of IT hardware and software will continue to improve—computers will keep getting better, faster, and, most importantly, cheaper to own and operate. They can facilitate an increase in the productivity of many business functions. But, as a consequence of implementing an IT solution, new and unexpected problems may well occur. In some cases, these new problems are either not severe or can be remedied over time. For example, on-line systems provide a mechanism that could allow anyone with computer know-how access to secret information. To fix this problem software security can be incorporated into the system. In other cases the new problems prove to be sizeable, and reduce the productivity gains that were anticipated. Customized systems that provide needed functionality but require programming support staffs to keep them operational represent one such problem. Estimates of the maintenance cost of existing software systems run from 60 to 80 percent of the entire data processing budget!

It can be argued that the unrealized promise of IT is not due to the failure of technology, but to the failure to integrate it with the existing work environment. If the new technology is introduced in a well-planned manner, the chances of its failure or a suboptimal performance can be minimized. More importantly, IT will be much more helpful if it is not seen as an electronic tool to do the same old things. Even today, many companies require vouchers with five and more copies that are needed to take care of the purchases and the related payments. When there are five or more copies of a document, five or more employees will be needed to handle the tasks. If a computer is being brought in to keep turning out the same vouchers with same number of copies, the productivity gains will not be all that impressive. The work may get done faster, but it will remain the same work with the same paper glut.

In other words, IT need not be seen as "snake oil" by controllers, but if it is to truly revolutionize work, then its introduction must be well planned. It will be even better if the work IT is meant to perform is redesigned.

## 5.4 DEMONSTRATING ITS USEFULNESS

The usefulness of IT as a tool to help solve the recurrent and common problems faced by accounting departments were examined in 1982 in an article by Vincent Giovinazzo.[1] In this article, Giovinazzo discussed a number of everyday problems that plague accounting departments in all sorts of organizations. The list includes problems having to do with the use of IT to help perform accounting functions as well as problems that predated the advent of IT. While past performance is no guarantee of future success, the progress made in addressing old problems would seem to be an appropriate reference point to show the effectiveness of IT in helping to provide solutions to operating problems.

Despite their universality, the problems listed by Giovinazzo lacked the chic or the glamour that might have allowed them to be discussed systematically in the professional literature. Such persistent, universally occurring but rarely discussed problems, according to Giovinazzo are:

- Source data vulnerability
- Involvement of non-accounting personnel
- Nonintegrated accounting systems
- Processing and reporting lags
- Uneven workload
- Underestimation of time requirements
- Human relations problems

For each of these issues, we examine (1) the problem in the IT climate of 1982, (2) Giovinazzo's proposed solution(s), and (3) the extent to which IT solutions exist today.

In order to avoid reliance on untested technology, we will limit the solutions to commercially available products. Finally, we will compare 1993 solutions with the 1982 proposals.

**(a) Source Data Vulnerability.**  *Problem*: The smooth working of accounting systems depends on accurate data entry. Incorrect data input tends to be the result of human error manifested in the form of miscalculations, typing errors, misreading of forms, sloppy handwriting, and the failure to follow directions. With early versions of IT, yet another source of potential errors was introduced as a result of the coding of data to make it computer-readable.

*Proposed solution*: Giovinazzo proposed to minimize the source errors by the use of well-designed forms, effective training, well-understood procedures, a clearly defined chart of accounts, better coding, manual reviews, establishing and meeting deadlines, and prompt feedback on errors. His technological recommendation was the use of intelligent terminals that code and input data at the point of entry. He alluded to such point-of-sale terminals (POS) being already at work in selected industries, such as retailing. He also argued the importance of on-line technology and saw it as becoming a part of the solution but did not see it as becoming universal.

*Current solution*: IT in the 1990s offers technological solutions that can minimize the impact of human errors and carelessness. Today, on-line systems are much more prevalent and allow for better (more accurate) data entry. Application software can instantaneously detect erroneous input of data; error detection capabilities are facilitated by modern programming environments. For example, modern database management systems provide for referential and range testing of data. Additionally, on-line systems allow entry of data at the source point by someone closer to the transactions, rather than a generic data-entry clerk who simply keypunched without knowing what was being entered into the system. Finally, much more is now known about human factors that were conducive to erroneous data inputs. As a result of such awareness, techniques in screen design have been developed that minimize the errors attributable to data input.

An even greater impact on the reduction of source errors is due to the explosion of alternative input devices. POS, bar code scanners, optical character recognizers, and custom built input devices permit direct input of data with minimal (and often complete) elimination of human involvement.

These devices not only improve the customer-service function, but also store and transmit transactions to various accounting subsystems. These functions were previously accomplished primarily by clerks. The inexpensive nature of such alternative input devices makes them cost-effective for small and medium-sized businesses.

During the last ten years, efforts have been made to facilitate the transfer of information between organizations. Electronic data interchange (EDI) allows data generated from the business process of one organization to be entered directly into the information system of another organization. With EDI, a customer can enter a purchase order in their system and have it transmitted to the appropriate vendor's information system. Once there, it is converted into a sales order on the vendor's information system. This technology increases data accuracy, as it reduces the number of people who are responsible for creating and/or processing documents. In addition, accuracy is increased as data from the document is mechanically entered into the accounting system rather than keyed in by a clerk.

**(b)  Involvement of Non-Accounting Personnel.**   *Problem*: Data enters accounting systems from within the firm and outside the firm. Almost all departments and operational areas must constantly provide source data to the accounting department. It is not uncommon for departments to resent this role. Besides such resentment, given the external origin of source data, the accounting department cannot control the clarity, accuracy, and timeliness of data entering the system. Thus, there are really two problems: resentment and the opportunity for data entry error (a potential cause of problem source data vulnerability, previously discussed).

*Proposed solution*: As with source data errors, Giovinazzo saw better forms as helping to minimize the errors in the data received from external sources. In addition, he argued the need for better communication between accountants and others within the organization to reduce resentment. The relationship between accountants and the rest of the organization could be improved if the reports provided by accountants were timely, error-free, and relevant to internal users. Such reports should have a managerial (rather than financial) accounting orientation.

*Current solution*: Increasingly, the direct entry of data from the "process" into the accounting system is being facilitated by input devices described previously. Since these devices increase the rate at which data is entered and the accuracy of the data, they reduce the data error problem. In addition, the devices generate the accounting data as a byproduct of the action performed by the non-accounting worker. For example, clerks at supermarkets use bar code scanners to compute the customers' bill and print a receipt. As a byproduct of this function, the accounting system will receive inventory and sales data that can be used to update appropriate ledgers. The generation of this (accurate) data required no additional effort by the employee computing the customers' purchases, so data accuracy is improved and resentment is reduced or eliminated.

**(c)  Nonintegrated Accounting Systems.**   *Problem*: The typical accounting system is comprised of a series of subsystems that are interdependent, but not fully integrated. Not all software subsystems integrate with all other applicable subsystems. That is, many firms have payroll, billing, and inventory subsystems that do not automatically integrate with the general ledger. It is common to find the various subsystems physically dispersed, with no communications mechanism to transmit and integrate the data between the subsystems. Not all subsystems are computerized; and often when they are, the software package is unable to handle many of the exceptional transactions. In addition, the application packages are so detail oriented that no single person could use the entire package. Each individual knows different parts but not the entire package.

*Proposed solution*: Giovinazzo argued that systems should be designed so that relevant data would be automatically transmitted among related systems. Accountants should insist that there be greater integration between the accounting subsystems. In general, he stressed that management accountants should have a larger role in defining specifications for accounting systems.

*Current solution*: Today, we have IT solutions that address these problems. The problem of subsystem integration had been addressed. Accounting software is obtained by developing it or purchasing packaged systems. For the custom approach, it is now a standard requirement that all subsystems be integrated. If the software cannot achieve the requirement, it will not be installed. For package systems, complete integration is a requirement.

The ability to communicate data generated by systems of this sort is standard within any hardware platform. In the past, the cost of such communications over long distances or to many locations was prohibitive. Competition and technological innovations have dramatically reduced this cost.

Finally, most package systems provide functions that address the problems of all departments. In the packaged solution area, much effort has been made to understand the entire processing of the subsystem so that a clear and streamlined process is programmed. Current tools such as help screens and graphical user interfaces (GUI) also help to make the software easier to use.

If a business is operating today without an integrated accounting system, it is not because of the unavailability of IT. Integrated accounting packages are now available to fit a variety of needs.

**(d) Processing and Reporting Lags.** *Problem*: A long-standing accounting problem is that of the time lag between when transactions actually occur and when they are reported in the form of accounting statements. This delay is caused both by procedural and mechanical deficiencies. The large amount of data to be entered, examined, verified, and processed takes time. Since most systems are batch oriented, delays occur as a consequence of gathering the data, ensuring its accuracy, coding it, and then entering it into the system. Additional delays are encountered in processing the data as a result of limited machine processing speed.

*Proposed solution*: This problem is seen as being primarily one of scheduling. Hence, the proposed solution is to improve the scheduling of work and then ensure that those assigned to various tasks met established deadlines.

*Current solution*: In the 1990s, faster machines have allowed the information systems to keep up with the increased transaction rates, even though the on-line systems require additional processing power. Notwithstanding the advances in technology, up-to-date reporting of transactions and their impact on accounting reports is not universally available. This is a consequence of continued difficulty in getting all relevant accounting data assembled and entered. As discussed, input device technology has progressed significantly towards the goal of complete on-line entry of data at the source. Unfortunately, it has not solved this problem for the myriad of sources that the general ledger requires in order to generate its standard periodic reports.

**(e) Uneven Workload.** *Problem*: Traditionally, accounting departments have had uneven workloads. During events such as the closing of the books, physical inventory, and/or audits the department suffers from an overload; the rest of the time the workload tends to be relatively slow. In large part, problems of the reporting lags and the uneven workloads are a direct result of the fact that the accounting department is not where transactions actually occur. The purchasing department orders the merchandise and it is received by the inventory department. Accounting gets the paperwork from both departments and must further process it (1) so that payment can be made for the goods ordered and (2) to update the accounting records. In organizations that do not have fully integrated systems, the occurrence of a transaction, the transmittal of paperwork relating to that transaction, and the updating of accounting records do not occur simultaneously. Prior to monthly closings, accounting records must be brought up to date. This requires accounting clerks to go hunting for the documentation elsewhere in the organization, process it, and then enter it so that accounting records are ready for the monthly closing. Sources of the accounting department logjams are often external to the department. Consequently, when the deadline of monthly closing approaches and/or when auditors arrive, the workloads of the accounting department increase considerably; this can lead to errors in the reports generated.

*Proposed solution*: Giovinazzo noted that possible causes of the buildup could be (1) late arrival of data from outside the accounting department or (2) inappropriate processing procedures within the accounting department. As a solution to the problem, he suggested analyzing how the work is done over a period of time and determining what factors cause the logjams. Measuring the quantity of work as well as measuring the error rates could help isolate and eliminate the logjams. Such work measurements could also support the argument for additional staffing, at least during peak work periods. In addition to analyzing how accounting work is done, he suggested that the problem may also be abated by convincing the operating departments to adhere to transmittal deadlines and data accuracy.

*Current solution*: With the computerized systems now available, there is no reason for lags between transactions and the updating of the accounting records. Since the system automatically updates the accounting records, there should not be a need to manually update the books before the monthly closings. Given the availability of such technology and the improved means of communication now available, work flow within the accounting department can be more evenly paced.

**(f)  Underestimation of Time Requirements.**   *Problem*: The problem of not estimating time requirements correctly results because accountants have to pay attention to many details, which can be time consuming and require close scrutiny. Additionally, verification and follow-up can take time because they involve personnel outside the accounting department, who may not cooperate. Finally, changes in plans are always being made.

*Proposed solution*: No solution was specifically suggested for this problem; however, the work measurement recommended to even the work flow within the accounting department can also lead to better estimates of the time required for various assignments.

*Current solution*: These are not IT-related problems, however even here E-mail, voice mail, and other IT-related means of communicating can reduce the time required to get discrepancies resolved and/or questions answered. In addition, robust systems should provide for much higher levels of accuracy, thus reducing the need for close review of the data.

**(g)  Human Relations Problems.**   *Problem*: Reference has already been made to the tensions between the accounting department and the operating departments. In addition, resentment between accounting and information systems (IS) personnel is also common. Friction between accountants and IS personnel is the result of factors such as:

- The perception by accounting personnel that their needs were not accorded the proper priority by the IS department.
- The failure of either group to remain on schedule.
- Accountants' perception that IS keeps changing the systems.
- Too much technical jargon in both departments makes communication difficult.
- Accountants lack IS expertise, so they fail to understand the technological constraints.

- IS is seen as taking over some of the functions previously done by accountants; this leads to "turf battles" between accountants and IS personnel.

*Proposed solution*: Giovinazzo suggested that accountants acquire better communication skills and show greater understanding toward the operating personnel and their work constraints. Efforts should be made to have accounting reports serve the needs of the operating managers. Relevant reports would improve the attitude of the operating managers toward accounting and make them more willing to work with accounting. Accountants should also be more involved in the design of the IS system(s).

*Current solution*: This particular problem has abated for a variety of reasons. Some of the accounting IS needs have been reduced because of software available to accountants. Principally through advanced use of spreadsheets, accountants have been able to fend for themselves in terms of data processing needs. In addition, database management system query languages provide a means for non-programmers to retrieve information from the database in an ad hoc manner. These two features are part of a general trend, called end-user computing; this approach encourages the user to retrieve, compute, and display information in any manner they choose without the need of the IS staff.

In addition, accounting personnel, by training and experience, have become more computer literate. This means they are able to use IT more readily and they are more familiar with the jargon. To a lesser extent, IS personnel are becoming more accounting literate. IS departments have become more concerned with having personnel that not only understand technology but also the business problems that they are being asked to address.

## 5.5 ASSESSING THE PROMISE OF IT

In general, it is difficult to assess the success achieved by using IT. Part of the problem results from the fact that many benefits associated with IT are intangible. We sought to get around the difficulties by using a 1982 article referring to commonly occurring accounting problems as our point of departure. The extent to which these problems can be resolved using the IT available in the 1990s is convincing evidence that IT's promises are not just marketing hype.

At the time Giovinazzo wrote his article, the technology was less advanced. Using the IT then available, Giovinazzo recognized the need for (1) the use of integrated systems, (2) the importance of on-line capabilities, and (3) a greater involvement of the accountants in the design and operation of the IS systems. As one looks at the current environment and technology, it is clear that the IT currently available has reduced the problem of accuracy and speed. Today, it is feasible to have on-line, integrated accounting systems with built-in controls available to ensure accurate data input. The price/performance ratios are improving and the systems are easier to use. In addition, accountants are becoming more knowledgeable about IT. The predictions made about technology have been greatly exceeded, and today it is entirely possible to find IT solutions for problems dealing with data acquisition, manipulation, speed of processing, and analysis. Using IT, source data vulnerability, nonintegrated accounting systems, and uneven workloads can be virtually eliminated.

The remaining four of the seven problems mentioned cannot be completely resolved through IT because they are people problems, but here too, IT has helped. This was also perceived by Giovinazzo. He expected the main thrust for solving the problems to come from nontechnical steps such as:

- Better forms to reduce data entry errors.
- Cooperation between information producers and information users to ensure that the information being reported is relevant.
- Better relationships between accountants and operating departments so that data concerning operations could be communicated expeditiously.
- Need to analyze and measure work done by accounting workers with a view toward the elimination of logjams.

Given this analysis, it's clear that technical problems have been greatly reduced, but the same is not true of other, nontechnical problems. In other words, problems that involve routine, repetitive work have been virtually eliminated. But problems that involve the system itself or deal with human behavior remain unsolved. If the procedures required by the system are cumbersome, they are not likely to become less so just by giving workers computers instead of hand-held calculators, and spreadsheets in place of twelve-column ledger sheets. Many operating departments also remain unconvinced of the importance of the accounting reports. Loud are the voices being increasingly raised to complain about the relevance of accounting reports to the needs of many of its users in the 1990s.

These problems are not likely to disappear as long as the current structures remain, and they cannot be made to disappear by IT alone. As long as an organization is structured in terms of departments, and performance is judged on a departmental basis, there are likely to be turf battles and struggles over getting the maximum share of the resources for one's own department. Furthermore, such departmentalization promotes partial answers rather than innovative, comprehensive solutions.

Techniques such as total quality management, activity-based costing, and, most notably, reengineering have been shown to have considerable success with the systemic problems encountered by accounting departments when they are used in conjunction with the IT now available. Using these proven tools, managers can help reevaluate the accounting procedures being currently used in the hope of eliminating procedural gridlocks and redefine the procedures documentation process.

## 5.6 CONCLUSION

Clearly, IT is useful and can be even more so. Indeed, the technology has almost always worked. Applications of IT fail because of a misunderstanding of the problems, improper application/implementation of the technology, or the embryonic nature of the IT itself.

What are the chances of a "misfit" between IT purchased today and the problems to be addressed? It's best to first identify the needs that IT is expected to satisfy. The likelihood of an IT failure can be greatly reduced by a clear understanding of the needs to be met, careful planning of the changeover, and proper training of the employees. Merely purchasing computers will not make a business operate like clockwork. When we buy clothes, we try them on for size; when doctors prescribe medication, they make it specific to the symptoms. The same holds true for IT and IT applications.

IT cannot help accounting information become more relevant or make the controller's department become more productive if the same old way of doing business survives. If this happens, then IT will only have replaced the calculators with a more expensive alternative that may not pay back the investment made to acquire it.

To make optimum use of the emergent technology, the accounting procedures must be redesigned keeping in mind the IT presently available. If one is buying IT to solve systemic problems, it is better to first address the systemic problem itself and, while investigating the problem, explore the potential use of IT to solve the problem or to design a way around it. To better address systemic problems, changes in accounting functions are needed, and such changes can be made by applying cross-functional approaches to many accounting functions.

Indeed, to facilitate changes in accounting functions, it may be better to think of accounting as a part of the information resource management function, rather than as a distinct entity. Audit and internal control constraints may dictate the separation of accounting and IS, however there ought to be greater coordination between the two departments in order to ensure effective management of information resources. The accounting department will never be trouble-free, but it can become a more optimally used resource for decision makers in the information age if its functions are redesigned with the help of emerging IT.

## ENDNOTES

1. "Accounting Problems Nobody Talks About." *Management Accounting*. (October 1982).

# CHAPTER 6

# TQE AND THE REPERTOIRE
# OF THE DEMING APPROACH

## 6.1 INTRODUCTION

This chapter seeks to advance the techniques that help organizations attain the evolution-ary competence described in Chapter 4. The focus in this chapter is on providing an overview of total quality environment (TQE), which will allow the readers to incorporate elements of it into the process governing the documentation of accounting procedures manual. Naming this chapter after Edward Deming is a recognition of the fact that the contemporary concern of managers around the world for implementing TQE—also referred to as *kaizen*, continuous improvement, and total quality management—within their respective organizations is largely owing to his efforts.

TQE has spawned other approaches to improve work, such as process redesign, activity-based costing, and the learning organization paradigm. These are discussed in Chapters 7, 8, and 9. There is ample literature describing success stories stemming from the use of these techniques. This evidence helps to substantiate the value and practicality of using TQE, process redesign, activity-based costing, and the learning organization paradigm to enhance organizational performance and processes. In Chapter 10, we will show how these various approaches can be used to redefine the procedure documentation process itself.

## 6.2 WHAT IS TQE?

The focus of TQE is quality, but it must not be taken to mean simply a freedom from defects. Instead, organizational TQE should be seen as being concerned with an ongoing self-examination on the part of an organization directed to improve all aspects of its performance. TQE is concerned on a continuous basis with doing things better, not just in a factory but in offices, schools, and even our homes. The starting point for such continuous efforts is self-knowledge: that is, an understanding of what organizational components do, why they do it, who does it, where they do it, and the resources it requires to get done. The organizational, departmental, and individual self-knowledge sought by TQE is not couched in abstract terms. In other words, statistics and quantitative perfor-mance measures serve as surrogates to facilitate an understanding of what, why, who, where, and how. To make organizational understanding tangible, performance measures drawn from statistics are brought into the picture. Using statistics, it is possible to portray and forecast how a given system operates and see to it that the system continues to operate in an efficient manner.

In effect, statistics help to isolate variances in managerial and production processes. They help to identify exceptions that are random and not subject to control, as well as variances that are due to a given source or a malfunction in the production or managerial process. This permits supervisors to practice management by exception and allocate their attention to problems that can be controlled, if not solved. In contrast to accounting standard cost variances, the TQE variances are not expressed in monetary amounts. Instead of translating production measures into the relatively unfamiliar jargon of accounting terminology, they describe production through terms such as labor time, time needed to complete a batch, material usage, number of units produced, and number of units rejected. Using the language of the production department makes the production-related exceptions more familiar to those responsible, and thus allows easier solutions for those exceptions.

Converting various production factors into monetary amounts does not provide any greater insight to those interested in those factors. The number of hours a process took is easier to represent than the number of hours times price per hour, because price is subject to a greater degree of fluctuation than are the hours needed to complete a process. For airlines, graphing a distribution consisting of fuel consumed for each mile flown is more predictable than a distribution of fuel cost per mile flown.

## 6.3 NOT BEING CONTENT WITH THE STATUS QUO

In addition to the concern with self-awareness, the TQE is also based on a belief that an organization and its various components cannot remain still. Instead of being content with the status quo, TQE seeks ongoing evolution, continuous improvement. Like self-knowledge, self-improvement is also meant to be tangible and measurable. To become tangible, improvements are described in statistical measures. Improvement in the safety record can be measured by the reduction in the number of actual accidents that occur.

The Deming-inspired gospel of TQE does not subscribe to the notion of maintaining the status quo, instead, it urges that the system should be improved continuously. Maintaining the status quo can only lead to obsolescence. Continuous improvement of the system is facilitated through a constancy of focus and an understanding of the mission.

In acquiring and constantly renewing the understanding of their mission, individuals and organizations must be able to go beyond the obvious. Universities may see their role as teaching accounting, when in fact they are teaching the means to measure and report about an organization's present performance and its future prospects in monetary terms. Those involved in the business of making horse buggies may have thought they were making buggies, when in fact they were providing people with means of transportation. Those who could not see beyond horses and buggies became obsolete, but those who saw themselves as being involved in the transportation business moved on to the horseless carriages of the twentieth century. Fiber optics may well be the horseless carriage of the twenty-first century, accomplishing business interaction without requiring the hassles and hazards of travel. By transporting work to the employees, instead of the employees to work, technology can and is redefining the focus of transportation.

## 6.4 CHANGE IS INEXORABLE

A major part of being focused in a continuously changing environment is keeping up with customers, knowing not just their current needs but also being in tune with their potential future wants. Customers' preferences are subject to redefinition with the changing of environment. A critical factor for TQE is the realization that nothing can or ought to remain unchanged. Instead, those responsible for doing the work should strive for constant improvement in the way things are done. To improve processes, it is important to first understand them. Secondly, within an organization there must be uniformity regarding how specific processes are performed. It is the function of management to see that all workers are following standard operating procedures. Through policies, rules, directives, training, and, if need be, disciplinary action, employees must be made to follow the standard procedures.

After such uniformity is achieved, managers and employees ought to focus on improving the process by doing things better. Improving means setting higher standards and then making sure they are met. Meeting the standards involves solving problems and overcoming obstacles. Furthermore, the attainment of set standards is not judged entirely on the basis of end results achieved. Managers seeking to implement TQE must not limit themselves to judging by the results, instead they should measure the efforts that go into improving the process. A sales manager may not limit himself to just evaluating the units sold, instead, he must also analyze the time spent on cultivating rapport with customers, developing new customers, and examining the sales approach of the individual salespeople. TQE does not limit itself to the summary reports, the proverbial bottom line, instead, it is also oriented to the underlying efforts.

Following the establishment and attainment of higher standards, it is necessary to ensure that the improvements are monitored on an ongoing basis. Such maintenance of the improved performance level is akin to catching one's breath before starting the upward spiral movement again. The ongoing climb up the spiral is done through what is known as the Deming cycle, which consists of the following steps:

Plan: Decide what needs to be done: define the problem, analyze it, identify its causes, and develop solutions

Do: Implement the plan agreed upon and decide on measures that will help judge and evaluate the implementation

Check: See that the implementation of the plan is having the desired impact

Act: Explore to see what could be improved upon next

In business parlance, the Deming cycle's four steps correspond to designing a product, producing the product, selling it, and then researching possible improvements in the product itself or in the way it is made.

Self-awareness and the perpetual search for improvement is the goal sought by the Deming cycle, calling for a Plan-Do-Check-Action approach on a continuous basis. Self-examination and the desire to improve leads to a plan that can be implemented. Such implementation must then be checked and measured to see if it follows the intended course of action. Once the desired improvement is firmly in place, an organization or its various components must once again subject itself to self-examination and a further search for improvement, which will also be implemented, checked, and measured to be followed by further self-examinations and quests for improvement.

## 6.5 THE SYSTEM AS A PROCESS

The quest for TQE is facilitated by seeing the organization as a system involved in interlinked processes, with beginnings and endings. Manufacturing or office systems are made up of several such interlinked and interdependent processes. Each process transforms input it receives from suppliers or an earlier process into something different by adding value to the input before sending it forth as an output. The problem in one process may have its origin somewhere upstream, and this origin may not be discovered, unless the interdependency of the processes is understood.

TQE's concern for organizational improvement is manifested in a realization that problems exist, but can be resolved by the cooperation of all those involved in the entire process, not just those directly responsible for problems. Problem resolution can be facilitated by addressing not just the symptoms, but also the underlying causes that are responsible for the existence of the problem. For example, 20 percent of the production may be unacceptable during a typical week. Following such diagnosis, steps can be taken to ensure that not only the symptoms but the underlying factors responsible for the unacceptable production are eliminated. Ideally, there should be no defects, but if they do occur, then the defective production must be eliminated as close to the source as possible. The further downstream a problem is caught, the more expensive it is to the business. Additionally, an organization wastes money and resources when it is forced to have a significant amount of the workforce devoted to inspecting and catching mistakes made upstream. Quality should be the concern of everyone in the system, not just the quality inspectors.

TQE proceeds under the assumption that the unacceptable production is merely a symptom of underlying problems that exist within the system. Problems occurring at one stage of production can have roots in another part of the system. If accounting clerks are making too many mistakes entering data in the journals, then it may well be due to faulty design of the source documents and not the carelessness or the incompetence of the clerks.

Addressing the underlying causes may seem like a course of action calling for additional expenses and a major commitment of resources. However, such a course of action directed at improvement in quality can actually lead to lower costs. Quality should be the focus throughout the system, but quality cannot be assured by just inspecting the finished product or spending more money doing things the same old way.

## 6.6 CROSS-FUNCTIONAL COOPERATION, NOT DEPARTMENTAL COMPETITION

TQE thinking does not promote the isolating of a department. Instead of niche obsessions and turf protections, it calls for cross-functional approaches to quality assurance, cost reduction, safety, product innovation, and vendor relations, as well as in meeting production quotas and delivery schedules. If the system is to work well and if it is to improve, the individual functions that comprise a process must cooperate and not compete. The cogs must move in unison if the wheel is to continue its movements. Activities such as individual performance evaluation or performance rating can direct attention away from the system. They can make workers compete with each other, instead of working for the good of the system. Management must lead, but should do so by getting all the components of the system to work together instead of competing with each other.

Such cross-functional cooperation is made easier when workers understand what they are doing and why they do it. They must not be simply trained, but educated—make everyone smarter, so that they will work smarter. Being educated means acquiring an expertise that allows employees to see not just the symptoms of a problem, but also gives them the motivation to learn the underlying causes responsible for the problematic exceptions and variances. Workers must be educated to see not just the twigs and the leaves on an individual branch, but to reflect upon the entire tree (of which the branch, with its twigs and leaves, is only a small part) and even the forest, together with its environment.

Cooperation applies not only to internal processes, such as different departments and responsibility centers within the systems, but also to the external processes linked with a given system, such as the vendors, who supply the raw material for a factory or consumer goods for a department store. Customers can be those buying the finished product or they can be the employees working on the product-in-process at the next stage of its manufacture. The customers, be they internal or external, should not be treated as adversaries but as an integral extension of the system. The vendors upstream, as well as the customers downstream, must be seen as part of the system, as partners, whose satisfaction should be the motivating force as well as the "weather vane" for the firm.

Customers' satisfaction is the most important factor a business must consider, notwithstanding its quirkiness and irritating unpredictability. A business's reason for being is making sure that its customers are satisfied enough to keep coming back for more. If anything can be described as representing the core of TQE, it would be the desire to satisfy the customers and to remove whatever inconveniences them.

## 6.7 WHO DOES WHAT IN TQE?

Typically, organizations are made up of layers of top managers, middle managers, supervisors, and employees. The numbers in each organizational layer follow an increasing order—one queen bee, presumably a few supervisor bees, and many workers. In bringing about continuous improvement and keeping the organizational hive functional, all of them have roles to play. In the context of TQE, the functions of each layer are as follows:

| | |
|---|---|
| Top managers: | Must decide a strategy for improvement and then let their plans be known throughout the firm. They must provide support and direction as well as conduct ongoing audits to see that the policy as established and plans as drafted are being implemented. Without their total commitment to its implementation, the structure of organizational TQE could never get off the ground. It is their responsibility to make sure that the middle management and supervisors are also dedicated to implementing TQE. |
| Middle managers: | Help to bring about a cross-functional perspective. They also establish, maintain, and ensure the upgrading of standards. This may require them to conduct skill audits to see that workers have the skills needed to implement plans. When TQE is getting off its foundation, the middle managers must learn that there is a culture change underway that empowers the employees to use their initiative, downplays organizational chain of authority, and fosters team efforts. The efforts directed at culture change, so crucial to the success of TQE, can be sabotaged if the middle managers insist on maintaining rigid organizational structures. |
| Supervisors: | They serve as the link between managers and workers. They may be called upon to maintain discipline, suggest improvements, and explain to workers what needs to be done. The supervisors, like the middle managers, must promote team spirit rather than insisting on playing the role of authoritarian taskmasters. |
| Workers: | They carry out established procedures, but at the same time are empowered and encouraged to use their work experience to suggest ways to improve the process. They also engage in self-improvement through education and training. |

## 6.8 IMPLEMENTATION LEVELS FOR TQE

Improvements cannot be implemented only at the management level, since they must filter down throughout the organization to be effective. The implementation of TQE requires tailoring the implementation to suit the organizational layer where it is being implemented. Some TQE experts argue that TQE's implementation at the management level, segment/process level, and individual level will require different perspectives. Implementation at the three levels will use similar tools and will be interdependent, however, the three respective implementations differ in their scope and duration. Management-level TQE implementation lasts for the entire duration of a project, which may be spread over several years, while the segment-level project may last only a few months, and is limited within the segment. The action at the individual level may be instantaneous, and is limited to a specific employee. At the management level, managers and professionals are integrally involved in the TQE approach; at the segment level, it is likely to be a group of people, such as those working in a quality circle; at the individual level, each employee works for TQE individually, albeit in a supportive environment, accompanied by others who are also working for TQE.

The costs of improvements at the individual level are relatively low, but are higher at the group and the management level. Changing the factory layout can only be authorized at the management level; decreasing the level of inventory maintained or adopting the just-in-time approach are also management-level decisions. Improving processes through quality circles is an example of segment-level TQE, while a suggestion system exemplifies individual-level TQE.

## 6.9 SETTING THE STAGE FOR TQE

Introducing TQE into an organization almost always involves a culture change. Such a change requires profound efforts. A survey of the literature describing implementation of TQE around the world suggests that the following factors should be considered prior to implementing TQE:

*Concern for the customers.*   TQE starts with the willingness to place customers' satisfaction as the motivation, the raison d'être, for all organizational efforts. Customer satisfaction must be measured in terms that reflect quality, price, and delivery. Efforts must be made to monitor customers' perceptions and to measure them on an ongoing basis.

*Recognizing the links.*   A fundamental tenet of TQE has been the knowledge that product or service quality downstream is best attained by helping to inspire it upstream. If the high schools do a poor job, then the colleges will have a much harder time ensuring that their graduates are properly educated. The concept holds not only for processes within the firm, but also applies to the relationships between the firm and its suppliers and customers. The efforts to implement TQE must include every link in the chain. To ensure that, a firm must not think of the vendors who supply the raw materials or those who actually sell its products as adversaries.

*Getting everyone involved.*   Recently, the College of Business of a mid-sized, privately owned American university sought to implement TQE. The dean, prodded by his associates, agreed that the initial step needed was to hold a faculty retreat to come up with ideas and to set the stage. For economic reasons, only 40 percent of the faculty were invited to the weekend retreat held at a resort off-campus. Unfortunately, this favoritism was seen negatively by those who were not invited. Many of the uninvited saw their exclusion as akin to setting up a caste system, in which some were "in" and others were "out." As a result, those particular efforts at implementing TQE at the College of Business were doomed. In addition to universal inclusion, getting everyone to subscribe to the tenets of TQE requires open communication throughout the organization. It also calls for a relationship based on trust, loyalty, and cooperation between managers and employees. In committees that bring together supervisors and workers, it is imperative that supervisors stop supervising and deal with their employees in an egalitarian spirit. Otherwise, they will inhibit trust and communication.

*Process orientation.*    Hitherto, the preferred organizational structure has been a departmental arrangement, but the focus of TQE is on processes that cross departmental lines. Accounting took care of the numbers that were reported in financial statements, while management information systems (MIS) tackled the machinery that processed the data into financial statements. The medium and the message went their separate ways, given the way accounting and systems guarded their respective areas under the prevailing departmental focus. However, a process focus can change the procedures and processes that comprise the accounting systems, as well as the relationship between accounting and MIS. Getting managers, who were trained as relatively narrow specialists, to adopt a cross-functional process orientation is a demanding task, and calls for changes in an organization's planning and control systems as well as alternative means to measure and reward managerial performance.

*Longer planning horizons.*    American managers have become notorious for their short-term focus. But reaping the potential of TQE does not occur overnight, it requires extended cultivation and care. To benefit from TQE, managers must extend their horizons beyond improvement in the bottom line for the next quarter and they must be allowed and encouraged to do so by the directors and the investors.

*Willingness to seek out problems.*    Upon confronting problems, the typical manager's tendency is to ignore its existence. Managers and employees fear that they will be blamed for the problems they uncover. In addition to pretending that problems do not even exist, procrastination as to their solution seems the preferred choice of many supervisors and workers. Under TQE, problems are seen as opportunities for improvement rather than as skeletons that must be locked up in closets. The organization must encourage, and even train, its workers to find problems and then help find solutions in order to make TQE become a reality.

*Taking time out to think through the work.*    In addition to being willing to continuously confront problems and exploit opportunities for improvement, managers have to provide occasions to do so. While employees are consumed with everyday routines, they don't have time to seek out exceptions. TQE can only be achieved by stepping back and looking at routines and standard operating procedures with a new perspective. This requires provision for reflective occasions, retraining, continuing education, and intermingling of specializations and diversity of cultures. Learning is facilitated by institutionalized occasions and structures for it.

*Top management's commitment.*   All the factors listed require substantial and sustained efforts throughout the organization. They require a focus away from adversarial relationships and short-term focus. But what they require more than anything else is the dedication and wholehearted subscription of the senior management to the idea and culture of TQE. Some questions that must be raised, and which require positive answers, before implementing TQE are:

- Is top management dedicated to TQE as a corporate strategy?
- Is it committed to spending enough resources to really understand the implication of TQE for itself and the firm?
- Is it committed to introducing cross-functional focus in place of a niche approach?
- Is it willing to adopt the controls and corporate structures that will support cross-functional process emphasis?
- Is top management dedicated to making TQE an ongoing, intrinsic part of the organizational culture rather than a one-shot, seasonal fad?

## 6.10 IMPLEMENTING TQE

No matter how well intentioned the members of an organization are, they cannot just will TQE into being. Sustained efforts are needed to implement TQE. In this section, we'll discuss what is required to implement quality management.

The improvement must start somewhere, and suggestions originating at any implementation level can set in motion the improvement process. Through suggestions provided by individuals, working on their own or in teams, almost every phase of work can be improved, among them:

- Ideas for new products
- Improvement in the use of tools
- Faster completion of processes
- Improvement in customer services
- More efficient use of resources
- Higher productivity and better safety records
- A more pleasant and less monotonous environment

While it is best to start with the problems, it is very common for workers to fail to see the problems. First, managers must educate workers to look for problems in their work routines.

The discovery of problems and/or opportunities can be made easier if daily routines are viewed with different perspectives. Putting groups of employees together makes it possible to look at old habits under a new light. Groups that are drawn from across the departments are even more likely to see old routines with new vision and think through the replacement of out-dated ways of doing things. Cross-functional teams can not only unearth problems but they can also help discover cutting-edge solutions.

Cross-functional management is central to the organizational implementation of a total quality program. Traditionally, companies are organized along functional lines such as design, production, marketing, and accounting. Each department works on a given, assigned function, and is, for all practical purposes, a separate entity unto itself. Such traditional hierarchies can be suboptimal when it concerns implementation of TQE. As a result, cross-functional management teams have evolved to help expedite policy implementation throughout the organization.

Goals having to do with quality assurance, cost reduction, and production scheduling cannot be limited to a given department: most, if not all, functional departments must get involved in order to attain them. This involvement can be accomplished through cross-functional teams. Functional departments can be assigned the responsibility of implementing tasks by the cross-functional teams. Cross-functional teams also help to minimize the impact of turf fighting among the functional departments.

Yet another reason for the use of cross-functional teams is to minimize the disruption that can occur if an organization is trying to continue with business while seeking to implement changes associated with TQE. Functional departments can maintain the current operations, while the cross-functional teams seek out ways to improve and to implement them in an optimum manner. The cross-functional team members can serve as the liaison for their respective departments, keeping others informed about the team's progress.

The following points also need to be mentioned in connection with cross-functional (CF) teams:

- Cross-functional teams are organized at the top-management level. They may be assigned specific tasks, such as new product development, product design, raw material acquisition, sales inspection, profit management, and quality assurance.
- Cross-functional goals should be established before the determination of department goals. Using the organization-wide CF goals, each level develops its own goals that will help reach the overall goals drawn up by the CF teams.
- Cross-functional policy should be formulated in terms of goals before it is deployed. The same applies to the goals developed at each level within the organization.
- The progress made in the implementation of the CF goals should be periodically audited to ensure that they are indeed being implemented. Such audits may start with the top management auditing the divisions, divisions auditing their segments, and so on until all levels are touched.
- Cross-functional teams also decide among various options. They are the ones who set priorities.

## 6.11  POLICY IMPLEMENTATION

The cross-functional teams develop the overall policies, which are then assigned to the respective departments for implementation. Prior years' performance serves as the yardstick for deciding new goals and before settling on new goals, potential obstacles to their implementation at each level are considered.

The policy and mission statements may begin in abstract terms at the top but become more concrete as they move downward. Policy statements are formulated differently at various levels:

Top
management: Issues general statement of direction for change, expressed qualitatively; sets priorities; allocates resources between divisions

Divisional
management: Elaborates the top management's statement in quantitative terms
Middle
management: Sets specific goals in quantitative terms
Supervisors:   See to it that specific actions are undertaken by the employees working for them

It is not sufficient to just verbalize policy and plans: they will remain empty rhetoric unless properly implemented. Managers and supervisors must work with their employees to make sure that the planned goals are being implemented. Implementation must also be seen as a collective act that involves the entire organization. It is critical that managers discuss plans with their supervisors as well as their employees, they do not do the planning in isolation from the levels above and below them.

Policy development can be facilitated by using set formats that help document the policy and the measures. Managers can use forms to standardize the work that goes into developing the plans envisaged by the policy. Such forms may include the following fields of information:

- Top managers' long-range strategy
- Top managers' short-range objectives
- Last year's departmental goals and objectives
- Last year's actual results for the department
- This year's goals and measures
- Schedule
- Task assignments

Management can facilitate policy implementation in two ways: first, it can make sure that goals are accompanied by specific control measures, actions that need to be taken at each level of the organization. Secondly, management must periodically conduct audits to see if the implementation of the policy is proceeding as planned. For example, top management may decide that delays in paying vendors ought to be eliminated. This would represent the policy. To implement it, managers must decide on measures that will indicate that the policy is being implemented. One such measure could be the number of bills that were paid after their payment date. The successful implementation of the agreed policy will mean that there were no late payments of invoices.

Audits to investigate that implementation is proceeding as planned are not meant to find scapegoats, rather they must be seen as diagnostic sessions. They are meant to help workers recognize where the process may be experiencing problems. They are meant to expose what is wrong, not to assign blame.

### 6.12  CONTROL POINTS AND CHECK POINTS

The control measures that are selected to monitor implementation of plans rely on the use of what are described as control points and check points. These points have their origin in statistical quality control (SQC) charts and the concept of manageable margin. In everyday language they may be translated as given effects and their probable causes. The SQC charts indicate the acceptable and unacceptable ranges. In a service station, management may set up 15 to 20 minutes as the time allowed to change the oil in customers' cars.

Somewhere in the shop, a chart will display the time it actually takes to change oil on customers' vehicles. As long as the time taken to perform the task remains between 15 and 20 minutes, no action is called for by the manager. If the time taken exceeds the acceptable range, it suggests an abnormal situation has arisen and corrective action is called for by the supervisor. Exceeding the acceptable range is an effect, and it prompts the manager to look for the causes of delay.

The likely causes for the delay in oil change may be due to any or all of the following:

- Availability of oil at the workstation
- Delay in getting the paperwork processed
- Poor performance of the workers

In the jargon of SQC, checking the time it took to get the job done will be a control point, while the factors listed above will be check points. At times the control points are called result (R) criteria and check points are called process (P) criteria. The control points are handled with the help of data, while the check points are attended to with the help of a process or the workers involved in the process.

It is important to remember that the control point at one level may be the check point for the level above. For the shop foreman, the time taken is the control point and the three possible causes are the check points. For the worker who does the paperwork, the control point may be the time it takes to do the paperwork and the check point may be knowing who will do the oil change itself.

## 6.13 TOOLS USED FOR TQE

Even though TQE can help an organization be at the cutting edge, the tools used for implementing TQE are by no means high tech. Beyond the right attitudes and shared perceptions, the tools and techniques of TQE consist of checklists and charts aimed at inducing desired employee behavior and improving organizational communication. Getting employees and managers to do the right thing requires educating them to notice dysfunctional factors in their work. Below are some commonly used TQE tools and techniques:

*Three-Mu checklists.* One popular technique used by Japanese companies to help expose dysfunctional aspects pertaining to their work is known as the Three-Mu checklist. The checklist has employees look for **Muda**, **Muri**, and **Mura** in the work they do. **Muda** means waste or unproductive use of resource, **Muri** refers to strain or extraordinary exertion of efforts, and **Mura** refers to variances or discrepancies. The Three-Mu exhort workers to look for unproductive use of resources, extraordinary exertion of efforts, and variances between expectation and outcomes. They are to look for **Muda**, **Muri**, and **Mura** in the environment around them. Specifically, they are to look for such symptoms in the specified control points. Below are some typical control points used, along with an example of waste, strain, or variance common to each particular control point:

1. Workers employed → Excess number of workers
2. Techniques used → Time allowed for production seems inadequate
3. Methods applied → Methods used seem too complicated

4. Time expended → Excess of idle time between setups
5. Facilities employed → Takes too long to get material delivered to production centers
6. Tools used → Inventory movements in and out of warehouse are tracked manually
7. Materials → Using material in excess of that allowed by specification
8. Output generated → High rejection rates
9. Inventory → Too much diversity in parts specification
10. Environment → Toxic substances are not controlled
11. Way of thinking → Discrimination based on gender or ethnic origin

*Questionnaires.*    Journalism students are instructed to get the who, what, where, when, why, and how of their news stories in the opening sentences of the stories they write for their newspaper. Japanese companies use the same questions to analyze processes and to discover problems and/or opportunities for improvement. Often, the questionnaires are used in conjunction with the Three-Mu checklist described previously. Questions may pertain to the entire process or to a small task involved in a larger process. Alternatively, they may refer to the operators, facilities, machines, and materials used, to operating techniques, or even to performance measures employed. They may be formulated as follows:

- Who is doing it? Who should be doing it? Who else can do it?
- What is being done? What should be done? What else can be done?
- Where is it being done? Where should it be done? Where else can it be done?
- When is it done? When should it be done? Can it be done at some other time?
- Why is it done? Why is it done a given way or at a given time? Can it be done in another way or at another time?
- How is it done? How should it be done? Can it be done in some other way?

Clearly, the questions asked of a control point, be it a task or the entire process, are meant to increase workers knowledge of it and to change their perceptions and perspectives.

*Visual Aids: Charts, Graphs, and Diagrams.*    In the first section of this chapter we briefly discussed the role of statistics in a TQE environment. Here we describe some of the ways in which statistics is brought to production and operating processes. Note that most of these are visual tools, which are invaluable in making statistics understandable for all employees. By using visual tools such as those described below, Deming's gospel becomes easier to implement in production and other organizational environments.

*Pareto diagrams.*    These diagrams classify problems according to cause and phenomenon. The problems are diagrammed according to priority, using a bar-graph format, with 100 percent indicating the total value lost. Value lost represents the phenomena which is plotted against different factors that are responsible for the value lost. A number of factors may be causing the units to be rejected at a particular process. By plotting them on a graph, one can learn which factors cause the most rejects. It may be that of a total of ten possible factors, three are responsible for up to 80 percent of the rejected units. Such graphic representations help isolate factors/problems responsible for most of the value lost.

*Cause-and-effect diagrams.* These diagrams are used to analyze the characteristics of a process or situation and the factors that contribute to them. Cause-and-effect diagrams may be accompanied with a Pareto diagram to help isolate and rank causes for most of the damage. Flow charts describing the steps involved in managerial processes are a variation of cause-and-effect diagrams that can aid employees' understanding of their work.

*Graphs.* There are many kinds of graphs employed, depending on the shape desired and the purpose of analysis. Bar graphs compare values via parallel bars, while line graphs are used to illustrate variations over a period of time. Circle graphs indicate the categorical breakdown of values, and radar charts assist in the analysis of previously evaluated items. Frequency distributions, histograms, and scatter diagrams are used to describe organizational processes, in addition to ordinary graphs.

*Control charts.* Perhaps the most important of the visual aids in use are control charts. Control charts serve to detect abnormal trends with the help of line graphs. These graphs differ from standard line graphs in that they have control limit lines at the center, top, and bottom levels. Sample data are plotted in dots on the graph to evaluate process situations and trends. These charts seek to differentiate between two types of variations: the inevitable variations that occur under normal conditions and those that can be traced to a cause. The latter variations, often called "abnormal," are the ones that set off remedial actions. Control charts also help in pointing out developing trends, which could become the reason for countermeasures by employees. These visual tools are widely used by quality circles and other small groups, as well as by staff engineers and managers, for identifying and solving problems. Since most of them use statistical analysis, employees have to be trained to use these tools in their routine activities. Such training must be provided to the employees by the firm, and indeed, should be a part of the efforts to help employees develop themselves.

*Slogans, signs, and ceremonies.* Human beings like to be reminded about things that need doing. The commercials blaring forth from radios are proof that timely reminders help in getting human beings to act in a given way. Similarly, slogans, signs, and ceremonies are used extensively by Japanese firms employing TQE to reinforce employees' willingness to improve the work they do. Slogans may be used to urge workers to organize their workstation, the tools they use, and even their work habits. Awards and recognition ceremonies are also very much a part of TQE implementation.

All these tools and others yet to be discovered do not help if they are not used. It is imperative to provide workers with time out from the daily grind to let them think through the work they do. Without these reflective intervals, whether they're in the form of informal meetings, in quality circles, or in focus groups, TQE cannot be attained. It is also best to initiate the search for TQE in good times, because once the bad times take over a company, its employees may be too immersed in putting out the brush fires to be distracted by the promises of TQE.

## 6.14 CONCLUSION

Terms like quality circles and continuous improvement and acronyms like TQE and TQM are very much a part of the current managerial landscape. Their spread attests to their efficacy. Quality improvement is not mere hype, it is useful. Indeed, what keeps it from becoming just another fad is the fact that TQE does work. It works well enough to put at risk of oblivion those firms that do not value continuous improvement in what they do and how they do it.

The survey of TQE concludes with the reiteration that it is concerned on a continuous basis with organizational improvements brought about by a commitment on the part of individuals, acting alone and in groups, to seek a better understanding of their work and to do it better. The starting point for such continuous efforts directed towards evolution is self-knowledge—an understanding of the organizational/departmental goals and objectives in terms of what, why, who, where, when, and how. Such awareness culminates in an unceasing search to evolve, not by means of expensive technology or through adoption of fads, and certainly not by mere wishful thinking, but through dedication, by the application of good sense, and by taking on the self-imposed obligation to provide quality service and value to customers, be they in another country or in the department located down the hall.

# CHAPTER 7

# BUSINESS PROCESS REDESIGN

## 7.1 INTRODUCTION

This chapter features a discussion of techniques that organizations can use to improve their evolutionary competence. We survey business process redesign (BPR), more popularly known as reengineering. Business process redesign has much in common with the Deming approach, the subject of the previous chapter, though it requires the use of information technology in addition to the tools of TQE. It has a different impact, and requires a somewhat different sort of implementation.

## 7.2 WHAT IS BPR?

Business process redesign, also known as reengineering and process innovation, is based on the premise that by exploring the what, when, where, why, who, and how of processes, an organization can achieve exponential productivity gains. In other words, the reengineering process starts by asking questions about the organization's activities. Following such inquiries into the nature of processes and procedures, BPR focuses on attaining the same goals by alternative, streamlined, more efficient, and presumably more technologically advanced means. Fundamental to such redesign is the desire to eliminate the non-value-adding steps from the process and to approach it from a cross-functional, rather than a departmental, perspective. Through focused reexamination and by using emergent information technology (IT), an organization can systemically change the way its work is performed and enhance its evolutionary competence.

A large number of current work practices exists because of the tradition that breaks work into its component tasks and then lets different specialists take care of their respective assignments, usually narrow segments of the overall process. Such thinking focuses on individual tasks much more than on the underlying goals or the entire process and the results of this approach are costly delays, lengthy paper trails, and an excess number of clerical employees whose function is to tend to clogged paper trails. The traditional philosophy also promotes departments to fight suboptimally for their own pieces of the procedural turf. Instead of focusing on the entire process, such as getting merchandise into the hands of the customer and collecting payment in the most efficient, cost-effective way, the current organizational philosophy focuses on smaller segments of the process, such as ordering, credit analysis, shipment, billing, collecting, and depositing. The various tasks are assigned to different individuals dispersed throughout the organization. By the time all the tasks get accomplished, customers may well be on their way elsewhere.

A similar compartmentalization governs purchasing and other organizational processes. Under the present setup, a decision to purchase an item is broken into its component tasks:

1. A department makes a formal request to the purchasing department for a specific item, after getting due authorization for the request (a complete, paper-laden process in its own right, often distinct from the purchasing process).

2. The purchasing department then selects the vendor. Such selections can involve solicitation of bids and other such measures.

3. Following vendor selection, the purchasing department orders the goods, with the help of a purchase order—a document that has as many as five copies to be distributed among not just the vendors and those making the requests, but also the receiving, payable, and cash disbursement departments. Because of such voucher outpouring, each of these different-colored siblings of the original document engenders the need for one or more employee to handle their respective copies.

4. When the shipment is received, yet another specialist verifies its accuracy with the help of a copy of the purchase order which, no doubt, has to be hunted down in the paper thicket.

5. Accounting waits to receive the pertinent pieces of paper from internal departments such as purchasing and receiving, as well as the external vendor. After all the matching copies are in the hands of accounts payable personnel, the accountants then issue a voucher, also in multiple copies, authorizing payment to the vendor.

6. This voucher and its technicolor multiples have to travel all over the organization to get approved and be finally paid.

Clearly, a lot of paper shuffling goes on in the paper-clogged trails, but what is worse, and much more costly to the firm, is the waiting between the various steps involved in the entire process. That waiting takes up most of the time spent by vouchers in the paper trail.

One of the most talked about instances of reengineering of the accounts payable process originally occurred at the Ford Motor Company, and has since been duplicated by others. Ford reengineered its process so that the delays were eliminated and cut 375 of the 500 employees who worked in accounts payable. What Ford did was this: first, they developed close relationships with a few vendors so that time was not wasted selecting the right vendor to supply a given item. Secondly, they stopped authorizing payments for purchases after the invoice was received; instead, payments are made when the goods are received. Paperwork is eliminated by creating the purchase orders on computer. These electronic purchase vouchers are available to those in receiving, so that when the order is received, it can be read directly into the voucher within the system. Gone is the waiting for and matching all the copies. The same voucher is available to those in charge of disbursements, so that upon receipt of the order, payments can be processed electronically. Vendors agree to and enter into such relationships because they have a much greater assurance of getting the business and also of getting paid sooner.

Wal-Mart, the retailing behemoth, went the same route as Ford, perhaps traveling even further when it did away with those involved with placement of orders. This way, the replenishment of inventory does not wait for a decision on the part of a narrow specialist who must rely on a paper trail. Instead, suppliers are beeped from Wal-Mart's point-of-sale information system, which is tied directly into their own information process. Inventory replenishment is triggered by the actual sales in real-time, is delivered by the manufacturer to the store where it is needed, and intermediaries such as buyers, distributors, and wholesalers are becoming extinct. The manufacturers themselves can plan production based on sales being reported in real-time, and not through guesswork.

## 7.3 THE IT CONNECTION

The agent of change for Ford, Wal-Mart, and others who have implemented similar systems happens to be the emergent IT. Using such technology, purchasing can do away with its multiple-copies purchase vouchers that require availability of different individuals in various departments to take care of their respective copies. Instead, while placing an order the purchasing agents can now enter it into the on-line system, also called the client-server system. In some cases, as we have seen with Wal-Mart, it does not even have to be entered by purchasing agents, but can be put into the system by the requesting department. When the ordered goods are received, the receiving agent checks for the order on the system (he may actually scan the packing slip with the help of imaging technology now available) and if he finds it, then with the push of a button, he lets the system know that the order was delivered. At this point, the system takes over and coordinates the processsesing as well as the issuing of payments.

BPR is for real, and what is allowing this reality is a revolution in software. Four key technologies are helping to make BPR a reality. They are graphical user interfaces (GUI), networking software, flexible databases, and imaging technology (IT).

*Graphical User Interface.* GUI allows an image, a picture on the screen, to represent the object in the real world. The image of a file folder serves the same functions file folders do in the world that surrounds the desktop client-server. GUIs such as Windows software or the system used by Apple Computer Inc.'s Macintosh makes computers easier to use.

*Networking software.* This software allows for geographically separated workers, even groups of workers, to jointly work on a project. Networking makes for easier and faster communication. It represents the information highway, which links schedules, permits electronic conferences, allows instant feedback, and facilitates distribution of reports.

*Flexible databases.* Flexible databases or relational databases assemble in one place all the data in an organization in a form that can be shared by many, if not all, departments and individuals working on their respective jobs. The information system permits the entry of data by anyone in the system, and instantly adjusts to the new information. Each time a sales clerk enters a transaction, the system updates its files. Relational databases make the system of tracking inventory truly perpetual.

*Imaging technology.* IT allows information to be entered and processed by means of electronic scanners. A photocopier can now be turned into a source of input into systems: words, numbers, and bar codes are scanned at cash registers, in controllers' departments, and at the suppliers' warehouses. Insurance claims are available to the adjusters and to their supervisors in the blink of an electronic eye. Tax returns can now be faxed to the IRS.

The IT software delivers information directly to those who need it when they need it. IT's advancements permit access to information across geographic distances. This permits coordination to a degree that was hitherto difficult, if not impossible, to achieve. BPR should be implemented with IT's capabilities in mind. This does not mean buy first and think later. It is still important to think through, and to justify strategically, the context within which IT will be used, but the possibilities made available through technology can now be considered in the early stages of BPR.

To illustrate the potential made available by IT for BPR, here is a brief list of some of the possibilities:

- It is possible to dispense with intermediaries by connecting buyers and sellers of goods, as well as the suppliers and the users of information.
- Tracking the status of a product, a document, a payment, or even the entire process can be done in real-time.
- Knowledge and expertise can be disseminated at the speed of light: the books in the Library of Congress or the British Museum will soon become available to someone waiting for a plane in Timbuktu.
- Vast numbers of details can be processed, analyzed, and portrayed in formats that are understandable and convenient. The preparation of budgets does not rely on filling out the bulky twelve-columns ledger paper with handwritten data received from various parts of the organization. As a result, what-if situations can be made tangible and studied. Activity-based costing, the subject of the next chapter, is possible because of the IT now available.
- The process does not have to progress in a step-by-step sequence. Steps can be worked on jointly or even simultaneously.
- Geography is no longer a barrier, thanks to IT. Distance does not keep information from moving rapidly and being accessible as soon as it is available.
- Transactions are handled in forms that were unimaginable a few years ago: cash withdrawals, test taking, and mortgage applications are commonly subject to such IT-inspired intermediation and structures.

Until recently, the information systems in use were set in place to meet departmental needs. It was not uncommon to find several systems being used by the different sub-unit of the same department, and it was very common to find different systems in different departments. The sales, credit check, and accounts receivable departments could all be using different systems. Such manifestations of the "Tower of Babel" syndrome were not conducive to process-oriented thinking. The emergent IT makes communicating across departments possible, and makes it more effective and efficient.

## 7.4 BPR VERSUS TQE

BPR has much in common with TQE, the subject covered in chapter 6. Both seek to furnish organizations with evolutionary competence and both are made possible with the help of frequent communication, deployment of sufficient resources, commitment of the leadership, and the cooperation of the employees. Both prefer a cross-functional perspective, and have been shown to be quite practical, yet the two must be seen as contrasting, though not mutually exclusive, approaches.

TQE requires a long-term perspective and delivers long-lasting impact in return for undramatic but steady commitment. It is slow-paced, but continuous. Adjectives that describe it are "gradual," "constant," and "incremental." To be implemented, it requires universal involvement and collective effort. It judges the efforts invested as much as the result obtained. It seeks to rebuild on the existing foundations and does not require breakthrough technology or major investments.

In contrast to TQE, BPR seeks its impact in short-term, dramatic steps. The adjectives to be associated with it are "abrupt," "intermittent," and "volatile." It does not seek maintenance, but instead opts for rebuilding. It depends on technological innovations and major financial commitments, and is implemented by a few with a trickle-down approach.

The literature on behalf of BPR has at times made it seem more desirable than TQE. Indeed, a few have wrapped BPR in patriotic colors, seeing it as more attuned to the American myth of rugged individualism. But setting up the two approaches as polar extremes is not at all necessary. The two can coexist. TQE can and will facilitate BPR. TQE could easily lead to reengineering if and when the need is felt, technology is available, and there is money to make the needed investments. Once the process is reengineered, its true potential benefit can best be realized with the help of a consciousness and work attitude borrowed from TQE. There is a case to be made for practicing TQE while waiting for BPR, and even after it has been put in place. In denying the organization the advantages of TQE while waiting for the dramatic dawn of reengineering, the company can subject itself to stagnation and a loss of business. While waiting for heart bypass surgery, the patient is urged to foreswear fat and cholesterol. Similarly, it is best for an organization to not entirely depend on the surgical solutions of BPR.

## 7.5 PREDECESSORS OF BPR

It is worth noting that even though the word is new, the concept of reengineering is not. The history of the rise of systems in American management at the turn of the century shows that such developments were really attempts in corporate reengineering using the emergent technology of the time. Frederick Taylor's philosophy of scientific management also anticipated today's reengineering, as did Henry Ford's assembly line.

BPR brings forward the earlier practices of disciplines such as task analysis, operational auditing, and work simplification in order to redesign the way work is undertaken. More recently, a relationship between improved, redesigned procedures and greater productivity was demonstrated by Intel Corporation. This correlation was documented in 1981 in an article in *Fortune* magazine several years before the word reengineering took on its mythic aura. By carefully reviewing its operating procedures, the firm was able to reduce the number of steps needed to hire an employee from 364 to 250. Its engineers no longer needed twelve pieces of paper and ninety-five administrative steps to order a mechanical pencil. The review of operating procedures cut the steps involved to eight and the forms used to only one. The cost savings were in the millions of dollars and were accomplished by using techniques that were not dependent on cutting-edge technology. By methodically analyzing administrative procedures, they were able to distinguish value-adding steps from those that were not. A breakdown of a process into its components permitted differentiation between those that were necessary activities and those that caused delays, were redundant, or were useless yet added to costs. Following the analysis, the activities that were necessary parts of the process were retained, while the non-value adding tasks were eliminated. Even though the 1981 article described the technique used as work simplification, it is quite similar to BPR.

## 7.6  THE BENEFITS OF BPR

Clearly, the question of BPR's historical origin is interesting, but for organizational well-being it is immaterial. What is of significance is the realization that reengineering can be helpful in ensuring organizational evolution. The advent of BPR and its increasing use is, in large part, a result of the growing need to improve the productivity of white-collar workers. This imperative is becoming increasingly desirable in the face of economic hardships plaguing banks, thrifts, insurance companies, accounting firms, airlines, and major retailers. Another factor that is causing the increasing use of BPR is the imperative to focus on a flexible, team-oriented approach that promotes interdependent activities throughout the organization. This leads to cross-functional problem solving rather than narrow specialization.

The improvement of white-collar productivity cannot be induced by firing the low-paid employees or by opening the corporate checkbook for high-tech automation. The billions spent on IT has often only led to employees and managers doing the same work but doing it somewhat faster. Computers were no more than expensive calculators, and they failed to revolutionize the process, though they expedited productivity. One can increase productivity more effectively by removing bottlenecks, avoiding mistakes, and focusing on customer service. Resorting to new technology should follow, not precede, removal of non-value-adding activities. Often, mistakes are made and procedural gridlocks are created not because of careless, indolent employees but because of the way workers have been told to do their work. Outdated, complex, and redundant standard operating procedures also contribute to low productivity.

Bell Atlantic, concerned about the time it took to fill orders for telephone lines from a long-distance carrier, undertook an analysis of its process. It found the delays were the result of its functional structure. Different tasks were handled in different locales and, given that the order passed through twenty-five different tasks, the average time it took to complete the process was measured in days and weeks. By means of a BPR, Bell Atlantic dispensed with less useful, non-value-adding tasks and assigned teams to follow through with an entire order instead of having different individuals performing different tasks. As a result, the time needed to fill an order by Bell Atlantic has been reduced to a matter of hours.

## 7.7  WHAT DOES BPR INVOLVE?

The *Random House Dictionary* defines "process" as a systematic series of action directed to some predefined end. This definition is a good starting point for understanding a business process. Here, the word "business" is used broadly by not limiting it to organizations involved in making money. While talking about a business process, it is best to add some additional attributes to its dictionary definition. A process is undertaken in order to meet some needs within an organization, and those needs are felt by and associated with those who work for or with the organization. These processes have outcomes that affect both those within and those outside the organization. Those affected by the process have a vested interest in the outcomes, and those with such interests can be either customers, owners, or stakeholders in the process. For the sake of convenience, we will use the term "customers" to refer to all who have a vested interest in the outcome of a process.

In addition to having customers, a process usually involves more than one organizational department. Most organizational structures do not reflect the processes undertaken by the organization. This is largely because the emphasis on process orientation is of relatively recent origin. The organizational structure was usually composed of functions such as accounting, finance, development, distribution, purchasing, production, personnel, security, and quality. Such a structure downplayed the fact that within each function, there are processes being performed that require inputs from throughout the organization. There are, however, processes that are unique to each function. Downplaying the cross-functionality of processes and its pervasiveness within organizations of all sorts and sizes ignores a basic truth of business. New product development, creation of a strategic marketing plan, the processing of an insurance claim, and the hiring of an employee, all cross-functional processes, are what organizations do all the time.

Most of the processes performed by the accounting department are cross-functional. Among them are ledger control, payroll, tax planning, accounts receivable, accounts payable, cash control, fixed assets control, cost accounting, labor distribution, budget preparation, pensions, benefits, and employee expense accounts. These processes affect others, and such impact often helps to shape the organization's behavior. At the same time, the processes performed by the accounting department are affected by others within and outside the organization. This is one reason that accounting cannot be seen as being solely the concern of accountants, just as an information system does not belong only to the systems department.

## 7.8 PROCESS MANAGEMENT

Good management can help make processes more effective by allowing them to be performed in an optimum, streamlined manner. Management can help see to it that a process is efficient and makes effective use of resources. Additionally, management will allow a process to have flexibility built into it. Customers change, and so do environments. The processes in place can best respond to changes if they are flexible.

If processes are managed well, they will display the following traits:

- Ownership of the process is well-defined. There is someone designated to manage the process who can be held accountable for it.
- The scope of the process can be identified.
- There exist measures that help evaluate the effectiveness and efficiency of the process.
- The process is flexible enough to deal with changes and exceptions.
- The process has built-in provision for feedback.
- Written procedures documentation exists, indicating effective process management. Such documentation should help new employees understand tasks and how to perform them.
- There is effective communication between those in different departments who perform the various tasks that comprise the process. There should also be an absence of turf battles between those who work on the different tasks of the same process.

Managed processes are geared to eliminate errors, are not subject to non-value-adding delays, make optimum use of resources, are adaptable, rely on cutting-edge technology, are documented, and are easy to understand. Well-managed processes can provide an organization with a competitive advantage, while poorly managed processes can be costly to an organization. But such mismanagement can be recognized and, given the importance of processes to an organization, must be eliminated. If the process is out of control, then it is time for process redesign.

## 7.9 HOW TO REDESIGN?

Process redesign is itself a process, and it can be more effective if it is implemented in a well-planned manner. A discussion of well-planned implementation of BPR follows.

1. Identify the process to be redesigned. Process redesign is a major undertaking and it ties up a considerable amount of resources. Given this, it is best to select processes that merit such a commitment of resources. Most managers have some perception of what processes are most crucial to their success, and this can help the selection. Alternatively or additionally, the selection may be based on how poorly an important process is performing. An additional impetus to redesign or select a process for redesign may be through benchmarking the process against industry standards. It may be advisable to measure the performance against firms that are rated high and are recognized as the industry pacesetters.

   Such identification of a process for redesign will be facilitated if the goals of the redesign have been defined. Reduction of cost, elimination of production delays, faster and improved customer service, and improved quality are objectives typically sought through reengineering of processes. The objectives should be specific. It is better if they are measurable. Above all, they should be an outgrowth of the firm's strategic vision, goals, and objectives.

2. Analyze the effectiveness and the efficiency of the process. Before any attempt is made to redesign a process, its current performance must be analyzed. Such analysis can use the attributes of well-managed processes discussed in the previous section to judge the current performance of the process in question. Or, specific outputs of the process can be measured. Additionally, benchmarking comparisons can be used to measure the effectiveness and the efficiency of the process and its outputs. Such comparative measures will not only reveal how much the process needs to be improved, they will also help to judge the subsequent success of the redesign. But such measures may not be of much help if radical change is desired. Redesign seeks to change things, not to improve them incrementally. Using old performance measures to judge the merits of the new process may be like comparing the speed of a mule with that of a four-wheel drive truck.

3. Analyze the relationship between BPR and the firm's strategic plans. The conventional wisdom requires that an organization consider its overriding strategic objectives before it makes a major investment of its resources. Since process redesign commitment calls for just such investments of resources, it is best to consider how the objectives sought through redesign fit with the firm's strategic plans.

4. Identify how the BPR can support strategic goals of the firm. This, in effect, continues the examination of the relationship between redesign and the firm's strategic plans. However, it should include an analysis to show that goals being sought from the redesigned process do not negatively affect other processes in use by the firm.

5. Explore IT that may be useful. IT is constantly changing, and it will certainly be worthwhile to explore how emergent technology can help realize the objectives sought by process redesign. Since knowing the technology available can have an impact on the objectives being sought through BPR, consideration of IT should occur early in the redesign process.

6. Design a prototype. It is best to test the new before giving up the old. Testing is necessary to reveal the redesign in action, and may help catch and correct unforeseen flaws. The new may not turn out to be as good as had been hoped and it could cause unexpected problems. It may have an adverse impact on other processes within the firm. Plans on paper often seem more attractive than they actually are in practice. Testing of a prototype will save costs and resources. Furthermore, such testing will ensure that the final results are satisfying. Implementation on a pilot basis can allow time to make changes in the plan to better achieve the objectives desired before the plan is implemented organization-wide.

7. Measure prototype for its effectiveness and efficiency. Once a prototype is operational, it should be tested against the original process for its effectiveness and increased efficiency. Such measurements may reveal a need for modifying the changes. The testing should rely not on anecdotal impressions, but on tangible measures.

8. Carry out full implementation. After testing the plan on a pilot basis and making any changes that may be called for, the project is ready to be implemented. Special efforts may be needed to smoothly manage the changeover. Employees and supervisors may be resistant to performing their work in a new way.

9. Measure how well the redesigned process performs. Like all plans, BPR should be tested periodically to see if it needs improvement. In theory, after the major change, the redesigned process should satisfy the owners' needs for some time. However, the changing environment within which business operates today will not permit such "resting." Holding on to the status quo may not be in the best interest of a firm. Indeed, it is best to follow up a BPR with a TQE to realize maximum benefits.

## 7.10 CONCLUSION

At its most basic, BPR means starting with a clean sheet of paper, a clean organizational slate, and drawing new sets of tasks to be carried out with new technology to fulfill organizational vision and goals. Reengineering is much in the news these days. According to one estimate in *The Wall Street Journal* (March 16, 1993), the potential loss of twenty-five million jobs in the United States due to process redesign is possible. Currently, the private sector employs about ninety million workers. Whether or not the predication in the press holds true, the fact remains that reengineering is becoming a popular management tool. But the benefits of BPR are likely to be greater if it is not touted as the only way; there is no such thing as the one and only way in the world of processes. Indeed, BPR in combination with other techniques may be more effective than BPR used as a make or break intervention.

Firms of all stripes are scrambling to let reengineering duplicate for them what it did for Motorola, Ford, Xerox, and Bell Atlantic, namely, exponential productivity gains. However, reengineering is not for everyone. It is a costly and extreme remedy, which should be used only when absolutely needed. A surgical remedy may not be the only solution. Furthermore, among the experts there are calls that would have BPR be accompanied by a restructuring of the networks and environments within which the firm operates. Fast-moving cars are of little use if the roads and highways do not support their optimum performance. Similarly, IT-inspired streamlining of the processes may be only the first step, which must be followed by a search for environments that allow the potential of such changes to be realized.

Despite the caveat about prescribing reengineering indiscriminately, the fact remains that it is a useful managerial tool. It is one more tool to help organizations acquire evolutionary competence. It can coexist with TQE and activity-based costing.

# CHAPTER 8

# ABC AND ABM

## 8.1 INTRODUCTION

Like TQE and BPR, activity-based costing (ABC) and activity-based management (ABM) are techniques that help organizations attain evolutionary competence. Like the former tools, activity-based techniques require organizational introspection. Such self-awareness is the starting point for organizational evolution. Armed with a thorough understanding of what they do, organizations are better able to create value for their customers. The various techniques used for organizational evolution all help managers better understand their organizations. ABC seeks to enhance such understanding in monetary terms. Both ABC and ABM can coexist and complement other approaches, such as TQE and BPR. But unlike TQE and BPR, ABC is accounting-oriented. The focus of ABC, in particular, is on cost tracking, but in the process of revealing the what, who, where, when, why, and how of cost occurrences, those charged with ABC analysis can also reveal task inefficiencies and suboptimal processes that drain away resources. Managers must know where the inefficiencies are before they can take steps to correct them. Since they help cost the inefficiencies, ABC and ABM give such knowledge to managers. ABC also satisfies the call for tangible measurements made by those implementing TQE.

## 8.2 COST TRACKING TO HELP MANAGERIAL DECISIONS

Tracking costs helps managers assess use of the resources needed to manufacture products or provide services for both internal and external customers. Effective use of resources is dependent on making production cost-effective. As part of their cost-tracking function, accountants prepare financial statements and other reports for management, which is then better able to make informed decisions pertaining to functions such as:

- Pricing products and bidding on projects
- Planning and control
- Performing strategic analyses, both short-term and long-term
- Evaluating performance

Cost tracking has always been critical to effective managerial performance, but the need for more accurate cost measurements is being driven by the growing global competition as well as the technological, regulatory, demographic, and behavioral changes in the business environment.

## 8.3 THE FOCUS ON ACTIVITIES

The need for greater accuracy through less aggregation and less distortion in accounting reports has led to the increasing adoption of ABC. It focuses on assigning the costs incurred by each of the various activities performed to manufacture products or provide services, rather than simply pricing aggregate production resources. Such a focus stems from the realization that costs occur because activities are undertaken.

Activities drive up costs, yet not all activities are equally capable of adding value to the outputs. Furthermore, the cost-benefit criteria applied to activities can actually enhance the quality of managerial decisions concerning the replacement of obsolete or suboptimal activities that drive up costs. Such decisions are the necessary ingredients for attaining evolutionary competence, and they can be facilitated by ABC.

## 8.4 ACCURATE REFLECTION OF RESOURCE USAGE

An incentive to management for actually using information provided through ABC is the presumption that it is a more accurate reflection of how resources are being used. In the 1980s, the cost systems in place in the United States, Canada, and Britain since World War II were widely accused by academicians and practitioners alike of being too aggregated and too distorted to be relevant in managerial decision making. The increase in the popularity of ABC owes a considerable debt to such disenchantment with outdated cost-tracking techniques. Cost accounting textbooks have since added chapters on activity-based costing, while retaining the traditional techniques of cost tracking.

## 8.5 DEFINITION OF ABC AND ABM

Seeing activities as a product's building blocks allows accountants to track costs by tracking activities. Recognizing such cause-and-effect relationships between activities and costs incurred can help to minimize aggregation and distortion, and leads to more accurate costing.

ABC has been defined as an activity-focused costing technique that seeks to calculate costs of products and services by assigning them costs in keeping with their use of resources. Such assignment of costs is tantamount to an assignment of resource usage among the various outputs. It is based on using activities as the vehicle for cost accumulation, instead of departments or even the plant itself, a more traditional means of collecting cost pools.

ABM is the approach management can use to ensure that activities necessary to produce goods and services are value-adding. Such an assessment of the value provided is done by means of activity-based costing techniques as well as information pertaining to revenue, customer demographics, quality measures, and market factors.

The relationship between ABC and ABM is explained by realizing that the initial step is activity analysis devoted to charting the relationship between inputs and outputs. The inputs are measured as activities. These activities are then costed, which gives rise to ABC. In the process of costing activities, the extent to which they add or do not add value to outputs is assessed. A number of activities may be found to not add any value at all. Such measurements of costs incurred and benefits realized are among the many factors that permit the management of activities for optimum results.

## 8.6  DIRECT VERSUS INDIRECT COSTS

To explain activity-based costing, it is best to start by explaining the meaning of terms like activity, cost center, and cost-driver, and by differentiating between direct and indirect costs. When accounts talk about an activity, they have in mind those activities that result in costs being accrued. Obtaining a college education is an activity that is responsible for costs occurring for students or their parents. The son or daughter attending college can be seen as a cost center or cost objective, which is another way of describing any activity for which a separate measurement of costs is desired by those who foot the bill. Parents like to know the costs involved in the undertaking. Attending college is not only an activity, it should also be seen as a cost driver, which, by definition, is the reason costs are being incurred.

All the costs involved in obtaining a college education are not alike. Money paid as tuition and to purchase textbooks are the direct costs of obtaining an education. The expenses involved in commuting to and from college, assuming the student lives at home, will be an indirect cost, because the car is very likely also used for other activities, or cost objectives. Hence, the expense of getting to college must be calculated by first obtaining the total costs associated with the car and then allocating the portion of such costs that pertains to obtaining a college education by our cost center. If the car is only used to commute to college, then the expenses involved in its purchase and upkeep would be a direct cost of attending college. In other words, an indirect cost is shared among more than one cost objective, and such sharing is what accountants call cost allocation or, more recently, cost assignment.

A direct cost can also be conveniently traced to the product or the output. If it cannot be traced easily or if it is not cost-effective to be so tracked, then it is classified as an indirect cost, making it a part of the overhead. It is for the sake of convenience that the wood used in making furniture is classified as a direct cost, while the glue used for the same sofa is seen as indirect. Even though both can be physically traced to the product, the reason for not treating glue used as a direct cost is because it is inconvenient to do so. Tracking the actual cost of the drops of glue that went into the sofa is not a job easily justified, since the amount is not material. Hence its classification as a part of the overhead, which will in turn be allocated to the final products.

Items such as the glue used on sofas are lumped together as overhead. Anything used in the production that was not specifically identified as direct labor or material is considered overhead. This includes the depreciation of the long-term assets, as well as the expenses of having staff members that work for all the cost centers for example, maintenance workers in a factory. Even the salaries paid to supervisors are classified as indirect costs, since their time is spent between different jobs, in contrast to the wages of assembly line workers, whose labor is more easily traced to specific jobs. Inventory-related costs such as those associated with receiving the raw material, getting it to the cost centers on the production floor, and keeping it secure are also seen as overhead items. Salaries of accountants and timekeepers are also overhead because their services are usually shared between all the cost centers.

In former times, overhead was not seen as a major item, judging from the use of the word "prime cost" to describe direct labor and direct material. However, overhead now tends to be much higher than direct labor costs. Moreover, support activities today make up a higher share of the total costs. For some firms, overhead can represent 90 percent or more of the costs incurred in getting their outputs into the hands of their customers.

Consequently, management of overhead is of prime importance. The upward creep in overhead must not be seen as being a phenomenon of the 1990s. Accountants have traced the upward trend back to the 1920s. Allocation almost always involves estimating. When overhead management was not given prime importance, the accuracy of the estimated overhead allocations was not critical. But with overhead representing an ever-higher share of the total costs, it is imperative to allocate it more accurately. Allocating common costs more correctly is what ABC is all about.

## 8.7 COST BEHAVIOR

It is generally known that different kinds of costs behave differently when the production volume changes. In the past, cost behavior emphasized the following differentiation of costs:

- Committed fixed costs. Changes in production volume do not result in changing committed fixed costs, as long as the changes in volume fall within a certain range. A factory built to produce ten thousand units presumably will not show changes in fixed costs, as long as volume produced was ten thousand units or less.
- Discretionary fixed costs. These remain fixed for a designated period. The salary of the controller remains the same for the year, at the end of which there may be a change in it. The budgets for training or advertising similarly can remain fixed until the new budget, when, at the discretion of managers, they may change.
- Variable costs. These vary directly in proportion to the volume. One battery per car and the hourly pay rate of direct labor are examples of variable costs.
- Step-variable costs. These vary in steps, hence their name. The costs for an automobile oil change every three thousand miles and the salaries of a mechanic and other maintenance workers are examples of costs that change in discrete blocks.
- Mixed costs. These have a fixed component as well as a portion that varies with the activity volume. A phone bill can have a fixed portion, such as the monthly fee for having a connection. It can also have variable charges based on the long-distance calls made.

## 8.8 COST BEHAVIOR UNDER ABC

Costs classification under ABC is on the basis of their association with activities at the following levels: unit level, batch level, facility level, and product level. None of these except the unit-level costs can be directly traced to the units of production, and from the perspective of units produced, they are fixed costs. But when linked to activity drivers, they become variable. These costs are first collected in a cost pool and then assigned/allocated using an appropriate basis. The allocation basis under ABC are labelled cost drivers. If an activity is to be considered a cost driver, there must be some relationship between it and the accrual of costs. A larger number of cost drivers are employed under ABC, compared with the traditional costing methods, which are content to use fewer drivers. Even now, firms using just one driver to allocate all the indirect costs are easy to find.

## 8.9 ACTIVITY LEVELS FOR ABC

The four levels of activities and the costs associated with each are described below:

**Unit-level costs** are the variable costs, and they correspond with the level of production. Direct labor or machine hours vary with the number of units produced.

An example of an activity that creates unit-level costs is assembly-line production. This activity results in the consumption of resources such as hourly wages of workers and direct material used such as tires and batteries for automobiles.

The cost drivers typically used to assign unit-level costs are the number of units produced, as well as labor hours and quantity of material.

**Batch-level costs** are those that correspond with the number of batches produced. A batch-level cost will accrue whenever a batch of units is made.

Examples of such costs are those associated with activities such as machine setup, order processing, and material handling. The resources used for these activities are labor costs of setting up machinery, processing job orders, and writing up purchase vouchers.

Typical cost drivers used at this level are the number of purchase orders issued, the number of times machines were set up, and the number of batches produced.

**Facility-level costs** are incurred to maintain the plant where the product is manufactured. These are the most difficult to assign.

The resources used at this level are for plant depreciation, payment of taxes, salaries of plant managers, and expenses incurred for the general training of employees.

The typical cost drivers used for facility-level costs are the number of employees in the activity centers, the volume of units produced, and the allocation percentages arrived at by management.

**Product-level costs** are those that pertain to activities such as product design, special handling, storage, and parts and product testing. Maintaining bills of materials and routing information are also seen as product-level activities.

Resources this level requires include the specialized equipment needed for a given product as well as the costs incurred to design and test it.

The cost drivers used at this level are the number of products developed or manufactured as well as the number of parts used by a given product.

## 8.10 THE SEARCH FOR MORE ACCURATE ALLOCATION

The search for more accurate allocation has been around for some time, even before the advent of ABC. Indeed, ABC itself was known at the end of the nineteenth century. Its failure to gain greater acceptance was largely due to a lack of computers that could perform the large number of calculations needed by an ABC system. Below are the various methods used to allocate indirect costs—they range from least to most accurate with respect to the cost assignments made.

1. The basic method involves all indirect costs incurred being added up and then assigned to units manufactured by means of a single basis such as direct labor hours

incurred in the production. Such a method is obviously most subject to aggregation and distortion of costs.

2. A greater degree of accuracy is produced when, instead of using just one basis for all the production departments, different bases are used for each of them. These drivers could be the cost of direct materials used in one department, direct labor hours in the next, and machine hours in yet another.

3. Even more accurate are systems that isolate from the total overhead the service departments' costs and assign them first to production departments and, through them, to units produced. Service departments are those that are not directly involved in the manufacturing process but facilitate the work of those departments that are so involved. These could be human resources, maintenance, cafeteria, and inventory warehouse. At times, the service departments will use different bases for assigning fixed and variable costs incurred by the service departments; for instance, budgeted machine hours for variable costs, but machine hours based on total capacity available for fixed costs.

All of the methods described seek allocations of common costs. The allocation is mandated by the generally accepted accounting principles (GAAP) for external reporting. However, such a mandate is self-defeating because it is willing to accept allocation methods that are questionable.

## 8.11 GAAP'S ACCEPTANCE OF IRRATIONAL ALLOCATIONS

One example where irrational methods are acceptable for allocating common costs is through examining the so-called joint product costing. Joint products are separate products that are obtained from a common source but then split into distinct products, which can in turn be further processed into yet more products. From an oil well, natural gas and crude oil are obtained. Both of them are joint products and both are further processed into yet more joint products.

In the meat products industry, the common source, such as a cow, turns into a number of different products, like steaks, ground beef, and bones. The costs that pertain to the purchase of the common source and getting it ready to be split into different products are called joint costs. The GAAP requires that such joint costs be allocated among all the products that result after the common source is split into its components.

A number of methods are acceptable for allocating joint costs. Among them is one that uses the net realizable value of each product as a basis for allocating and another that uses the quantities by weight of separate products obtained at the split-off as the basis. Under the latter, steaks, ground beef, soup bones, and hide and entrails will all be allocated a share of common costs based on their respect weights at the split-off. The values assigned to bones, hides, and chitterlings will be higher, because by weight they represent considerably more than they do in terms of final profits realized. Such cost allocations will make them appear to be the losers in financial reporting because the revenue they provide is less than their allocated costs. But dropping them from the portfolio of products sold will lower the total profits. Dropping them causes the loss of whatever revenue they realized, but the common costs will not change in the total. The share that was being allocated to these less fashionable products must now be split between those that remain—the ones remaining will be allocated a higher share of the common costs. Notwithstanding the problem described, this method can be used for financial reporting.

Under each method used for joint cost allocation, the cost of goods sold and the inventory values for each joint product is different, often vastly different. However, such distortions are offset when they are aggregated for the final accounting statements. So, for the external reporting using cost and revenue aggregates, it does not matter if the costs and the profits reported for individual joint products are inaccurate, since the bottom line will take care of the inaccuracies. But in decisions concerning individual products, for instance those that involve product pricing, such distorted costs are not to be believed. Pricing decisions for joint products are best made by treating joint costs as common costs that need not be allocated.

## 8.12 TRYING TO DO WITHOUT ALLOCATIONS

Not allocating common costs is against GAAP, but for decisions that are of concern only to internal decision makers, such a departure from GAAP is acceptable, and is indeed quite widespread. Pricing decisions for joint products are not the only ones to treat indirect costs thus by ignoring allocations of the common costs. The contribution approach to costing is applicable and useful for a wide variety of policy decisions, and it will not allocate common fixed costs to respective products or even divisions. Instead of cumbersome allocation, common costs are treated as a corporate period expense. But the danger in such a treatment is that product managers will ignore the common costs, since they are not being charged for them. It is acceptable to ignore common fixed costs for some short-term policy decisions, but the same does not hold for strategic policy decisions.

Unfortunately, a firm will not long survive if its strategic policies do not seek to recover its common costs. In recent years, the airlines in the United States have shown just such an obliviousness towards common costs, judging from the frequent price wars that occur between them. The bankruptcy of Pan American and others can be partly attributed to their failure to recover the common fixed costs of operating an airline through their pricing policies. Their pricing policies incorrectly relied on the application of a short-term remedy for long-term, structural problems such as too many airlines competing in the same markets.

## 8.13 TRACKING ABC

Tracking cost under an ABC system will occur through the following steps:

1. Assemble related actions into activity groups
2. Track and pool costs by activities
3. Select cost drivers that link cost pools for activities with outputs (cost objectives), such as units produced
4. Calculate rates to help assign the cost of activities to outputs in keeping with their activity usage
5. Assign common costs or rather the cost of activities to units produced or other cost objectives

These steps are briefly explained below.

**(a) Assemble Activity Groups**   A large number of activities are performed in any organization. The cost accounting ideal is to cost each activity on an a la carte basis, however, doing this outside of a restaurant menu is very likely not cost effective. Hence, actions are arranged in activity groups to help make cost-effective tracking more manageable. These groupings are often done in terms of their relationship to different activity levels.

Care must be taken to arrange the costs in a given pool only if they are driven by the same or at least correlated activities. Set-up costs are actually a mixture of labor and supplies, and the two of them are not necessarily correlated. But if they are pooled together and then assigned using just one driver, like the set-up time in a combined pool, distortion will result. Similarly, if costs arranged in the pool do not have the same degree of proportionality with the driver used, there will be distortion. An example of such distortion is seen when the variable and fixed portions of utility costs are combined and a single rate is used to assign the combination to the output. As previously mentioned, the activity levels of use in ABC are unit, batch, facility, and product. Some activity groups may be contained within one department, while others can be spread across several departments.

**(b)  Track Cost by Activities**   Following the development of activity groups, costs must be grouped by the activities. In order for costs to be so tracked and pooled, changes in the typical charts of accounts may be necessary.

**(c)  Select Cost Drivers**   Cost drivers are the connecting link between costs, activities, and outputs. A cost driver can link a cost pool associated with an activity to units produced, or it can link costs in one activity center to activities in another center. One way to help classify cost drivers is in terms of first-stage and second-stage cost drivers. First-stage drivers link the cost of resources consumed in an individual activity to other activity centers. Such assignments are similar to the way a service department allocates its costs to other departments. The second-stage drivers link costs from the activity center to actual units of production.

**(d)  Calculate Rates to Assign Cost**   Once costs are tracked to the pools of activities and cost drivers have been agreed upon, then the rates per cost driver unit can be calculated. This rate could be based on either the planned or actual activity level.

**(e)  Assign Costs of Activities to Outputs**   Using the rates calculated above, the final step can be undertaken. This consists of distributing costs to units of output or rather the cost objectives. All costs involved in the manufacturing process are eventually assigned to products.

## 8.14  CONCLUSION

To gain maximum benefits from ABC, it must not be limited to costing products. Evidence presented in accounting journals such as *Management Accounting* shows that ABC's use is increasing, that those using it represent a great diversity, and that most users have found it extremely helpful.

ABC is a technique that provides economic information about the firm. Such an organizational map can be useful to managers because it provides specific, tangible information about the deployment of the firm's resources. As discussed in Chapter 3, TQE stresses measurements and facts, and it seeks these facts by asking who, what, where, when, why, and how processes and performances work. ABC answers those same questions in cost-related terms.

Judging from the literature on ABC, its use can help point out the suboptimal nature of production-based operations. It translates the organization into tangible, monetary terms. The view provided by ABC is a cross-functional, integrated view of the firm and its activities and processes. The information provided by such a system can be of interest both for operational and strategic planning and control. Such information enhances decisions pertaining to product lines, market segments, and customer relationships. ABC can not only help measure the costs of activities and processes, it can even provide economic information about the measures being taken to implement TQE, such as improved quality, reduced process time, and the elimination of inventories and working capital. It can coexist and indeed complement other approaches to organizational growth, making it a vital component in efforts aimed at TQE and BPR. The information ABC provides by itself does not facilitate attaining evolutionary competence, or for that matter, improved profits and enhanced performance, but it can assist in their attainment. To help attain evolutionary competence, managers assisted by ABC must rely on setting in motion conscious, planned efforts towards that goal. In other words, to ensure continuous improvement, a mechanism that promotes organizational learning must be put in place.

Since this is not a book about cost techniques, the interest in ABC is not for the sake of the specific cost-related information. The discussion here pertains to the process that underlies ABC, as well as to the diversity of organizational environments that seem able to benefit from ABC. The documentation process can become an illuminative occasion by incorporating within itself the approach used by ABC, alongside the concepts and techniques of TQE and BPR.

## ABC AND ABM BIBLIOGRAPHY

There has a great wealth of material written about ABC and ABM. To help readers who wish to read more about the topic, a bibliography of material dealing with the topic follows.

What should be of special interest to readers is the literature pertaining to ABC and ABM in the service sectors and support services, since they have direct bearing on accounting function.

## SOURCES AND SUGGESTED REFERENCES

"Activity-Based Cost Accounting." *Management Accounting* (UK) 67, no. 8 (September 1989): 84.

Beaujon, George J., and Vinod R. Singhal. "Understanding the Activity Costs in an Activity-Based Cost System." *Journal of Cost Management* 4, no. 1 (spring 1990): 51–72.

Berliner, Callie and James A. Brimson, eds. *Cost Management for Today's Advanced Manufacturing: The CAM-I Conceptual Design.* Boston: Harvard Business School Press, 1988.

Borden, James P. "Review of Literature on Activity-Based Costing." *Journal of Cost Management* 4, no. 1 (spring 1990): 5–12.

Campi, John P. "Total Cost Management at Parker Hannifin." *Management Accounting* (January 1989): 51–53.

Capettini, Robert and Donald K. Clancy. "Cost Accounting, Robotics, and the New Manufacturing Environment." Florida: American Accounting Association, 1987.

Cooper, Robin. "Implementing an Activity-Based Cost System." *Journal of Cost Management* 4, no. 1 (spring 1990): 33–42.

Cooper, Robin. "The Rise of Activity-Based Costing—Part One: What is an Activity-Based Cost System?" *Journal of Cost Management* Vol. 1 (summer 1988): 45–54.

Cooper, Robin. "The Rise of Activity-Based Costing—Part Two: When Do I Need an Activity-Based Cost System?" *Journal of Cost Management* Vol. 1 (fall 1988): 41–48.

Cooper, Robin. "The Rise of Activity-Based Costing—Part Three: How Many Cost Drivers Do You Need, and How Do You Select Them?" *Journal of Cost Management* 2, no. 4 (winter 1989): 34–46.

Cooper, Robin. "The Rise of Activity-Based Costing—Part Four: What Do Activity-Based Cost Systems Look Like?" *Journal of Cost Management* 3, no. 1 (spring 1989): 38–49.

Cooper, Robin. "The Two-Stage Procedure in Cost Accounting—Part One." *Journal of Cost Management* (summer 1987): 43–51.

Cooper, Robin. "The Two-Stage Procedure in Cost Accounting—Part Two." *Journal of Cost Management* (fall 1987): 39–45.

"Cost Management in the 1990s." *Management Accounting* (UK) 67, no. 11 (December 1989): 16–17.

Drury, Colin. "Activity-Based Costing." *Management Accounting* (UK) 67, no. 8 (September 1989): 60–63, 66.

Dugdale, David. "Costing Systems in Transition: A Review of Recent Developments." *Management Accounting* (UK) 68, no. 1 (January 1990): 38–41.

Eiler, Robert G., and John P. Campi. "Implementing Activity-Based Costing at a Process Company." *Journal of Cost Management* 4, no. 1 (spring 1990): 43–50.

Eiler, Robert G. "Managing Complexity." *Ohio CPA Journal* (spring 1990): 45–47.

Greene, Alice H. and Peter Flentov. "Cost Management Practice: Managing Performance Maximizing the Benefit of Activity-Based Costing." *Journal of Cost Management* (summer 1990): 51–59.

*Harvard Business School Case Series*. Boston: Harvard Business School Publishing Division. (For a description of these case series see "Review of Literature on Activity-Based Costing" by James P. Borden, cited earlier.
  1. "Camelback Communications." 185–179. 5 pp. 1985.
  2. "Fisher Technologies." 186–188. 46 pp. 1986.
  3. "Ingersoll Milling Machine Company". 186–189. 25 pp. 1986.
  4. "John Deere Component Works." 187–107. 18 pp. 1987.
  5. "Mayers Tap, Inc." 185–111. 59 pp. 1986.
  6. "Mueller-Lehmkuhl Gmblt." 187–048. 12 pp. 1986.
  7. "Schrader Bellows: A Strategic Cost Analysis." 186–273. 82 pp. 1985.
  8. "Siemens Electric Motor Works (A-B)." 189–089/090. 9 pp., 12 pp. 1988.
  9. "Tektronix (A-C)." 188–142/143/144. 1988, 18 pp./12 pp./18 pp.
  10. "Union Pacific" (A, B, Introduction) 186–176/177/178. 1985 14 pp./19 pp./20 pp.
  11. "Winchell Lighting." (A-B). 187–074/075. 1987, 23 pp. 12 pp.

Jeans, Mike and Michael Morrow. "The Practicalities of Using Activity-Based Costing." *Management Accounting* (UK) 67, no. 10 (November 1989): 42–44.

Johnson, H. Thomas. "Activity-Based Information: Accounting for Competitive Excellence." *Target* 5, no. 1 (spring 1989): 4–9.

Johnson, H. Thomas. "Activity-Based Information: A Blueprint for World-Class Management." *Management Accounting* 69, no. 12 (June 1988): 23–30.

Johnson, H. Thomas. "Activity Management: Reviewing the Past and Future of Cost Management." (Part 1), *Journal of Cost Management* 3, no. 4 (winter 1990): 4–7.

Johnson, H. Thomas. "Managing Cost Versus Managing Activities—Which Strategy Works?" *Financial Executive* 6, no. 1 (January/February 1990): 32–36.

Kaplan, Robert S. *Measures for Manufacturing Excellence*. Boston: Harvard Business School Press, 1990.

Lee, John Y. "Cost Driver Accounting: A Case Study." *The Ohio CPA Journal* (Spring 1990): 15–18.

"Management Accounting: Evolution not Revolution." *Management Accounting* (UK) 67, no. 9 (October 1989): 5–6.

McNair, C.J., et. al. *Meeting the Technology Challenge: Cost Accounting in JIT Environment*. Montvale, NJ: National Association of Accountants, 1988.

McNair, C. J. "Interdependence and Control: Traditional vs. Activity-Based Responsibility Accounting." *Journal of Cost Management* (summer 1990): 15–24.

Mecimore, Charles D. "Regaining Competitiveness Through Activity-Based Costing." *Corporate Controller* 2, no. 4 (March/April 1990): 10–14.

Morrow, Mike and Peter Scott. "Accounting Issues: Easy as ABC." *AA: Accountancy Age* (September 1989): 44–49.

O'Guin, Michael. "Focus the Factory with Activity-Based Costing." *Management Accounting* (February 1990): 36–41.

Ostrenga, Michael R. "Activities: The Focal Point of Total Cost Management." *Management Accounting* (February 1990): 44–49.

Peavey, Dennis E. "Battle at the GAAP?: It's Time for a Change." *Management Accounting* (February 1990): 31–35.

Pryor, Tom E. "Activity Accounting: The Key to Waste Reduction." *The Accounting Systems Journal* 1, no. 1 (fall 1989): 32–38.

Pryor, Tom E. "Designing Your New Cost System Is Simple (But Not Easy)." *Journal of Cost Management* 3, no. 4 (winter 1990): 43–47.

Romano, Patrick L. "Trends in Management Accounting: Activity Accounting." *Management Accounting* (May 1988): 73–74.

Romano, Patrick L. "Activity Accounting—An Update—Part One." *Management Accounting* (May 1989): 65–66.

Romano, Patrick L. "Activity Accounting—An Update—Part Two," *Management Accounting* (June 1989): 63–65.

Romano, Patrick L. "Where is Cost Management Going?: Activity-Based Costing and Other Approaches." *Management Accounting* (August 1990).

Rotch, William. "Activity-Based Costing in Service Industries." *Journal of Cost Management* (summer 1990): 4–14.

Roth, Harold and A. Faye Borthick. "Getting Close to Real Product Costs." *Management Accounting* (May 1989): 28–33.

Schubert, John K. "The Pitfalls of Product Costing." *Journal of Cost Management* (summer 1988): 16–26.

Sephton, Marcus and Trevor Ward. "ABC in Retail Financial Services." *Management Accounting* (UK) (April 1990): 29, 33.

Shank, John K. and Vijay Govindarajan. "Transaction Based Costing for the Complex Product Line: A Field Study." *Journal of Cost Management* (summer 1988): 31–38.

Sharman, Paul. "A Practical Look at Activity-Based Costing." *CMA* 64, no. 1 (February 1990): 8–12.

Shields, Michael P. and S. Mark Young. "A Behavioral Model for Implementing Cost Management Systems." *Journal of Cost Management* (winter 1989): 17–27.

Staubus, George G. *Activity Costing for Decisions: Cost Accounting in the Decision Usefulness Framework.* New York: Garland Publishing Co., 1988.

Troxel, Richard B. and Milan G. Weber, Jr. "The Evolution of Activity-Based Costing." *Journal of Cost Management* 4, no. 1 (spring 1990): 14–22.

Turney, Peter B.B. "Activity-Based Costing: A Tool for Manufacturing Excellence." *Target* 5, no. 2 (summer 1989): 13–19.

Turney, Peter B.B., ed. *Performance Excellence in Manufacturing and Service Organizations.* (Proceedings of the Third Annual Management Accounting Symposium, San Diego, CA, March 1989), American Accounting Association, Sarasota, FL, and NAA, Montvale, NJ, 1990.

Turney, Peter B.B. "Ten Myths About Implementing an Activity-Based Cost System." *Journal of Cost Management* 4, no. 1 (spring 1990): 24–32.

Turney, Peter B.B. "Using Activity-Based Costing to Achieve Manufacturing Excellence." *Journal of Cost Management* 3, no. 2 (summer 1989): 23–31.

Turney, Peter B.B. "What Is the Scope of Activity-Based Costing?" *Journal of Cost Management* 3, no. 4 (winter 1990): 40–42.

Turney, Peter B.B., and James M. Reeve. "Cost Management Concepts and Principles: The Impact of Continuous Improvement on the Design of Activity-Based Cost Systems." *Journal of Cost Management* (summer 1990): 43–50.

Vining, G. William. "Managing Causes and Costs with Activity-Based Process Analysis." *The Journal of Corporate Accounting and Finance* 1, no. 2 (winter 1989/1990): 107–114.

# CHAPTER 9

# ORGANIZATIONAL LEARNING

## 9.1 INTRODUCTION

This book began by comparing traditional and current views of procedures documentation, a business communication genre. Seen as archival afterthoughts, procedures manuals have turned into neglected tools that are rarely used even though well-written, current documentation can play many useful roles. Given the lack of respect accorded procedures manuals, it is not enough to merely advocate a greater respect for them, instead, it is better to change the very perception of the procedures documentation process. The documentation process must involve not only the end product, namely, the finished procedures manual, but also the steps that create it. When we create a valuable process, the product that results is inevitably worthy of respect.

## 9.2 RETHINKING THE DOCUMENTATION PROCESS

It is time to change the role currently assigned to procedures documentation. The justification for such a shift lies in the changes taking place in the global environment, in the introduction of more effective management practices, and in the emergent information technology. But an even greater imperative to change the perception of the documentation process is the current redefinition of authority and managerial leadership.[1]

Today, leadership need not be an omniscient manager at the top of the organizational pyramid. Instead, managers are being urged to exercise their leadership by acting as designers, teachers, and stewards who seek to build shared visions, challenge prevailing work patterns, and foster systemic patterns of intellectual exploration. This type of leadership encourages employees to expand their contributions for shaping the organization's future and allow them to learn. Such learning is crucial to developing an evolutionary competence—the ability of an organism to survive in a rapidly changing environment.

Changing the perception of how the documentation process is viewed can help make it an occasion for institutional learning. Managers should see to it that the process governing procedures documentation serves as an occasion for sharing mental maps, for challenging work habits, and for exploring and improving. That such a shift in perception is practical is evidenced by the success of techniques that help organizations attain evolutionary competence. Techniques such as TQE, BPR, ABC, and ABM (discussed in previous chapters) have shown themselves capable of changing the way organizations perform. If it is possible, practical, and profitable to redesign and continuously improve procedures and activities, why not do the same for the process involved in their documentation? Indeed, why not redesign the documentation process so that it will serve as an occasion for institutional learning and not be seen as a neglected afterthought?

## 9.3  WHAT PREVENTS ORGANIZATIONAL LEARNING?

As we stressed in the previous chapter, organizations need evolutionary competence if they are to survive, let alone flourish and prosper. To acquire it, they must provide their employees with occasions where institutional learning and TQE, BPR, ABC, and ABM may occur, where personnel can step back from daily chores and reflect on their work.

One of the most common obstacles that prevents personnel from attending to organizational evolution is their workload. The cause for such absorption in work may not be the slowness of the employees, or even the workload itself, but rather the work design. When daily "firefighting" overwhelms personnel to the extent that they can no longer maintain normal operations, let alone keep up with the changing environment or technology, organizational and accounting standard operating procedures have passed the point of being useful. Such organizations must rethink what work they do, why they do it, when they do it, who does it, and where it is done. Indeed, knowing when organizational routines and standard operating procedures are about to lose their effectiveness may well be the key to preventing obsolescence.

## 9.4  RETHINKING THE PERCEPTION OF WORK

To facilitate organizational learning, we need to rethink how work is viewed. In the closing decades of the nineteenth century, Frederick Taylor observed that the emphasis in the industrial environment needed to shift from the end product to the processes that help create the product. He also realized the importance of employee motivation, even though he saw employees as playing a passive role. But Taylor's ideas emphasized reductionism: breaking things down into isolated parts in order to better control them. Scientific management emphasized finding the one best way to perform and then enforcing it. The enforced standardization that was gospel for Taylor promoted efficiency in the first half of the twentieth century. But in a shrinking, unpredictable world dominated by volatility and ever-accelerating change, reductionism and enforced standardization may become a one-way ticket to oblivion. Today, the call is for self-management, learning through feedback, and for systems that are able to redesign themselves.

Some management experts are also calling for institutional structures that incorporate built-in mechanisms to respond to change. Indeed, these experts see the ability to change faster than the competition as a cardinal virtue and a strategic advantage.

It is becoming the norm for organizations to force change, to periodically examine the work they do and justify to themselves and their stakeholders the viability of what they do and what they know. Journals are busily writing about cutting-edge organizations that are able to "bail out" before the markets for their products reach maturity or saturation. These companies keep reinventing their products and services. Such organizations prosper because of their willingness to cultivate organizational learning.

## 9.5  THE LEARNING PARADIGM

Notwithstanding the current resurgence of the term "learning organization," the concept has been around for some time. The paradigm also underlies Edward Deming's crusade for improved productivity. The term "learning" should not be equated with simply digesting information and taking in details. Instead, learning organizations allow workers to expand their ability to deal with current problems and visualize future opportunities. A learning organization is concerned with things as they could be, not just as they are; it seeks to anticipate opportunities, not simply deal with crises when they occur. As noted by management expert David Garvin, a learning organization has skills that not only allow it to create, acquire, and transfer emergent knowledge, it also is able to modify itself in keeping with its newly acquired knowledge and experiential insights. Employees in learning organizations are skilled at the following activities, according to Garvin: systematic problem solving, experimentation with new approaches, learning from their own experiences and best practices of others, and transferring knowledge quickly and efficiently throughout the organization.

**(a) Promoting the Learning Attitude.**   To foster learning in an organization, leaders of organizations must ensure that their employees are able to question currently-held stereotypes and abandon them if they are unable to meet changing needs. Left to itself, organizational learning is likely to be very slow. Personnel are likely to find learning difficult when they are harried or rushed. Organizational learning, however, can be expedited by fostering an environment conducive to learning. A learning environment allows employees freedom for reflection, analysis, and creativity, and permits them to invent better ways of working and to ask questions about customer needs and current work processes.

To promote an environment suitable for learning within an organization, a number of techniques are available. Garvin lists the following as the catalyst for learning: frequent strategic reviews of a firm's competitive environment, product portfolio, technology, and market positioning; periodic audits of various systems, in particular those involving cross-functional processes and delivery systems; internal and external benchmarking; and study missions and symposiums to share and learn from each other. In a similar vein, management expert Daniel Kim has proposed a model for promoting organizational learning, which he calls the OADI-SAMM model. The acronym means "observe, assess, design, implement—shared mental models." Kim argues that by creating systems and processes that support these activities and integrate them into the fabric of daily operations, companies can enhance institutional learning.

It is worth noting that the advice given by Garvin and Kim echoes the tools used by those implementing TQE, as well as the ideas of Deming. Indeed, TQE and organizational learning are sometimes seen as identical, however, we would like to see the latter as representing the attitudes that ensures that an organization will not only seek out and foster learning opportunities for its employees, but will see to it that the lessons learned and the knowledge acquired are put to use. TQE is the set of techniques that allow learning. It is the implementation of the learning paradigm. Both are needed—the willingness and motivation on one hand, the tools and the knowledge on the other—for learning to occur.

This book seeks to add the procedures documentation process to the techniques suggested to enhance learning. The role being argued here for the procedures documentation process presupposes that it will incorporate elements from the various evolutionary competencies being surveyed in this book. By incorporating these techniques within the documentation process, organizational learning can be instituted in the accounting department as well as the firm.

In the rest of this chapter we will talk about benchmarking, frame breaking, and process consultation. The first two are techniques being widely used to promote organizational learning, and as such they are worth a special mention. Organizational learning can be enhanced if those seeking to promote it are familiar with techniques that facilitate group efforts. Such behavioral expertise is as important as knowledge of emergent technologies in inculcating evolutionary competence in an organization. A review of two real-life learning organizations follows the discussion of the above topics.

## 9.6 BENCHMARKING

Institutional obsolescence can result when standard operating procedures (SOPs) become embedded. Over time, the institutionalization of SOPs can delay corrective actions in the face of change. It is crucial to know when organizational routines such as SOPs are appropriate and when they are not. The practice of benchmarking can help make corrective action timely.

Benchmarking should be seen as learning from others in order to accomplish similar objectives. Benchmarking attempts to identify success, excellence, and superiority, and then imitates them. Such imitations borrow from those at the cutting edge in terms of their products, services, processes, and procedures. The objective is to keep pace with the best with regard to costs, cycle time, satisfaction and value, and working capital on hand, in particular inventory.

The rationale for using benchmarking is not unlike the philosophy of professors who curve grades. Telling students that, in order to receive an A in the course, they must get 90 percent of the total points possible or 95 percent of the score compiled by the best student in the class is a form of benchmarking. Such a standard provides students with expectations against which they can judge themselves and, more importantly, decide how hard they want to work. Without standards and expectations, one is deprived of the means to judge performance, to differentiate between good, bad, and unwanted.

Judging from the published accounts, benchmarking permeates all aspects of organizational behavior. Businesses of all types and sizes are using it, including Xerox, AT&T, and IBM. Ford Motor used it to design its Taurus model and to redesign its accounts payable department. Firms borrow from those in the same business as well as in different areas. Xerox learned a few lessons in warehousing from L. L. Bean. Granite Rock Company learned about on-time delivery from a pizza chain. Some have tried benchmarking in small, incremental steps, others have gone for major remaking of their organizational structure and products through benchmarking studies that last months, sometimes years. In many circles, benchmarking is seen as a continuous effort.

To help firms perform benchmarking, two resource centers have been set up to act as clearinghouses for information and case studies. The American Productivity and Quality Center, located in Houston, has organized the International Benchmarking Clearinghouse, while the Strategic Planning Institute has set up the Council of Benchmarking in Cambridge, Massachusetts. Consultants on benchmarking abound. Periodicals continue to provide a steady stream of accounts about cutting-edge companies that could serve as the trigger to launch benchmarking in areas such as changes in method of production, product design, order entry, invoicing, accounts payable and receivable, payroll, and human resources.

(a) **How to Use Benchmarking.**   Given the resource being devoted to the practice and propagation of benchmarking, it is no surprise to see the number of approaches being put forward. According to Jeremy Main, writing in *Fortune*, October 19, 1992, IBM has a four-phase approach, AT&T has used nine steps, and Xerox has ten steps.[2] No matter what the number used, it is best to remember that benchmarking ought not to be a shopping spree.

Treat it not as an impulse to borrow indiscriminately, but as a process subject to thinking, planning, measuring, analyzing, implementing, and reassessing. Each of these steps in the benchmarking process (BMP) may be briefly explained thus:

1. **Thinking** involves the initial impulse or the motivation to undertake the BMP. It may be triggered by a customer survey, a chance encounter between two purchasing managers representing different firms, or even the readings of a manager.

2. **Planning** means deciding upon what it is that needs to be benchmarked and agreeing on who will be used as the role model, the pacesetter, for the BMP.

3. **Measuring** is what keeps the process from turning into an expedition based on impressions. Tangible measurements of how the firm is performing when compared with the performance of others will reveal to what extent change is warranted.

4. **Analyzing** means evaluating the changes being contemplated on the basis of the measurements obtained in step 3. Such measurement-based analysis helps to justify the need for change and also reveals, during the implementation phase of the project, whether the change is effective.

5. **Implementation** involves putting in place the changes planned with the help of benchmarking.

6. **Assessing** a project undertaken is always a necessary ingredient. Assessments show whether the change(s) are working out as planned.

Benchmarking provides an organization with models for improvement. Since BMP reveals processes that are in place and being implemented, it should ensure that changes are practical, realistic, and attainable. Through BMP, it is possible to keep pace with the industry norms if not its cutting edge, which helps a firm improve its performance. In short, BMP is an antidote to the status quo, and even obsolescence; it is one form of "stealing" that is not unethical.

## 9.7 FRAME BREAKING

Benchmarking is, in effect, a special form of the practice known as environmental scanning. Another variation of environmental scanning is frame breaking, an exercise that helps managers think about scenarios that are unlikely and yet could somehow come to pass—like the price of oil falling to ten dollars a barrel at a time when the rate was more than thirty. Such a scenario was far from the minds of oil barons. Yet the managers at Shell were planning strategies for dealing with just such an event, were it to come about. When it actually did happen, they were ready, having planned for it as a learning exercise.

Thinking through and planning such scenarios can help disrupt the familiar mental models, which can lead to enhanced creativity. Being in a rut is not the route to becoming imaginative. The key that turns frame breaking into learning is being able to visualize the unusual and develop plans for strategic response in the face of the unlikely. Such planning, even without deployment of the plans, expedites learning, according to its practitioners. One of the more famous practitioners of scenario planning is Shell. Another firm that goes to considerable lengths in planning scenarios, according to Peter Senge, is the insurance firm, Hanover.[3] The approach is also used in Japan. Given all this, it is fair to assume that frame breaking is not mere science fiction; instead, it is an effective tool in training managers as creative planners. Such exercises in creative learning translate to enhanced evolutionary competence as well as the all-important profits.

Certainly, a role for frame breaking exists within accounting contexts. Some years ago the thought that a firm could operate without working capital would have ranked as managerial heresy, if not outright lunacy. Yet today there are Fortune 100 firms that are doing just that.[4] Twenty years ago, a few accountants could have enriched themselves considerably if they had planned for a world without working capital and convinced their top managers of the benefits of traveling down such a strategic highway.

### 9.8  PROCESS CONSULTATION

Process consultation is familiar to students and professionals in the field of organizational behavior. Edgar Schein, who helped develop the approach to group facilitation, calls the approach "a general philosophy of helping" intergroup processes.[5] The role of helper is not limited to consultants and group facilitators, indeed, one's relatives and friends, as well as supervisors and managers, could be seen as helpers. Consultants find themselves playing various roles in dealing with their clients. According to Schein, at various times a consultant may find himself playing the role of an expert, a doctor, and even a process consultant. The focus of process consultation is on getting the clients to become more helpful to their fellow employees.

One of the considerations process consultants need to be concerned with is ensuring that they do not impose their preconceptions on the group with which they are working. They must make sure that they understand the objectives of the group as seen by its members. They should not form any conclusion about what the group ought to be doing.

Schein differentiates between three approaches consultants can use while working with their clients. One approach has them playing what he describes as the role of an expert. At the functional level, this means the clients want to know how best to approach the problem they confront. Underscoring that assumption is the knowledge that the individual and/or the group has thought through the problem and has communicated it correctly to the expert. It also assumes that the expert has the knowledge with which to resolve the issue faced by the client. The problem such a role presents is knowing when the clients are using the expert as a crutch, avoiding the work required to come up with a solution. In such situations, experts must know how to say no.

The second approach consulting can take is that of a doctor. Clients expect consultants to tell them the nature of the problem they face and to provide a cure. Theoretically, this seems very practical, but the approach may go awry because the clients may not be communicating the symptoms of the problem accurately. Alternatively, the consultant may not know any more than the client about the cure for the problem. Furthermore, playing the role of a doctor assumes the client will accept the diagnosis, as well as the cure suggested. In practice, this is often not the case.

**(a)  Assumptions Behind Process Consultations.**   These two approaches—of acting out an expert or playing a doctor—are not as effective as the approach used in process consultation. The following are summaries of the assumptions Schein believes are required for using this approach:

- The consultants assumes that clients do not really know or understand the problem. If they knew the problem, they would not be in need of an expert. They need help in figuring out how to label, define, and understand their symptoms.

- It is best to assume that clients do not know what sorts of resources are available to them to help in solving the problem or how those resources may relate to their problems. .

- It is also necessary to assume that clients will benefit from involvement in the process that will lead to uncover the problem, particularly if they themselves are part of the problem.

- It is also helpful to keep in mind that the clients know what kind of remedies will help resolve the problem. They have the knowledge to figure out what works and what does not in their context, but they need help in unlocking the solution.

- It is best to remember that clients expect to benefit by learning how to solve future problems more effectively.

These assumptions underline that Schein is seeking to teach clients how to help themselves.

Schein also suggests that consultants start out in the process-consulting mode rather than in the expert or doctor roles. The latter two roles may be useful down the line, but at the start it is best to use an explorative approach. The initial intervention should be aimed at discovery, although the consultants should already be thinking through the problem. It's essential to let the group feel that it is they who own the problem. The initial intervention should convey that the consultant and the group will work as a team to solve the problem. Schein calls for the consultant to have a true sense of curiosity aimed at finding out more about the client's situation. There may be times when intervention, even at the start, can be actual advice, rather than simple inquiry. If the group members are not listening to each other or if the meeting format is wrong, it is best not to wait to make a correction.

Clients should be actively leading the quest for solutions on their own. They should be provided with new information in a form that will make it more acceptable. Such acceptance is facilitated when the new ideas are presented as alternatives and possibilities, and not as executive orders or even expert advice.

Schein also has some important advice regarding the use of surveys and feedback. He is not in favor of the approach in which a survey is undertaken and results are passed down to the employees, who are asked to work on the part of the problem that their level can resolve, letting the next level take care of the next stage. He is for feedback that identifies problems in such a way that the organizational level specifically concerned with that problem can address it directly. Schein recommends that the feedback be aggregated by group from the lowest level up. Each level is given its own results, but not in the presence of their supervisors. They are given the feedback for two reasons: first, to resolve the problem or clarify the data; second, to separate the problems reported into those they can solve themselves and those that need external action, either by managers or other departments, to be solved. This way, each level identifies the problems that are of concern to the employees at that level, allowing the problems to be worked on by that specific group. The senior managers do not have to be told about variances.

## 9.9 ACCOUNTANTS AND PROCESS CONSULTATION

Accountants should see themselves in the role of helpers rather than as experts or doctors. Their position within the organization is akin to that of a consultant giving clients help in helping themselves. The function of information accountants provide should be similar—to help managers manage better.

Such an orientation leads to information being delivered in forms most relevant to those receiving it. Variances will not reveal how the labor time spent on a given batch affected the income statement, rather they will be in forms that are of interest to those who can do something about it in the future. Variances will be reported not for the sake of reward or punishment, but for corrective action.

Information could also be provided that is of specific interest to the given users. Accountants might ask the information users themselves what performance measure they need, rather than supplying what accountants think they need. Clearly, far too much of the accounting information provided is based on the notion that all users have the same needs. It is time to learn that one size does not fit all.

Once such a user-orientation for accounting information is achieved, the procedures used to collect and report accounting information can be adjusted accordingly. Even if the orientation does not change at the outset, the procedures documentation process can gain from the assumptions that underscore process consultation. Groups called together to review current accounting procedures can work in keeping with the process consultation approach espoused by Schein. This will lead to a greater degree of employee ownership in procedures. The group processes dealing with procedures review and documentation will itself work in a more optimum manner and will also facilitate the implementation of the procedures redesign.

## 9.10 LEARNING ORGANIZATIONS ILLUSTRATED

Before concluding this chapter, we would like to show the learning paradigm in action with the help of two organizations where it seems to be in place: a steel mill in Texas and a California business that produces material for residential, commercial, and highway construction.

**(a) Chaparral Steel.**    This case is adapted from "Chaparral Steel: Unleash Workers and Cut Costs" by Brian Dumaine, published in *Fortune* magazine.

Executives from all across the United States are driving into Midlothian, Texas (pop. 5,141), home of Dee Tee's coffee house, seeking to discover how little Chaparral Steel has turned itself into the lowest-cost steel producer in the world. Its employees work in flawless unison, and they exude the elements that are considered attributes of effective management—customer service, empowerment, quality, training, and even more importantly, the ability to work as a team for the good of the organization.

It produces steel in 1.6 hours of labor per ton, versus the 2.4 hours it takes other mini-mills and 4.9 hours for integrated producers. In making products like beams for building skyscrapers and rods for concrete reinforcing, Chaparral is "in a down and dirty commodity business." Focusing on its philosophy of lowering costs, it has made money in a so-called mature industry, even in a sluggish economy.

Chaparral's success is being directed by CEO Gordon Forward, a Canadian with a PhD in metallurgy. In 1975, Texas Industries, which owns 81 percent of Chaparral, asked Forward to leave his job at a Canadian steel company to help found a mini-mill. In order to make his steel mill into the world's lowest-cost producer, he focused on three ideas: the classless corporation, universal education, and freedom to act.

In Chaparral's classless organization, top management treats workers like adults—"management by adultery," as Forward likes to joke. This attitude goes beyond gestures such as free coffee and casual executive dress, since workers in this egalitarian society receive a salary and bonus based on individual performance, company profits, and new skills learned. There is no need to punch a clock, and the workers can set their own lunch hours and breaks. After they come to work, they can park next to the CEO—if there's a spot—and then walk through the executive offices to get to their locker room, which is as clean and bright as any in a country club. Says Forward: "We figured that if we could tap the egos of everyone in the company, we could move mountains." In return for extraordinary freedom and trust, workers are expected to show initiative, use their common sense, knowledge, and intelligence, and see that the job gets done. To ensure an educated workforce, Chaparral makes sure that at least 85 percent of its 950 employees are enrolled in courses, cross training in such varied disciplines as electronics, metallurgy, and credit history.

Such adherence to the learning organization paradigm lowers cost in assorted ways: While designing its new mill for making wide-flange steel beams used in bridges and buildings, its workers developed patent-pending technology that manufactures a final product with just eight to twelve passes through the system, versus traditional methods that require up to fifty passes. In another part of the mill, two maintenance workers invented a machine for strapping bundles of steel rods together that costs only $60,000, versus the $250,000 required for the old machines, and the new machine did the job faster and more flexibly.

Such savings are not limited to production. In Chaparral's customer service center (where the goal is "to be the easiest steel company in the world to deal with"), the sales, billing, credit, and shipping departments sit under one roof, with all workers trained to handle one another's jobs. If a customer calls sales and has a credit question, chances are the salesperson will know the answer; that saves time and uses employees more efficiently.

Can this kind of entrepreneurial culture work in a big organization? Forward thinks so, although it wouldn't be easy. Managers must learn to abjure the traditional trappings of power and let people do their jobs. Forward says, "Real motivation comes from within. People have to be given the freedom to succeed or fail." His results speak for themselves.

Chaparral is remarkable because, in the words of an associate professor at Harvard University's Business School, it has turned itself into a "learning laboratory."[6] Professor Dorothy Leonard-Barton defines " a learning laboratory" as an organization dedicated to "knowledge creation, collection, and control."

Such organizational structure does not occur by chance; it has to be visualized, designed, planned, created, and maintained by communicating the underlying vision throughout the firm. According to Leonard-Barton, organizational learning requires that four subsystems exist. They do in Chaparral Steel. They are:

- Owning the problem and solving it
  a. Willingness to solve problems through one's own initiative: independent problem solving
  b. Sense of egalitarianism and respect for individuals
  c. Rewards that recognize skill accumulation by individuals as well as performance
- Garnering and integration of knowledge
  a. Integrating knowledge available internally

  b. Valuing shared knowledge
  c. Rewarding acquisition of knowledge
- Challenging the status quo
  a. Continuous experimentation; benchmarking
  b. Valuing risk-taking
  c. Hiring employees on the basis of their ability to challenge the status quo, willingness to learn, and enthusiasm, as well as their specific skills.
- Promoting research through networking
  a. Integrating external knowledge
  b. Openness to knowledge from outside
  c. Developing alliances and networks with customers, competitors, and suppliers

**(b) Granite Rock Company.**  The Granite Rock Company won the Malcolm Baldrige National Quality Award in 1992. The following discussion of the techniques and strategies it used to merit the Baldrige Award is based on the Granite Rock Company's in-house publications provided to the author.

**(i) Defining the Mission.**  Vying for customers in a commodity industry that typically buys from the lowest-bid supplier, Granite Rock Company, a California producer of construction materials, is expanding the terms of competition to include high quality and speedy service. By emphasizing the hidden costs associated with slow service and substandard construction materials, such as rework and premature deterioration, the company is convincing a growing number of contractors of the value of using Granite Rock's high-quality materials and unmatched service.

This strategy is working for the Watsonville-based firm. Since 1980, the regional supplier to commercial and residential builders and highway construction companies has increased its market share significantly. Productivity has also increased, with revenue earned per employee rising to about 30 percent above the national industry average.

Most of the improvement has been realized since 1985, when Granite Rock started its total quality program. The program stresses satisfying two types of customers: the contractor, who normally makes the purchasing decisions, and the end-point customer, who ultimately pays for the buildings or roads made with Granite Rock materials. To spread its quality message, Granite Rock sponsors seminars for contractors, developers, architects, and suppliers.

Granite Rock can assure customers that its materials exceed specifications as a result of its investments in computer-controlled processing equipment and widespread use of statistical process control. Customers also can be confident that the materials will arrive when they need them. Granite Rock's record for delivering concrete on time, a key determinant of customer satisfaction, has risen from less than 70 percent in 1988 to 93.5 percent in 1991. That record tops the on-time delivery average of a prominent national company that Granite Rock benchmarked to improve its process.

**(ii) Granite Rock: A Snapshot.**  Founded in 1900, Granite Rock produces rock, sand, and gravel aggregates; ready-mix concrete; asphalt; road treatments; and recycled road-base material. It also retails building materials made by other manufacturers and runs a highway-paving operation. Granite Rock has a wide range of products and services available for any construction job, large or small:

- ready-mix concrete
- asphaltic concrete
- emulsions, road oils, spreader truck fleet
- slurry seal aggregates
- sand, clay, and baserock fill materials
- truck and rail transport services for building materials: brick, block, stone, masonry accessories and tools, drywall, and much more

In addition, the company's general engineering construction division, Pavex Construction Company, specializes in road construction and restoration projects, and supplies cement treated bases and roller compacted concrete.

Granite Rock competes in a six-county area extending from San Francisco southward to Monterey. Most of its major competitors are firms owned by multinational construction material companies. A vertically integrated company, Granite Rock employs four hundred people, who are distributed among branch offices, several quarries, fifteen batch plants, and other facilities. Approximately 250 of the employees are members of five unions.

**(c) Total Quality.**    Nine complementing corporate objectives, as shown in Exhibit 9–1, distilled from analyses of customers' requirements, are the cornerstone of Granite Rock's quality program. During the annual integrated business and quality planning process, senior executives systematically evaluate company-gathered data and develop measurable "baseline goals" to help the company advance toward each objective.

**CUSTOMER SATISFACTION AND SERVICE**
To earn the respect of our customers by providing them in a timely manner with the products and services that meet their needs and solve their problems.

**SAFETY**
To operate all Graniterock facilities with safety as the primary goal. Meeting schedules or production volume is secondary.

**PEOPLE**
To provide an environment in which each person in the organization gains a sense of satisfaction and acomplishment from personal achievements, to recognize individual and team accomplishments, and to reward individuals based upon their contributions and job performance.

**PRODUCTION EFFICIENCY**
To produce and deliver our products at the lowest possible cost consistent with the other objectives.

**PRODUCT QUALITY ASSURANCE**
To provide products which provide lasting value to our customers, and conform to state, federal or local government specifications.

**FINANCIAL PERFORMANCE AND GROWTH**
Our growth is limited only by our profits and the ability of Graniterock people to creatively develop and implement business growth strategies.

**PROFIT**
To provide a profit to fund growth and to provide resources needed to fund achievement of our other objectives.

**MANAGEMENT**
To foster initiative, creativity, and commitment by allowing the individual greater freedom of action (in deciding how to do a job) in attaining well-defined objectives (the goals set by management).

**COMMUNITY COMMMITMENT**
To be good citizens in each of the communities in which we operate.

**Exhibit 9–1.    A chart of Granite Rock Company's baseline goals.**

After annual improvement targets are set, the executive committee expects branches and divisions to develop their own implementation plans. Coordination across divisions is fostered by ten corporate quality teams that oversee and help align improvement efforts across the entire organization. Although committees are chaired by senior executives, members include managers, salaried professional and technical workers, and hourly union employees. Teams carry out quality improvement projects as well as many day-to-day activities and operations. In 1991, nearly all workers took part in at least one of the company's 100-plus quality teams.

In 1987, the company introduced the "Individual Professional Development Plan," a voluntary program in which 74 percent of Granite Rock employees now participate. At least once a year, workers meet with supervisors to define their job responsibilities, review accomplishments, assess their skills, and set skill- and career-development goals.

Granite Rock encourages all employees to continue learning and sponsors a series of classes and speakers on technical topics. In 1991, Granite Rock employees averaged thirty-seven hours of training at an average cost of $1,697 per employee, three times more than the mining industry average and thirteen times more than the construction industry average. Granite Rock University courses range from heavy equipment maintenance skills for mechanics and concrete mix design seminars for mixer truck drivers to courses in health and wellness, personal growth, and basic business law.

Training is a core part of Individual Professional Development Plans, or IPDPs, which each Granite Rock worker can develop annually. These plans highlight job accomplishments, identify individual skill strengths, and form a commitment from the company to invest in the individual. The aim of the IPDP is to support total quality through involvement and growth of employees.

Responding to customer concern over rising trucking costs, the company developed GraniteXpress, the construction industry's version of an automatic teller machine.

With the automated system for loading aggregate, a driver inserts the equivalent of a credit card into a terminal, keys in the type and amount of aggregate, and proceeds to the loading facility where the truck is accurately filled over an electronic scale. The service, which operates twenty-four hours a day, seven days a week, has reduced the time a trucker spends at the quarry to nine minutes, as compared with twenty-four minutes before GraniteXpress was installed.

As part of Granite Rock's effort to reduce process variability and increase product reliability, many employees are trained in statistical process control, root-cause analysis, and other quality assurance and problem solving methods. This workforce capability helps the company exploit the advantages afforded by investments in computer-controlled processing equipment. Its newest batch plant features a computer-controlled process for mixing batches of concrete, enabling real-time monitoring of key process indicators. With the electronically controlled system, which Granite Rock helped a supplier design, the reliability of several key processes has reached the six-sigma level (six sigma is a statistical term indicating a defect rate of 3.4 per million). The new system and real-time data collection will be adopted by the company's other concrete plants, where control of product variability either approaches or exceeds the three-sigma level.

Applying statistical process control to all product lines has helped the company reduce variable costs and produce materials that exceed customer specifications and industry- and government-set standards. For example, Granite Rock's concrete products consistently exceed the industry performance specifications by 100 times.

Innovative applications of technology have helped the company enhance its service offerings. Granite Rock's Arthur R. Wilson Quarry may be the most advanced aggregate production facility in the country. Heavy investments in recent years have improved production efficiency, quality control, and customer service.

Charts for each product line help executives assess Granite Rock's performance relative to competitors on key product and service attributes, ranked according to customer priorities. By 1995, the company aims to build a 10-percent lead over its nearest competitor for each indicator of customer satisfaction.

Granite Rock uses an annual survey that allows buyers to rank the company against its competitors. Every three to five years, more detailed surveys are conducted. Customer complaints are handled through product/service discrepancy reports that require analysis of the problem and identification of the root cause. Ultimate customer satisfaction is assured through a system where customers can choose not to pay for a product or service that doesn't meet expectations. Dissatisfaction is rare, however. Costs incurred in resolving complaints are equivalent to 0.2 percent of sales, as compared with the industry average of 2 percent.

Today, Granite Rock Company is the largest American owned and managed construction materials supplier in Northern California. It focuses on customers by applying experience and innovative technology to creatively satisfy their needs. Company co-presidents, Bruce W. and Stephen G. Woolpert, A.R. Wilson's grandsons, and Granite Rock employees continue to strive for the same high standard Wilson established more than ninety years ago.

**(d) Quality and Service.** Granite Rock's nine corporate objectives are the cornerstone of the total quality program and service. The following quotes from Granite Rock's company literature illustrates why they may be called a learning organization.

- Customer satisfaction:
  "We are serious about customer satisfaction. So serious, that any customer needn't pay their invoice unless completely satisfied. We regularly survey our customers, track and respond promptly to comments and complaints, and make sure our products arrive quickly and on time. Our goal is to achieve 100% satisfaction of our customers' needs."
- Safety:
  "Granite/Rock people know that guaranteeing job safety is the most important thing they do. This awareness in reflected in our safety record, two times better than the state's average for our industry."
- People:
  "Recognition and promotion from within have helped Granite Rock strengthen its quality efforts. Both are supported by the Individual Professional Development Plan (IPDP) for skills training and development. The average training investment for each Granite Rock person is nearly four times the industry average."
- Product quality assurance:
  "Our A.R. Wilson Quarry processing plant is a world class facility, designed to assure that our products meet and exceed specifications with tight uniformity. Our Materials Research and Testing Laboratory oversees product formulation and quality assurance for aggregates, concrete and asphaltic concrete. And each Granite Rock person has the power to stop delivery of a product that isn't just right. All this results in products of exceptional quality, according to our customers, who report high product satisfaction."

- Community commitment:
  "Granite Rock people are encouraged and recognized for their community activities. Granite Rock produces an annual Fourth of July concert to benefit community organizations, is involved in "Adopt-A-School" programs, and has a long list of other charity activities, sponsorships and contributions. Committed to responsible use of its natural resources, Granite Rock's work in quarry reclamation, water conservation, and plant beautification has received national and local recognition."

- Management:
  "At Granite Rock, the manager's role is to be a teacher, leader and coach, and to stimulate new ideas, fresh thinking and interdepartmental cooperation. Impeccable integrity, customer commitment, and a constant focus on quality are a priority for all Granite Rock managers."

- Training
  "Granite Rock subscribes to the belief that quality products and exemplary service can only be accomplished with highly skilled and motivated people who take personal ownership of company goals. Driven by that belief, it focuses on training and expecting each Granite Rock employee to set individual goals toward doing the best job possible."

**(e) Some Specific Steps.**    Below are some specifics that describe Granite Rock's organizational culture and performance.

*Egalitarian Systems.*    The firm used a lottery to ensure that all employees had an equal chance to attend the ceremony at the White House when the firm was awarded the Baldrige Award.

*Listening to Customers.*    The firm makes it a point to find out how customers feel about the service they receive. In one of their periodic surveys, the company learned about the trouble customers had finding the Granite Rock plant. Working in teams, they came up with solutions, such as signs to the plant along the way. In another innovation, an automatic clock stamped the time the customer entered the plant on a ticket. This ticket was returned by the customers when they left the plant. This system allowed them to measure the amount of time customers had to spend at the plant.

Customers are also able to load aggregate after hours using an ATM-like card. Granite Rock has set up an information center at one of their branch offices to help customers get answers to their questions.

Unhappy customers are allowed to pay less on their invoice. Such short payments, serving as a form of feedback, are then tracked to measure customers' satisfaction.

*Employees Take Initiative.*    When it comes to service, employees do not wait. They take it upon themselves to provide the service requested or do what needs to be done.

*Emphasis on Measurements.*    In a variety of ways, Granite Rock tries to measure the quality of both their product and their service. They rely heavily on statistical quality controls for their measurements. Surveys are also used, and such surveys are designed in-house.

*Making Accounting More Accountable.*   To better serve internal and external customers, accounting expedited its processing of information and the time it took to notify managers of the short payments made by unhappy customers.

*Being Better Than the Average.*   Granite Rock customizes products for the particular customer, instead of relying on industry-supplied specifications.

*Planning For the Unexpected.*   The company learned to plan for contingencies after careful study. This way they are ready for the unexpected.

*Making Communication Important.*   Part of Granite Rock's organizational culture is open communication, yet another aspect of their adherence to the learning paradigm.

The company has done well with its customers even when the economy of the region has been very slow, which lends support to the argument that learning to provide quality pays off in tangible ways.

## 9.11 CONCLUSION

It is undoubtedly true that procedural documentation manuals are archival instruments, but they need to be more than that. They must not be seen as merely the documented reflection of the way things are; instead, the preparation and maintenance of procedures and operating manuals ought to be seen as an occasion for analyzing how the organization is carrying out its tasks. Such analysis must be conducted with a view towards eliminating procedural gridlocks, incorporating new technologies, and enhancing organizational productivity. The analysis should become an occasion for learning from employees' experience, for frame breaking, for benchmarking, and for continuous improvement. When this philosophy is applied to procedures documentation, corrective action is not left to chance when once-appropriate standard operating procedures become obsolete.

While writing the manuals, it is essential to be able to apply the methodology of activity management, process redesign, and the paradigm of learning organizations to the task of procedures review and documentation. Writing procedures documentation should be an occasion for organizational self-examination, ensuring that deadwood and redundancies are not allowed to accumulate. This will ensure that procedure documentation is not an afterthought but an illuminative act, an inoculation against obsolescence.

## ENDNOTES

1. The following works present a sampling of what leading-edge experts are saying about organizational learning. In developing a perception of organizational learning, the author has relied on these works.

Argyris, Chris *Knowledge for Action*. San Francisco: Jossey-Bass, 1993.

Drucker, Peter. "The New Society of Organizations." *Harvard Business Review*. (September/October 1992): 95–105.

Garvin, David. "Building a Learning Organization." *Harvard Business Review*. (July/August 1993): 78–92.

Kim, Daniel. "The Link between Individual and Organizational Learning." *Sloan Management Review*. (fall 1993): 37–50.

Nonaka, Ikujiro "The Knowledge Creating Company." *Harvard Business Review*. (November/December 1991): 96–104.

Normann, Richard and R. Ramirez. "From Value Chain to Value Constellation: Designing Interactive Strategy." *Harvard Business Review*. (July/August 1993): 65–77.

Quinn, James. B. *Intelligent Enterprise*. New York: Free Press, 1992.

Senge, Peter M. *The Fifth Discipline: The Art and Practice of the Learning Organization*. New York: Doubleday, 1990.

Stata, Ray. "Organizational Learning—The Key to Management Innovation." *Sloan Management Review*. (spring, 1989): 63–74.

Tyre, M. & W. Orlikowski. "Exploiting Opportunities for Technological Improvement in Organizations." *Sloan Management Review*. (fall 1993): 13–26.

2. Main, Jeremy. "How to Steal the Best Ideas." *Fortune*. (October 19, 1992): 102–108.

3. Senge in his book *The Fifth Discipline* devotes considerable space to discussing mental models and the various means some firms are using to train their managers in breaking out of the familiar.

Another book that treats the subject is Peter Schwartz's *Long View*. A short introduction is provided in Arie de Geus, "Planning as Learning," *Harvard Business Review*, March/April 1988): 70–74.

4. Tully, Shawn, "Raiding a Company's Hidden Cash," *Fortune*, (August 22, 1994): 82–89.

5. "A General Philosophy of Helping: Process Consultation." *Sloan Management Review*. (Spring 1990): 57–64. Edgar Schein has lot more to say about the theory and practice of process consultation in a two-volume book, *Process Consultation*, 2nd edition. Reading, Massachusetts: Addison-Wesley, 1987.

6. Leonard-Barton, Dorothy. "The Factory As a Learning Laboratory." *Sloan Management Review*. (fall 1992): 23–38.

# CHAPTER 10

# DECONSTRUCTING ACCOUNTING PROCEDURES: A TOOL FOR ORGANIZATIONAL LEARNING

## 10.1 INTRODUCTION

In this chapter, we will apply the various competencies, TQE, BPR, ABC, ABM, the organizational learning paradigm, and process consulting to the process involved in the documentation of accounting procedures. Through the application of such techniques, the documentation process becomes broader. It becomes not a simple reflection of the status quo, but an occasion to ascertain the value of the existing procedures, as well as to evaluate their efficiency and cost effectiveness.

In order to understand the value of a given procedure, it must first be deconstructed to see its objectives and its value, and whether it can be modified and improved in any way. Once such questions are answered, the procedure may then be reconstructed with modifications and improvement. If the existing procedure is not worth redesigning, an alternative approach is implemented.

We have deliberately chosen "deconstructing," a term used in the literary circles to describe what must be done to a given procedure before it is documented. By using it we seek to demonstrate that labels, while useful, must not turn into straitjackets. One must be free to borrow from any source if such borrowing helps the realization of the final objectives.

## 10.2 DECONSTRUCTING A PROCEDURE

We will analyze an accounting procedure to help demonstrate the approach proposed in the book. Our analysis uses tools that are being used in various approaches to organizational improvement discussed in chapters 6 through 9. Using the tools in this manner is in keeping with the argument that various approaches to organizational improvement can coexist and complement each other.

The procedure will be analyzed by deconstructing the documentation provided by a well-regarded book about accounting policies and procedures manuals. The procedure to be examined is one used to reimburse travel expenses incurred by the faculty at a state university. The complete documentation for the procedure chosen is provided in the appendix for this chapter, and we will refer to it throughout as we review it.

We are using the documentation of an actual accounting procedure from the book *Design and Maintenance of Accounting Manuals* for two reasons.[1] Perhaps the more important of the two is the fact that the procedure illustrated is quite representative of the practice at a large number of academic institutions. While the amounts available for faculty travel at different universities may vary, as would the actual processing as well as the extent to which the system in place is computerized, the procedure as described is typical. In contrast to some procedures, it is relatively uncomplicated. It will also serve to illustrate the differences between what may be called the archival versus the illuminative approach to procedures documentation.

Archival documentation reflects the procedures, and satisfies the accepted standards for accounting documentation. But even though the documentation is dated September 2, 1986, it could just as easily have been written in 1976, or even 1966. The major difference over the three decades would have been in the amount allowed for daily meals. However, the procedures that will be written for 1996 are bound to be different, given the changes taking place in the environment within which accounting functions are performed.

## 10.3 RAISING SOME QUESTIONS ABOUT THE PROCEDURE

As we evaluate the procedure documentation included in the Appendix, a number of questions come to mind. Merely raising the questions is a fact-finding exercise; it does not imply a negative judgment about the merits of whatever is being questioned. The judgment concerning the value provided by the procedure can only be exercised after the facts are available. There may be very valid reasons for undertaking a certain task within a procedure in a given manner. If so, they can be allowed to remain in effect. If this is not the case, then different, more effective procedures could be put in place.

Below are questions and observations that ought to be considered in a review of the procedures such as the one being analyzed:

- The nature of the approval process
  1. Is highly centralized?
  2. Is it cost-effective to have financial affairs be a part of the approval process?
  3. Does financial affairs have the resources to devote to such approvals for the entire university?
  4. Do the resources devoted to the approval of such travel take time and attention away from other tasks?
  5. What does it take in terms of costs and resources for financial affairs to undertake such approvals?
  6. Can the approval be delegated to a lower operational level and still be cost-effective?
- The nature of the paperwork required
  1. The procedure seems task-intensive; a faculty member wanting to travel may have to fill out rather lengthy forms with multiple copies as many as four times: to get the initial approval, to get the advance, to get reimbursed, and, if he uses a university vehicle, to get authorization.
  2. There seems to be considerable duplication between the different forms. A considerable amount of data is being required afresh with each form.
  3. What is the justification for requiring such detailed, itemization for the expenses incurred for meals and lodging?
- The cost benefits of the procedure
  1. What is the clerical and administrative costs associated with such a procedure?

2. Who pays how much for the cost of administering the procedure. Those to be considered are the faculty member applying for permission and reimbursement; their department; the college; the controller's department; the university; and the tax payer.

3. What is the ratio between the total amount allowed for such travel and the costs associated with this accounting procedure? Has it changed over the years?

- The positive attributes of the procedure
  1. The procedure has attempted to minimize the paperwork by not requiring documentation for all meals. Meals for less than $25 do not need validation with receipts.
  2. The procedure specifically points out the time it will take to get the reimbursement processed.
  3. The instructions for filling out the forms, as well as the documentation for the procedure itself, are very understandable and well-written.
  4. The procedure is quite specific, but is that extent of detail really needed? Why not trust the faculty's judgment for using the most economical and convenient means of traveling?

- Given the quantity of paperwork being generated, how can emergent information technology be of use in improving the process?

## 10.4 TRACKING THE COMPONENTS OF THE PROCEDURE

In addition to raising the questions listed in the previous section and having them answered, those seeking to review the procedure prior to documenting it will also need to track the steps required for its completion. In addition to listing the individual elements of the process, we indicate the following for each element: we will use the letter "V" to indicate whether the step adds value to the institution; the letter "P" will be used to show that it helps to move the process forward; the letter "N" will be used to suggest that it hinders the completion of the process—in other words, is non-value-adding. If the analysis were to be actually performed at the university, then we would also indicate the estimated time and costs incurred for each of the various elements. This is an essential part of the actual review process. Those estimates provide tangible measures about the value of the procedure being used. Of necessity, such estimates will be left out of the analysis being done here. It is also possible that some of the elements listed are not part of the process at a given school. If so, that institution has already embarked on a quest to improve the procedure.

| # | Category | Element |
|---|---|---|
| 1. | V | Professor's paper is selected for presentation at a conference. |
| 2. | P | Professor fills out a travel request form and gives it to his department chair. |
| 3. | N | The travel request waits for the approval of the department chair. |
| 4. | P | The department chair approves the request. |
| 5. | N | The form goes to the dean of the college and waits for his approval. |
| 6. | P | The dean approves the request and sends it over to financial affairs (FA). |
| 7. | N | The request waits for action at FA. |
| 8. | P | FA checks the availability of funds. |
| 9. | P | FA approves and informs the professor. |
| 10. | P | The four copies of the form are distributed. Two go back to the professor, one goes to the controller's office for their records, and the last one goes back to the college where professor belongs. |
| 11. | V | The professor attends the conference and makes the presentation. |

| 12. | P | The professor has to ensure that documentation for his expenses is collected, and he records his expenses as they occur or soon afterwards. |
| 13. | P | The professor fills out the form called employee travel voucher and includes the necessary documentation in the package. It then is sent to the department chair. |
| 14. | N | The voucher waits for the department chair's approval. |
| 15. | P | The chair signs it and sends it to the controller's office. |
| 16. | N | It waits for processing in the controller's office. |
| 17. | P | One or more staff members there will take care of the tasks required to approve the reimbursement. |
| 18. | P | The approval will be verified. |
| 19. | P | The check will be written and sent to the professor. |
| 20. | P | The professor will deposit the check in his bank. |

As the result of analysis and as a part of the review, the steps involved in the process should be examined to see if they must remain. Certainly those labelled as N ought to go. Items categorized as P may remain, be altered, be dropped, or be replaced, depending on the costs incurred in processing them.

A number of the steps categorized as P really duplicate the work. For instance, the professor's department chair approves the travel voucher. This is to ensure that the correct and approved amount is being requested as reimbursement by the faculty member. The amount allowed is already on file in a number of places, including accounting. If the professor is submitting an excess amount, it will be noticed. Within the accounting departments, there is the processing, which is followed by the verification. Such care—given that the amount to be reimbursed is already approved, may not be cost-effective; indeed, some will say, such redundancy amounts to costly overkill!

It should also be obvious that emerging information technology could help alter the procedure. There is technology available to help improve the process. Having the approval of the request on file in the system could save the professor from submitting a good deal of data twice: first in connection with the request itself, and second with the submission of the employee travel voucher. A good deal of paper, not to mention labor, could be dispensed with if the procedure were handled through a system that relies on client servers rather than old-fashioned forms with four copies of different colors.

The review suggested above will not produce identical answers for all universities. Each academic institution is likely to have its unique arrangements for the implementation of the procedure, even though the procedure on paper is fairly uniform. Most colleges have faculty members fill out the request for permission to travel; these are reviewed by their chair and the dean of the college where the department is located. The dean may either provide the final approval, or, in some institutions, the level above the dean may also be involved. The rest of the steps involved in the procedure are also quite typical.

## 10.5 EXPLAINING THE ILLUMINATIVE APPROACH

Using the illuminative approach to accounting procedures documentation implies more than simply writing down the status quo. The documentation process should begin with a review of the procedure currently in place to see the extent to which it is efficient, effective, and value-adding. The impetus for such an approach comes from the organization's commitment to the TQE as well as the learning organization paradigm. The actual procedures review and redesign should use all or some of these nine steps.

- Identify the process to be redesigned
- Analyze the effectiveness and the efficiency of the process
- Analyze the relationship between the procedure and the organization's strategic plans
- Identify how the procedure supports the mission and strategic goals of the organization
- Explore IT that may be useful
- Design a prototype
- Measure prototype for its effectiveness and efficiency
- Carry out full implementation
- Measure how well the redesigned process performs

In this case, by being selected for analysis, the procedure has been already identified. However, were it a situation in a real university, the selection itself could have been the result of a process devoted to selecting procedures for review. Alternatively, the procedure could have been chosen for review after learning about how industry is handling its travel and expense reimbursement procedures.

An article describing how such reimbursements are being reengineered in a recent issue of *Management Accounting* could very well have led to such a review.[2] The article notes that cutting-edge firms are streamlining their travel reimbursement policies. Given the redesigned streamlining of the policy, its documentation is accomplished in one page instead of the twenty or more pages that were sometimes used. A number of requirements pertaining to management approval have also been abandoned, presumably for reasons of cost-effectiveness. Those approval tasks were costing more in time and resources than they were saving in unauthorized expenses. Reimbursements are being provided by means of electronic fund transfers. Managers are provided credit cards that even allow for cash transfers, thus cutting down on yet another source of paperwork, namely, cash advances for travel. Giving credit cards has not led to unauthorized uses of funds. Such a redesign of reimbursement process has meant lower costs and greater planning and control, as well as enhanced satisfaction on the part of the corporate travelers, the customers for the reimbursement process.

## 10.6 THE ELEMENTS OF THE ILLUMINATIVE APPROACH

The illuminative approach to documention would include an analysis of the effectiveness and efficiency, as well as the cost-effectiveness of the procedure under review. It could be done by an individual charged with the assignment or by a group of employees who are already involved with the various elements of the procedures. Such a group would need less time in order to acquaint themselves with the process, however, they would still need to orient themselves because their knowledge of the process may well be limited to their particular slice of the action, and not the entire process.

## 10.7 FIRST UNDERSTAND, THEN DECONSTRUCT

Before trying to change things for the better, the reviewer(s) must understand the existing procedure well and develop an appreciation for the procedure's current performance and its resource usage. Such an understanding could be obtained by:

- Talking with those involved in administering the procedure
- Surveying the boundaries of the procedures
- Reading up on the currently available documentation pertaining to the procedure
- Talking with those performing the tasks involved
- "Walking through" the procedure
- Detailing the process on a flow chart, describing not just activities, but also the time involved in getting from one task to another and the time spent waiting for a given task to be performed
- Calculating, or at least estimating the cost of the procedure
- Comparing the expense of the procedure with the actual budget for faculty travel, in order to put the costs into perspective

To help focus the review of the procedure, look for the following during the procedure review:

- Non-value-adding tasks
- Duplication; unnecessary administrative paperwork
- Potential for simplification/improved form design
- Opportunity for standardization
- Chances to improve relationships with customers
- Alternatives to the procedure
- Use of information technology to streamline, upgrade, or even reduce costs.
- Proper implementation of the procedure

In looking for the above points it is helpful to remember some of the tools used by TQE such as "Three-Mu checklists," as well as the "who, what, where, when, why, and how" approach.

In order to ensure that the review is effective, it should not be carried out in isolation of the organizational mission and strategic objectives.

### 10.8 CONCLUSION

This chapter illustrates the illuminative approach to procedure documentations. The use of the illuminative approach to documentation can help to bring learning to the accounting department. Procedures documentation can become the occasion to question the status quo in the accounting department. The ripple effect can even spread the learning to the rest of the organization.

### APPENDIX: AN EXAMPLE OF PROCEDURES DOCUMENTATION

The sample documentation of a policy/procedure dealing with the reimbursement of faculty travel at a public university, which follows, is adapted from:

Brown, Harry. *Design and Maintenance of Accounting Manuals*, Second Edition. New York: John Wiley & Sons, 1993.

## Administrative Procedure

THE UNIVERSITY OF SOUTHERN MISSISSIPPI

ACCOUNTING SERVICES

- RETRIEVAL NO.    10,015
- PAGE    1 of 2
- ISSUE DATE    January 10, 1983
- ORIGINATOR    Accounts Payable
- SUPERSEDES

SUBJECT:  NEW REMITTANCE VOUCHER (ACC 14)

Many expenditures for services do not require a purchase order and authority to pay has been processed through the use of a memo, a purchase requisition sent directly to Accounting Services, or by a remittance voucher.  The new Remittance Voucher form is to be used by anyone authorized to approve disbursements for an account.  The use of the purchase requisition for payment authorization is no longer acceptable.

Use of the Remittance Voucher

Items which do not normally require purchase orders and for which this Remittance Voucher is to be used are:

Seminar and conference registration fees for employees
Honorariums to non-employees
Speaker's fee
Travel expense reimbursement for non-employees (recruiting, speakers,
    professional fees and so forth)
Stipends
Agency accounts
Items on contract (leases, land purchases, entertainment groups, Cablevision, etc)
Legal, audit and other professional fees
Payroll taxes and amounts withheld for payment to others
Insurance
Postage (Post Office only)
Advertising
Utilities
Telephone
Freight when shipper is USM
Band and music awards (not trophies or plaques)
Registration workers
Officials - Athletic and Intramural Departments
Sales Tax
Refunds - housing deposits and Continuing Education fees

The form is used as a cover document for form ACC 14A - Multiple Vendor-Payee Attachment.

Supporting documents are to be attached to the accounting copy of the form. The preparing department retains the pink copy for comparison to the monthly budget activity report.

**Appendix 10–1.  An example of procedures documentation.**

Completing the Form

Department enters:

    A.  Department name
    B.  Account number
    C.  Telephone number of preparer
    D.  Name and address of vendor.  If used as a cover form for the Multiple
        Vendor-Payee Attachment, enter "See Attached Forms".  Social Security
        number must be supplied for payment to individuals.
    E.  Description of charge to be paid, including name of registrant(s) at
        a seminar or conference, purpose of payment and so forth.  If
        registration form, document or letter is to be sent with check,
        indicate that enclosure is attached.
    F.  Amount to be paid.  Attach supporting document, if any.
    G.  Total to be paid on this voucher.
    H.  Signature of person requesting payment and date signed.
    I.  Approval signature and date.  Payments to individuals require two
        approval signatures if an invoice is not available.
    J.  General Ledger Code, Object Code and Department to be charged.  If
        department is not entered, the charge will be made to the account
        shown at top of the form.
    K.  Amount to be charged to this account.  Form provides distribution for
        up to eight accounts.

Accounting Services enters:

    L.  Vendor code number.
    M.  Voucher number.
    N.  Voucher date.
    O.  Purchase Order Number, if any.
    P.  Name of person processing the voucher and date processed.
    Q.  Name of person reviewing or verifying information.
    R.  Preprinted number of special check used to pay voucher.
    S.  Special handling required or enclosure to be mailed.
    T.  Amount of encumbrance to be liquidated, if any.

This form is available from Accounting Services, 201 Forrest County Hall.  The
use of unauthorized forms should be discontinued.

Any questions concerning this procedure should be directed to Accounting Services,
Box 5143, Telephone 4084.


Attachment:  Copy of Remittance Voucher

**Appendix 10–1.  Continued.**

**REMITTANCE VOUCHER**
**UNIVERSITY OF SOUTHERN MISSISSIPPI**
**HATTIESBURG, MISSISSIPPI   39406-5143**

THE UNIVERSITY
OF SOUTHERN MISSISSIPPI

| DEPT NAME | (A) |
| ACCT NO | (B) |
| TEL NO | (C) |

VENDOR ____(D)____

VENDOR CODE

(L)

| DESCRIPTION | AMOUNT |
|---|---|
| (E) | (F) |

| REQUESTED BY (H) | DATE | APPROVED BY (I) | DATE | TOTAL | (G) |

| ACCOUNTING USE | GL (J) | OBJECT (J) | DEPARTMENT (J) | LIQUIDATION (T) | EXPENDITURE (K) |
|---|---|---|---|---|---|
| VOUCHER NUMBER (M) | | | | | |
| VOUCHER DATE (N) | | | | | |
| PURCHASE ORDER NUMBER (O) | | | | | |
| PROCESSED BY: DATE (P) | | | | | |
| VERIFIED BY: (Q) | SPECIAL CHECK NO. (R) | ACCOUNTING USE (S) | | | |

ACC 14 (REV 1-83)         WHITE — ACCOUNTING         CANARY — VENDOR         PINK — DEPARTMENT

**Appendix 10–1.   Continued.**

# Administrative Procedure

THE UNIVERSITY OF SOUTHERN MISSISSIPPI

ACCOUNTING SERVICES

- RETRIEVAL NO.   30041
- PAGE            1 of 11
- ISSUE DATE      September 2, 1986
- ORIGINATOR      Controller
- SUPERSEDES      30023

SUBJECT:   TRAVEL PROCEDURES

Employees of the University are reimbursed for reasonable and necessary expenses incurred while in the performance of approved travel.   The following forms are used in the procedure.

| Form | Source |
| --- | --- |
| Permission to Travel Form (ACC 1) | Printing Center |
| Travel Voucher (ACC 2) | Printing Center |
| Remittance Voucher | Financial Affairs |

INDEX

Advance payment, registration fees
Approval for travel
Authorization for Travel
Cancellation, Travel Request
Change, Travel Request
Common Carrier
Domestic Travel
Employee Business Expense
Foreign Travel
High Cost Cities
Lodging, shared, spouse, family
Meals, day trips, overnight trips
Motor Pool
Other expenses
Parking
Permission to Travel Form and Instructions
Personal Travel Log
Private Vehicle
Pro Travel
Receipts
Registration fees
Reimbursement Procedure
Rental car
Spouse accompanying
Tips
Tolls
Travel Advance
Travel Voucher and Instructions
University vehicle, credit card

**Appendix 10–1.  Continued.**

## AUTHORIZATION FOR OFFICIAL UNIVERSITY TRAVEL

Each employee required to travel in performance of official duties and entitled to reimbursement for expenses incurred shall have prior authorization from the department chairman and/or other designated officials.  The Permission to Travel Form must be submitted at least two weeks in advance of expected departure date to conferences, conventions, and meetings.  In the case of out-of-country travel, the request must be submitted at least 90 days in advance of the requested departure date.

The originating department should be certain that the travel request form is properly completed.  The four-part Permission to Travel Form is prepared by the applicant and routed as follows:

A.  Domestic Travel:
    (1)  Department Chairman
    (2)  Dean of College of School or Division Chairman
    (3)  Financial Affairs (funds available)

B.  Foreign Travel, including Hawaii and Alaska, add:
    (4)  Vice President
    (5)  President (after obtaining State approval)
    (6)  Financial Affairs (for distribution)

When Financial Affairs receives the request for Permission to Travel, it will determine if funds are available.  If there is a problem and the request cannot be further processed, it will be returned to the Dean's or Director's Office.  If funds are available, Financial Affairs will retain the white and canary copies and return the other copies to the applicant.

## Distribution

White - Financial Affairs (for encumbrance)
Canary - Employee (used to obtain travel advance, if needed)
Pink - Employee (to file with travel voucher)
Goldenrod - Departmental file

If a travel advance is requested, the employee will obtain the canary copy from Financial Affairs and take it to the Business Office to obtain cash advance.  Money will not be advanced earlier than two weeks prior to the meeting or conference.  No cash advance for trips within the State.

The pink copy of the travel request must be attached to travel voucher when it is submitted for payment.  The goldenrod copy of the travel request should be maintained in the department's file.

**Appendix 10–1.  Continued.**

## Required Approvals

The approval requirements for travel are as follows:

A.  In-State and Out-of-State Business trips require approval of immediate supervisor. No permission to travel form required. Categories 1 and 2 on Travel Vouchers.

B.  In-State and Out-of-State trips for conferences, conventions, associations and meetings require approvals of the Chairman, Dean, or Division Chairman and Financial Affairs. Permission to Travel Form required. Categories 3 and 4 on Travel Voucher.

C.  Travel outside the Continental United States requires the approvals in B above, the Vice President, President, Trustees, the Executive Director of the Commission of Budget and Accounting and the Governor of Mississippi. Category 5 on Travel Voucher.

## Changes and Cancellation of Travel Request

An amended Permission to Travel request form must be submitted to change place or dates of meeting, or to change department account number.

When a Permission to Travel request has been canceled, a copy of the request with the word "Canceled" written across the face should be sent to Financial Affairs to release the encumbrance of funds. Funds will remain encumbered until the travel voucher is paid or until notice is received of cancellation.

## Advance Payment – Conference Registration Fees

To pay registration fees in advance of the conference, these procedures must be followed:

>   The request should be made at least 20 days in advance of the start of the conference.

>   The request should be made on a Remittance Voucher form with both the literature concerning the conference and photocopy of the Permission to Travel Form attached.

## University Vehicles

To obtain a vehicle from the Motor Pool, the request is made to the Physical Plant Division as outlined in the University Motor Pool operating procedures and regulations.

The University of Southern Mississippi uses the State Credit Card for vehicles assigned to the Motor Pool and USM Gulf Park.

The Motor Pool will charge the account through a Motor Pool Invoice. The amount charged is based on mileage with a $15.00 per day minimum. Motor Pool charges should never be reported on a Travel Voucher.

**Appendix 10–1.  Continued.**

REIMBURSEMENT OF TRAVEL EXPENSE

GENERAL

Travel Vouchers received in the Financial Affairs Office by 5 P.M. Thursday will be paid the following Thursday, provided there are no problems with the voucher.

Expense Reimbursement

Immediately upon returning from a trip, the traveler should submit a voucher for reimbursement of travel expenses.  If not filed within 30 days of the return, any advance received will be deducted from the traveler's pay check for the next full pay period.  The first two copies of the Travel Voucher, with required receipts attached, should be completed as outlined in the attachment hereto and submitted to Financial Affairs.  For other than business travel, the pink copy of the Permission to Travel must be attached to the Travel Voucher when it is submitted to Financial Affairs for payment.  The departmental account number shown on the Permission to Travel will be charged regardless of the department shown on the Travel Voucher.  Vouchers submitted with errors other than mathematical errors will be returned.  Travel expenses will be reimbursed only for the employee's own expense.  One employee cannot submit expense for reimbursement on the same voucher with another person.

If members of the family accompany the University representative, request hotel clerk to note the single room rate on the bill; otherwise, reimbursement will be made for one-half cost of the room.  Expenses as a result of unofficial stopovers, side trips, telephone charges, or any other items of a personal nature should not be submitted for reimbursement.  Long distance telephone charges must be documented to show place, party called, and purpose of call.

Reimbursement for Meals

Reimbursement for meals will be for actual expenditure (plus tips) at a reasonable amount with the following maximum daily limits for three meals:

| | |
|---|---|
| In Mississippi | - $18 per day |
| Out of Mississippi | - $24 per day |
| High-cost cities (specific) | - $30 per day (receipts required) |

Day Trips

No reimbursement is authorized for meals when travel is confined to the vicinity of your home campus.  For other travel, reimbursement for meals shall be:

Breakfast - When travel begins before 6 A.M. and extends beyond 8 A.M.
Lunch - When travel begins before 12 P.M. and extends beyond 2 P.M.
Dinner - When travel extends beyond 8 P.M.

No reimbursement is authorized for a dinner meal when the meeting is completed in sufficient time for the traveler to return by 6 P.M.

**Appendix 10-1.  Continued.**

## Lodging Expenses

Reimbursement will be made for lodging expense incurred in a hotel or motel on the presentation of a paid original bill.  When a room is shared with other employees on travel status, reimbursement will be calculated on a pro rata share of the total cost.  The other traveler must submit a copy of the lodging receipt indicating that the room was shared.  An employee on travel status, if accompanied by spouse who is not an employee on travel status, is entitled to reimbursement at the single room rate.  Request the hotel clerk to note the single room rate of the bill.

Normally if the order of business for which the travel is authorized begins after 3 P.M., reimbursement will not be made for lodging prior to the first day of business.  If the order of business begins prior to 3 P.M., reimbursement is made for lodging and meals for the preceding day if the lodging is necessary for the traveler to be present prior to the first order of business.  Reimbursement is made for lodging the final evening of the trip if the traveler is not able to return home by 9 P.M.

## Modes of Transportation

Transportation authorized for official travel include University vehicles (Motor Pool), private vehicles, common carriers, and rental cars.

If travel is by other than the most direct route between points where official University business is conducted, the additional cost must be borne by the traveler.  No traveler can claim transportation expense when he is gratuitously transported by another person, or when he is transported by another traveler who is entitled to reimbursement for transportation expense.  Private vehicle mileage reimbursement cannot exceed cost of round-trip air coach fare.

## Private Vehicles

The employee shall receive the legal rate established by the State of Mississippi for each mile actually and necessarily traveled in the performance of official duties.

The following situations justify the use of private vehicles for travel:

1.  When travel is required at such time or to such places that common carrier transportation may not be reasonably available.

2.  When one or more persons travel to the same destination in the same car and total mileage claimed does not exceed the total airline tourist fares for transporting the same number of people.

**Appendix 10–1.  Continued.**

## Common Carriers

Employees are reimbursed for actual airline and train fares.  All airline tickets must be purchased from Pro Travel, the official travel agent of the University.

The employee will purchase his own common carrier transportation and claim reimbursement.  Employees are <u>not</u> <u>allowed</u> <u>to</u> <u>charge</u> the transportation to the University.  Ticket cost reimbursement <u>must</u> be handled through the travel voucher.

Travel by airline shall be at the tourist rate.  Certification that tourist accommodations are not available will be necessary when travel is first class.

## Rental Cars

Limit the use of rental cars as much as possible.  There are times when common carriers, private cars, and University vehicles are not available and rental cars are the only means of transportation.  Below are examples:

1. When a destination has been reached by common carrier and several locations in the same vicinity must be visited.

2. When transportation between airport terminal and destination is needed and taxi or limousine service is not fesible or available.

3. When a schedule cannot be met through the use of common carrier.

Before you accept the rental car, examine the vehicle for any prior damage. If prior damage is discovered, it should be reported immediately to the rental agency to prevent improper claims against the University.

At the time you return the rental car, report any accident involving the rental vehicle and file an accident report with the rental agency.

## Other Expenses

Registration fees paid by the employee at the conference will be reimbursed on a travel voucher when supported by a paid receipt.  The portion of the fees applicable to meals shall be reported as meal expense.

Tips for meals and taxi should be included as part of those charges.  Tips reported here include baggage handling tips when arriving and departing a hotel or at airports.

Actual parking fees while away from home, and road and bridge tolls are reported here.

**Appendix 10–1.  Continued.**

Receipts

Major expense incurred by an employee while on official travel for the University require receipts.  The receipts must be originals and not copies.

Expenses that <u>require</u> original receipts:

1.  Lodging
2.  Rail, plane, or bus
3.  Registration fees
4.  Car rental (including gasoline tickets)
5.  Telephone expense (only as listed on motel receipt)
6.  Gas for University vehicle (see reimbursement procedure)

Expenses that <u>do</u> <u>not</u> require receipts:

1.  Meals, including tips (receipt required for any meal if over $25)
2.  Mileage of personal vehicle
3.  Tips
4.  Taxi/Limousine, including tips
5.  Parking/Tolls

Reimbursement Procedure

Payment checks for travel expenses are sent to Business Services and applied to the employee travel advance receivables.  If reimbursement exceeds advance, Business Services will send a check for the excess to the traveler.  If advance exceeds expenses, Business Services will notify the traveler of the amount applied and amount due.  Monthly statements are sent also.

**Appendix 10–1.  Continued.**

COMPLETING THE PERMISSION TO TRAVEL FORM

a.  Date submitted.

b.  Employee name, title and social security number.

c.  Name of convention, association or meeting. If for normal business travel out-of-state, enter "Business Travel."

d.  Enter city and state where meeting is being held.

e.  Enter dates of meeting.

f.  Enter department name and account number to be charged for the travel.

g.  Enter purpose of the meeting.

h.  Estimated total cost of attending.

i.  Signature and Southern Station Box Number.

j.  Required approvals of domestic travel.

k.  Additional approvals for travel outside continental United States.

l.  Amount of advance when requested.

m.  Signature of employee.

n.  Date advance received.

o.  Completed by Business Services when advance is issued.

p.  Completed by Financial Affairs when Permission is approved.

All costs of conventions, associations and meetings are reported annually to the State Peer Committee.  This report shows the convention name, place, dates, employee attending and total cost.

**Appendix 10–1.  Continued.**

## PERMISSION TO TRAVEL

Submit at least two weeks prior to date of proposed trip (90 days prior to foreign travel).

THE UNIVERSITY OF SOUTHERN MISSISSIPPI

**a** _____ , 19_____

Name: **b** _____ Title **b** _____ S.S. No. **b** _____

In compliance with Section 25-3-45 Mississippi Code 1972, request is made for authorization to attend the following convention, association, or meeting:

**c** _____  **d** _____

Complete Name of Convention, Association or Meeting (Do Not Abbreviate)     Place of Meeting

**e** _____  **f** _____  **f** _____

Dates of Meeting     Department Name     Department Code

Purpose of convention, association, or meeting (If an advance is needed, but cost of trip will be reimbursed by an outside organization, please explain).

**g**

**Domestic Travel**

Estimated Cost  $ **h** _____

1. Chairman **j** _____

I acknowledge that I have read and that I understand the summary of travel policies on the back of this form.

2. Dean or Division Chairman **j** _____

3. Funds Available—Accounting (5143) **j** _____

**Foreign, Hawaii, Alaska Travel**

4. Vice President **k** _____

**i** _____

Signature of Applicant

5. President **k** _____

Southern Station Box No. **i** _____

---

**FOR ACCOUNTING AND BUSINESS OFFICE USE ONLY**

**ADVANCE (Cannot exceed estimated cost above)**

Amount of advance  $ **l** _____ Signature **m** _____ Date **n** _____

I hereby certify that the above trip has been properly approved. The amount advanced will be repaid from reimbursement check for travel expenses, and it is expressly understood and agreed that unless this amount is repaid by me before the next full pay period after the date of my return, it may be deducted from my next salary check.

**ACCOUNT RECEIVABLE . o**

| Date MMDDYY | F/GL | OBJ. | SOCIAL SEC. NO. | AMOUNT | | DR |
|---|---|---|---|---|---|---|
| | 1014 | 1169 | | | | 4 |

**ENCUMBRANCE p**

| P.O. CONTROL | F/GL | OBJ. | DEPARTMENT | SOCIAL SEC. NO. | AMOUNT | | DATE MMDDY |
|---|---|---|---|---|---|---|---|
| | | | | | | | |
| | | | | | | | |
| | | | | | | | |

White Copy—Accounting  •  Canary Copy—Employee (For Advance)  •  Pink Copy—File with Voucher  •  Goldenrod Copy—Dept. File Copy

ACC 1 (Revised 7/84)

## Appendix 10–1.  Continued.

COMPLETING THE EMPLOYEE TRAVEL VOUCHER

a. Employee name, social security number and address.

b. Purpose of trip and city and state.

c. Department name and account number to be charged. Must be same as Permission to Travel.

d. Names of others on trip, whether traveling together or not.

e. Date of travel, departure and arrival times.

f. Meals -- Daily meal maximums are:

| | |
|---|---|
| In Mississippi | $18 |
| Outside Mississippi | $24 |
| Listed high-cost cities | $30 |

g. Lodging -- Original hotel or motel bill required. If bill shows two persons, indicate single room rate for this room. Do not report charges other than room charges as lodging (telephone, room service, etc.)

h. Travel by personal vehicle -- Indicate if University Motor Pool vehicle was used. If so, do not enter any miles here. If personal vehicle is used, enter departure and arrival location and miles. Enter current approved rate (20¢ mile).

i. Travel by Public Carrier -- Date, city leaving from, destination city, mode and ticket amount. Airline ticket coupon must be attached.

j. Registration fees -- If paid by employee, receipt or copy of program stating fee must be attached.

k. Tips -- Only for baggage handling or valet parking.

l. Taxi/limousine -- Actual taxi fare plus tip or airport limousine charge.

m. Parking/Tolls -- Actual parking charges or road or bridge tolls paid.

n. Car rental -- Receipt required.

o. Check travel category. If category 3, 4 or 5 are checked, approved copy of Permission to Travel form must be attached.

p. If travel advance was received, enter amount and date received.

q. Enter total of all expenses on Travel Voucher.

r. Maximum Reimbursement Allowed -- If chairman approving enters a smaller amount here than the total amount, then the smaller amount will be paid. Some departments have a travel limit on each trip.

s. Signature of employee and account director and dates.

t. For Accounting use only.

**Appendix 10-1. Continued.**

# EMPLOYEE TRAVEL VOUCHER

THE UNIVERSITY OF SOUTHERN MISSISSIPPI

**IMPORTANT—SEE INSTRUCTIONS ON BACK**

| Employee **a** | | S.S. No. **a** |
| --- | --- | --- |

| Address To Which Check Should Be Sent | **a** |
| --- | --- |
| | **a** |

| Purpose and Place of Visit **b** | Dept. **c** | Dept. No. **c** |
| --- | --- | --- |
| | Others On Trip **d** | |

## MEALS AND LODGING

| **e** | Date | | | | | | | | Total |
| --- | --- | --- | --- | --- | --- | --- | --- | --- | --- |
| | Departure Time | AM-PM | AM-PM | AM-PM | AM-PM | AM-PM | AM-PM | AM-PM | |
| | Arrival Time | AM-PM | AM-PM | AM-PM | AM-PM | AM-PM | AM-PM | AM-PM | |
| **f** | Breakfast | | | | | | | | |
| | Lunch | | | | | | | | |
| | Dinner | | | | | | | | |
| **g** | Lodging | | | | | | | | |
| | | | | | | | Total Meals and Lodging | | |

## **h** TRAVEL BY PERSONAL VEHICLE (Did you use University vehicle? ☐ Yes ☐ No)

| Date | From | To | Miles | |
| --- | --- | --- | --- | --- |
| | | | | |
| | | | | |
| | | | | |
| | | Total Miles | X Rate = | |

## **i** TRAVEL BY PUBLIC CARRIER

| Date | From | To | Mode | Ticket Amount |
| --- | --- | --- | --- | --- |
| | | | | |
| | | | Total Travel By Public Carrier | |

## OTHER EXPENSES

| | Item | Date | Place Where Expense Incurred | Amount |
| --- | --- | --- | --- | --- |
| **j** | Registration Fees | | | |
| **k** | Tips (baggage handling) | | | |
| **l** | Taxi/Limousine | | | |
| **m** | Parking/Tolls | | | |
| **n** | Car Rental | | | |

**o** Check Category of Travel (3 Through 5 Require Permission to Travel Form)

1 ☐ Business Trip In-State/No Advance

2 ☐ Business Trip Out-of-State/No Advance

3 ☐ Business Trip Out-of-State/With Advance

4 ☐ Conventions, Conferences, Associations/In-or Out-of-State

5 ☐ Out-of-Country

Travel Advance For This Trip
$ **p** Date **p**

| Total Other Expenses **q** |
| --- |
| **Total Expenses** |
| MAXIMUM REIMBURSEMENT ALLOWED ▶ **r** |

Employee Signature **s** Date **s**

Approved By **s** Date **s**

| Accounting Distribution | F/GL | Object | Department | Liquidation | Expense |
| --- | --- | --- | --- | --- | --- |
| Voucher No. **t** | | | | | |
| Voucher Date **t** | | | | | |
| P.O. No. **t** | | | | | |
| Verified By Date **t** | | | | | |

ACC 2 (4/85)          White Copy—Accounting  •  Canary Copy—Accounting  •  Pink Copy—Department  •  Goldenrod Copy—Emplo

**Appendix 10–1.   Continued.**

## ENDNOTES

1. Harry L. Brown, *Design and Maintenance of Accounting Manuals*, Second Edition. New York: John Wiley & Sons, 1993).
2. R. W. Gunn, et al. "Shared Services: Major Companies are Re-engineering their Accounting Functions." *Management Accounting*. (November 1993): 22–28.

# CHAPTER 11

# INTERNAL CONTROL
# AND RISK EXPOSURES

## 11.1 INTRODUCTION

Control is a concept relevant to every system. This chapter, as well as the following chapter, will examine many aspects of control within the context of a business organization. Our survey starts with a historical review of the concept and then moves on to discuss control processes, control objectives, control systems, risk exposures, feasibility considerations, and the varied controls that are needed to counteract risk exposures. An understanding of control and how it functions enhances the documentation process.

This chapter will enable you to:

- Understand the historical evolution and the nature of control and control processes in information systems.
- Describe the objectives of the internal control structure, as well as its facilitating elements and various control systems within an organization, including the management control system, operational control system, and accounting control system.
- Identify the exposures to risk that a firm faces with respect to assets and data, including those involving computer fraud.
- Discuss the control problems caused by computerization of the accounting information system of a firm.
- Describe the considerations that affect the feasibility of controls and control systems within firms.
- Discuss the forces that are instrumental in the improvement of control systems within firms.

## 11.2 NATURE OF CONTROL

The process of control is one of management's basic functions. Management must establish and maintain control over its operational system, organization, and information system. The decisions made by management in doing so are crucial to the organization's success. Effective control decisions enable a firm to employ its resources efficiently, to fulfill its legal responsibilities, and to generate reliable and useful information. It is important to distinguish between "control" and "controls." The term "controls" refers to various measures and procedures put in place to counteract exposure to certain risks. An integral part of a firm's control is a "framework of controls." The framework of policies and control procedures is also called a firm's internal control structure. Composed of

119

underlying management policies and a wide variety of control procedures, this framework spans all of the firm's transactions. The overall purpose of this framework is to provide reasonable assurance that a firm's objectives will be achieved. Thus, it also encompasses the firm's operations and system of management.

This structure is the means through which the process of internal control functions. If the internal control structure is strong and sound, all of the operations, physical resources, and data will be monitored and kept under control. Information outputs will be trustworthy. However, a weak and unsound internal control structure can lead to serious repercussions. The information generated and made available for managerial decision making by accounting departments and others in the organization is likely to be unreliable, untimely, and perhaps unrelated to the firm's objectives. Without control, the implementation of policies may be dysfunctional. Furthermore, the firm's resources may be vulnerable to loss through theft, carelessness, sabotage, and natural disaster.

Internal control, as both a process and a structure, is of great concern to accountants. As key users of the accounting information system (AIS), they need to know which control procedures are in effect. Therefore, accountants often take active roles in developing and reviewing control frameworks. They work closely with system designers during the development of information systems to ensure that the planned control procedures are adequate and auditable. For instance, accountants/auditors make sure that totals will be balanced and reconciled properly and that audit trails are clearly established. During audits they determine the adequacy of measures and procedures that facilitate the internal control process, so that they can assess the reliance to be placed on the internal control structure when performing subsequent auditing program steps. Developing and evaluating control frameworks are skills at which accountants can and are expected to excel.

## 11.3 HISTORICAL DEVELOPMENT OF CONTROL

In *Accounting Terminology*, published in 1931, the American Institute of Accountants (now known as the American Institute of Certified Public Accountants) defined internal control as an accounting device whereby a proof of the accuracy of figures can be obtained through the expedient of having different persons arrive independently at the same result. In 1936, the same group in another book, *Examination of Financial Statement*, defined "internal check and control" as those measures and methods adopted within the organization itself to safeguard the cash and other assets of the company as well as to check the clerical accuracy of the bookkeeping.

In 1958 and again in 1972 the American Institute of Certified Public Accountants (AICPA) defined the difference between what it called administrative controls and accounting controls. The administrative controls focused on authorization by management, while the accounting controls pertained to policies governing the safeguarding of assets and the reliability of financial statements.

Investigations by governmental agencies such as the Security and Exchange Commission (SEC) have revealed illegal activities within American firms that were not detected by their internal control structures. As a result, Congress passed the Foreign Corrupt Practices Act (FCPA) in 1977. In addition to prohibiting certain types of bribes and hidden ownership, this act requires subject corporations to devise and to maintain adequate internal control structures. Managers of these corporations are legally as well as ethically responsible for establishing and maintaining adequate internal control structures. They are expected to establish a level of "control consciousness," so that employees are not inclined to subvert the control structures. Those managements that do not comply with the act's requirements are subject to large fines and imprisonment. It is not surprising that many managements have taken significant action. For instance, many corporations now maintain sizable internal audit departments.

The FCPA in 1977 changed the focus on internal controls by establishing legal requirements for firms that sold their securities to the public. The act was primarily designed to prevent bribery of foreign officials in order to obtain business, but its effect has required companies to have good systems of internal accounting control. American firms were using slush funds to bribe foreign officials as well as to make political contributions domestically. The act sought to make this more difficult. Specifically, the FCPA requires companies to make and keep books, records, and accounts, that, in reasonable detail, accurately and fairly reflect the transactions and dispositions of the assets of the issuer of securities for sale to public. Such firms must devise and maintain a system of internal accounting controls sufficient to provide reasonable assurances that:

- Transactions are executed in accordance with management's general or specific authorization
- Transactions are recorded as necessary to allow preparation of financial statements in accordance with generally acceptable accounting principles and to permit accountability for assets
- Access to assets is permitted only in accordance with management's general or specific authorization
- Periodic inventorying of the firm's assets occurs, and action is taken when there are differences between the recorded and the actual quantities of assets

In a related context, in 1986 Congress passed the Computer Fraud and Abuse Act. Under this law it is a crime to knowingly, and with intent to defraud, gain unauthorized access to data stored in the computers of financial institutions, the federal government, or computers used for interstate or foreign commerce. This law applies to corporations as well as to individuals.

Another milestone was reached in 1992, when after several years of study, a Committee of Sponsoring Organization report was issued which sought to define internal control as well as practical guidance to companies to help them assess and improve their internal control. The Committee of Sponsoring Organization (COSO) was set up by the American Accounting Association, the AICPA, the Institute of Internal Auditors, the Financial Executives Institute, and the Institute of Management Accountants. The COSO asked the firm of Coopers & Lybrand to conduct the study on its behalf.

The COSO report looks at internal control in much more than purely financial terms, it attempts to unify the views on internal control held by the various groups that sponsored the committee. It defines internal control as a process put into effect by people (management, the board of directors, personnel) to provide reasonable assurance but not absolute guarantee of the attainment of objectives concerning: effectiveness and efficiency of operations, reliability of financial reporting, and compliance with applicable laws and regulations. In addition, the COSO report describes internal control as consisting of five interrelated components:

- Control environment—the setting in which people operate
- Risk assessment—necessary to design controls

- Control activities—policies and procedures to manage risks
- Information and communication—necessary to identify and manage risks
- Monitoring—to evaluate performance

As one might expect, there are those who feel the COSO report did not go far enough in making management more responsible to its shareholders.

## 11.4  THE CONTROL PROCESS

In brief, the control process consists of measuring actual results against planned accomplishments and taking corrective actions when necessary. These corrective actions are the decisions that keep the firm moving toward its established objectives.

The best way to understand the nature of the control process is by seeing it in terms of a household heating system. The overall purpose of a heating system is to control the temperature, which is the characteristic or performance measure. A homeowner begins the process by setting the thermostat reading for the desired temperature (the bench mark). He or she chooses a setting that is determined by personal objectives such as comfort and economy. Then an automatic process takes over. As the furnace (the operating process) generates heat, the thermostat's thermometer (the sensor element) detects the actual temperature. Next, the thermostatic mechanism (the regulator element) compares the actual temperature (fed to it by the sensor element) with the bench mark temperature. When the actual temperature rises above the preset temperature, the control element notifies the activating mechanism in the thermostat to shut down the furnace. Later, when the actual temperature drops below the preset temperature, the information feedback leads to the furnace being turned on again.

What allows a thermostat to function is feedback. As the foregoing heating system example shows, feedback is an information output that returns ("feeds back") to a regulator element and then to the operating process as an input. Feedback therefore provides the means for deciding when corrective action, such as turning off the furnace, is necessary. Like the temperature control system, a control process also both monitors and regulates. What makes this possible is feedback obtained through various information systems.

## 11.5  ELEMENTS OF A CONTROL PROCESS

A typical organizational control process consists of six elements:

- A factor being controlled, called the characteristic or performance measure
- An operating process that gives rise to the characteristic
- A sensor element that detects the actual state of the operating process
- A planned accomplishment, or bench mark, against which the actual state of the characteristic is to be compared
- A planner who sets the bench mark
- A regulator element that compares the actual state of the characteristic against the bench mark and feeds back corrections to the operating process

An entity such as a business firm may employ a variety of control processes. They may be preventive, designed to keep problems from occurring in the first place. Wearing seatbelts or locking one's car are examples of preventive control. Controls could also be detective, aimed at discovering problems as soon as possible. A smoke detector is an example of a detective control. Alternatively, they could be corrective, meant to solve problems after they are discovered. Corrective controls work in association with detective controls. Hotels have automatic sprinkler systems meant to go on after the smoke detectors set off alarms.

As a firm grows in size and complexity, the number of needed control processes (and controls) are likely to increase. Moreover, these processes may vary in sophistication and could take the form of feedback or feedforward. The heating system is an example of a first-order feedback control process or system, if the bench mark is assumed to remain unchanged after being set. In a business firm a budgetary control system is an example of a first-order feedback control process. By contrast, a second-order feedback control system employs a bench mark that adjusts to meet changing environmental conditions. An example is a flexible budgeting system, in which the budget values are adjusted with changing levels of activity. A third-order feedback control system attempts to predict future conditions and results. Based on such predictions, the system anticipates future problems and suggests corrective actions before the problems occur. Several third-order feedback control systems, also known as feedforward control systems, are typically used by business firms. One example is a cash planning system. Based on periodic cash forecasts, the treasurer of a firm may anticipate seasonal cash shortages by arranging for bank loans before the shortages actually occur.

## 11.6 ILLUSTRATING THE CONTROL PROCESS

The accounting information system (AIS) is central to a firm's control processes, since it can serve as both a monitor and a regulator. We can see how it functions through the example of inventory control systems. The inventory control process begins when an inventory manager decides how much to carry of each inventory item.

Let us assume that the only item carried by a firm is a wheelbarrow, and that the manager decides to carry at least 100 wheelbarrows for sale. When the quantity on hand drops to 100 (the reorder point), the manager acquires fifty more wheelbarrows from a wholesale store across town the same day. The AIS keeps track of the inventory level by deducting the quantity of each sale from the on-hand quantity shown on an inventory record. Thus, the AIS serves as a monitor and provides feedback information to the manager via a first-order feedback control system.

Now assume that the firm recognizes that the sales of wheelbarrows face seasonal fluctuations. More wheelbarrows are sold each month in the spring when gardeners and contractors intensify their labors. Thus, the manager adjusts the level of minimum inventory by referring to sales for the same month last year and adjusts the reorder quantity by reference to the current on-hand quantity. Now the AIS provides added feedback information that is affected by changing conditions; thus, it acts as a second-order feedback control system.

Finally, assume that wheelbarrows can be obtained more cheaply from a wholesale firm in the next state, but the lead time is two weeks. Also, the firm's sales have grown substantially but fluctuate more significantly from week to week. The manager now employs a reorder point that is affected by the expected demand in the weeks ahead as well as by the lead time. The AIS automatically informs him when the reorder point has been reached. Then, the AIS computes the economic reorder quantity, which also depends in part on the expected sales demand for the coming weeks. Because it provides inventory information that anticipates future needs, the AIS now acts as a feedforward control system.

Another illustration of the working of internal control may be seen in the recommendations for effective payroll procedures:

- A system of advice forms should be installed so that new hires, terminations, rate changes, etc., are reported to the payroll department in writing. Such forms should be approved by the employee's superior.
- Before an applicant is hired, his background should be investigated by contacting references to determine that he is not dishonest and has no other undesirable personal characteristics. There are some legal limits on what may be asked of an applicant's references.
- The supply of blank timecards should be controlled. At the beginning of each week the payroll department should provide each worker with a timecard stamped with his name.
- Install a timeclock and have the workers "punch" in and out. A responsible employee should be stationed at the timeclock to determine that workers are not "punching" the time cards of other workers who may be late or absent, or who may have left work early.
- The foreman should collect the timecards at the end of the week, approve them, and turn them over to the payroll clerk. All timecards should be accounted for and any missing cards investigated.
- If the company has a cost system that requires the workers to prepare production reports or to account for their time by work tickets, the timecards and the production reports or work tickets should be compared.
- Payroll clerks' work should be arranged so that they check each other. As an alternative, one clerk may make the original computations for the full payroll and the other clerk do all the rechecking.
- The payroll checks should be prenumbered to control their issuance.
- The payroll checks should be distributed to the workers by a responsible person other than the foreman. Unclaimed checks should be held in safekeeping by the payroll department until claimed by the worker. Direct deposits in employees bank accounts should be encouraged.
- A responsible person other than the chief accountant and the payroll clerks should reconcile the payroll bank account.
- From time to time, an officer of the company should witness a payroll distribution on a surprise basis.

An illustration of internal control is also seen in the following recommendations to help improve control over cash receipts.

- All sales tickets should be prenumbered and all sales ticket numbers should be accounted for daily. All sales tickets stamped "paid" should be reconciled to the duplicate deposit slip receipted by the bank. This should be done by a responsible employee other than the cashier or a member of the credit department.

- An employee other than the cashier or a member of the credit department should open the incoming mail and prepare daily a list in triplicate of remittances received that day. The original of the list should accompany the checks (and cash, if any) turned over directly to the cashier; one copy of the list of remittances should be routed to the responsible employee mentioned above; and one copy should be routed to the person responsible for posting to the accounts receivable ledger.
- The responsible employee who received the copy of the list of remittances should compare the remittances shown thereon with the duplicate deposit ticket at the same time he reconciles the cash sales tickets to the deposit ticket. Any checks or cash not deposited the day received should be investigated.
- Different forms (or colors) of sales tickets for cash, C.O.D. and credit sales would facilitate the daily accounting for sales tickets used.
- The cashier, who should have no duties connected with accounts receivable, should prepare bank deposits, and forward the deposit to the bank. Someone other than the cashier or a member of the credit department should establish agreement of the list of remittances and daily collections with the daily deposit ticket.
- Remittances should not be held; those for each day or for each batch of mail, whichever is the more practical, should be deposited intact. The credit department may make whatever record it needs for further follow-up on the remittances that are not in the correct amount.

## 11.7 INTERNAL CONTROL STRUCTURE

Since the internal control structure is the vehicle through which the AIS employs an array of controls, it deserves prominent attention. In this section control objectives that the structure fosters, its key elements, and the three control systems that it incorporates are examined.

**(a) Internal Control Objectives.** According to the American Institute of Certified Public Accountants (AICPA Committee on Auditing Procedure's "Internal Control—Elements of a Coordinated System and its Importance to Management and the Independent Public Accountant." Statement on Auditing Standards no. 1, section 320. New York: AICPA, 1973) the internal control structure of a firm has the following objectives:

- To encourage adherence to management's prescribed policies and procedures
- To safeguard the assets of the firm
- To assure the accuracy and reliability of the accounting data and information

The control objectives above pertain to the operational system of a firm. In essence, their purposes are to assure that operations are performed effectively in a manner that best achieves the firm's broad goals; that operations are performed efficiently in a manner that does not waste resources; and that resources are provided with adequate security. (Security consists of safeguarding and protecting a firm's tangible and intangible assets via physical measures and control procedures.)

The last control objective, concerning accuracy and reliability, is dependent on the transaction processing portion of the AIS. In order to fulfill this objective, such control sub-objectives as the following must be accomplished:

- Ensuring that all transactions entered for processing are valid and authorized
- Ensuring that all valid transactions are captured and entered for processing on a timely basis
- Ensuring that the input data of all entered transactions are accurate and complete, with the transactions being expressed in proper monetary values
- Ensuring that all entered transactions are processed properly to update all affected records of master files and/or other types of data sets
- Ensuring that all required outputs are prepared by appropriate rules to provide accurate and reliable information; for example, that financial statements such as income statements are prepared from complete and up-to-date records and in accordance with generally accepted accounting principles

In striving to achieve these objectives, the internal control structure performs a variety of established procedures. For instance, the actual cash on hand and in the bank is matched to the amount reflected in the cash ledger account; the customer number on a remittance advice is matched to the number of an account in the accounts receivable master file.

Control objectives are not easily achieved, however. One difficulty arises from the changes faced by a modern firm, ranging from ever-changing tax laws to rapidly developing technology. Another difficulty derives from the array of risks to which the internal control structure and its firm are exposed. A third difficulty, related to the first, concerns the use of computer technology within the control structure. A fourth difficulty may be traced to the human factor, since control objectives are accomplished through people. Managers may establish policies, procedures, and bench marks that are unclear or unwise. Employees may not follow procedures consistently or may make incorrect comparisons. A final difficulty relates to the costs of controls.

**(b) Elements of the Internal Control Structure.** The internal control structure may be viewed in different ways. In this section we present a view that has been proposed by the American Institute of Certified Public Accountants in a statement entitled "Consideration of the Internal Control Structure in a Financial Statement Audit," issued as Statement on Auditing Standards (SAS) No. 55 in April 1988. In this audit-oriented view, the internal control structure consists of three elements: the control environment, the accounting system, and control procedures. In the next chapter we survey control from other perspectives. During this survey we should be aware that the three elements overlap each other.

One of the elements of the internal control structure is control environment. According to SAS No. 55, the control environment is the effect various factors exert collectively on establishing, enhancing, or mitigating the impact of specific management policies and procedures. The factors that make up an organization's control environment include the following: management's philosophy and operating style; the entity's organizational structure; the functioning of the board of directors and its committees, particularly the audit committee; methods of assigning authority and responsibility; management's control methods for monitoring and following up on performance, including internal auditing; personnel policies and practices; and various external influences that affect an entity's operations and practices, such as examinations by bank regulatory agencies.

**(c) The Control Environment.**    The management's philosophy and operating style is the most critical component of the control environment. If managers act like internal control is important, so will the employees. Management's philosophy and operating style can be defined by answering the following types of questions:

- Does management take undue business risks to achieve objectives, or does it act prudently in assessing risks and rewards before committing the organization to a course of action?
- Does management attempt to manipulate measures of business performance?
- Does management put pressure on employees to achieve results regardless of the methods required, or is it concerned that employees behave ethically? Is the organization dominated by one or a few individuals, or is responsibility distributed among a team?

Organization structure defines the lines of authority and responsibility in an organization. The structural factors that in turn affect internal control are:

- Relative degree of centralization or decentralization within the organization.
- The use of functional or divisional organization structure.
- The organization of the accounting and information systems functions.
- The roles played by the audit committee of the board of directors. An audit committee is responsible for overseeing the corporation's internal control structure, its financial reporting process, and its compliance with related laws, regulations, and standards. It works closely with both the internal and external auditors. All companies traded on the NYSE must have an audit committee consisting solely of outside (non-management) members of the board of directors.
- The methods used by management to assign authority and responsibility.
- The existence of a formal code of conduct, a policy and procedures manual, and job descriptions can also affect organizational structure by assigning responsibility and stating policies in effect.
- Methods of monitoring performance (supervision, internal audit and responsibility accounting).

**(d) Accounting Control System.**    The second element of internal control, the accounting control system, is designed to achieve two objectives: to safeguard the firm's assets and to ensure the accuracy and reliability of the data and information. Assets refer to resources, including data and information; accuracy means freedom from errors and reliability means consistency in results from processing like data. To achieve these objectives, the accounting control system employs groups of accounting controls. One group of accounting controls, called safeguards or security measures, protects information from unauthorized users. Another group of accounting controls focuses on accuracy and reliability within the records and procedures (methods) of the accounting system. These controls represent screening mechanisms over the flows of transactions where data errors and losses can occur. In effect, the accounting control system provides a series of controls throughout transaction processing activities, from the points where authorizations occur to the points where financial outputs are provided to users.

Examples of transaction-oriented accounting controls are:

- Charts of accounts
- Double-entry bookkeeping and trial balances
- Audit trails
- Batch control totals
- Prenumbered and well-designed source documents

The accounting control system will be discussed at greater length elsewhere, as it is a significant consideration while designing and documenting accounting procedures.

**(e) Control Procedures.**   This element includes all of the procedures that are not covered by the control environment and accounting system. Among the control procedures categorized by SAS No. 55 are:

- Sound design and use of documents and records. Thus, source documents and other forms should require transactions to be recorded and established procedures to be followed. For instance, prenumbered shipping documents should be prepared, and shipping registers should be used to monitor the use of the documents. To encourage adherence to prescribed procedures, adequate and up-to-date documentation should be maintained.
- Adequate segregation of employees' duties to ensure organizational independence. In order to achieve adequate segregation, different persons or organizational units should be assigned the responsibilities of authorizing transactions, recording transactions, and maintaining custody of the assets to which the transactions pertain. Organizational independence through the separation of functions reduces the opportunities for a person to be able to perpetrate and then conceal errors or irregularities in the normal course of duties. More will be said about this topic in a subsequent chapter.
- Proper authorization of transactions and other activities. Unauthorized transactions can lead to abuses and errors. As noted earlier, authorizing transactions is closely related to adequate segregation of duties.
- Adequate safeguards and security measures for assets, computer facilities, records, and data sets. Safeguards and security measures include locked safes and doors, fenced-in areas, alarm systems, and passwords.
- Independent checks on performance. In addition to assigning one employee to check the work (e.g., counting cash) of another, independent checks include reconciling a control account balance with the total of balances in a subsidiary ledger and comparing on-hand physical inventory levels with the recorded inventory balances.
- Proper valuation of recorded amounts. Control procedures are needed to ensure that assets are assigned proper quantitative values based on sound measurement methods. These procedures often include reviews by knowledgeable persons. For example, credit managers frequently review aged trial balances of accounts receivable, just as customers generally review their monthly statements of amounts owed.

## 11.8 IMPEDIMENTS TO EFFECTIVE CONTROL

Effective control requires ongoing efforts. Impediments to effective managerial control can be caused by a number of factors, individually or in combination. Among them are:

- Exposure to risks
- Technological problems
- Behavioral problems
- Cost-related considerations

Some examples of potential impediments or risks to control are:

- Clerical and operational employees, who process transactional data and have access to assets
- Computer programmers, who prepare computer programs and have knowledge relating to the instructions by which transactions are processed
- Managers and accountants, who have access to records and financial reports and often have authority to approve transactions
- Former employees, who may still understand the control systems and may harbor grudges against the firm
- Customers and suppliers, who generate many of the transactions processed by the firm
- Competitors, who may desire to acquire confidential information of the firm, such as new product plans
- Outside persons, such as computer hackers and criminals, who have various reasons to access the firm's data or its assets or to commit destructive acts
- Acts of nature or accidents, such as floods and fires and equipment breakdowns

Every entity, such as a business firm, faces impediments like those discussed above, which reduce the chances of it achieving its control objectives. For the controls provided by the internal control structure to be adequate, they should counteract all the significant risks to which a firm is exposed. Risk exposures may arise from a variety of internal and external sources, such as employees, customers, computer hackers, criminals, and acts of nature. In order to design sound control systems, accountants and system designers should be able to assess the risks to which a firm is subject. Risk assessment consists of identifying the risks, analyzing the risks in terms of the extent of exposure, and proposing effective control procedures.

## 11.9 TYPES OF RISK EXPOSURE

Among the system-related risks that confront the typical firm, other than poor decision making and inefficient operations, are:

*Unintentional Errors.* Errors may appear in input data, such as in customer names or numbers. Alternatively, they may appear during processing, as when clerks incorrectly multiply quantities ordered (on customers' orders) and unit prices of the merchandise items. These errors often occur on an occasional and random basis, as when a clerk accidentally strikes the wrong key on a terminal keyboard. However, errors may occur consistently. For instance, an incorrectly written computer program may produce computational errors each time the program is executed. In any of these situations, the erroneous data will damage the accuracy and reliability of a firm's files and outputs. Unintentional errors often occur because employees lack knowledge owing to inadequate training; they may also occur when employees become tired and careless or are inadequately supervised.

*Deliberate Errors.*   Deliberate errors constitute fraud, since they are made to secure unfair or unlawful gain. These irregularities may appear in input data, during processing, or in generated outputs. For instance, a clerk may increase the amount on a check received from a customer or underfoot a column of cash receipts. Either type of error damages the accuracy and reliability of files and/or outputs. Additionally, deliberate errors may also conceal thefts (and, hence, losses) of assets. For example, a manager may enter a misstatement in a report or financial statement. This type of error could mislead and thereby injure stockholders and creditors.

*Unintentional Loss of Assets.*   Assets may be lost or misplaced by accident. For example, newly received merchandise items may be put into the wrong warehouse bins, with the result that they are not found by pickers when filling orders. Data, as well as physical assets, may be lost. For instance, the accounts receivable file stored on a magnetic disk may be wiped out by a sudden power surge.

*Thefts of Assets.*   Assets of a firm may be stolen by outsiders, such as professional thieves who break into a storeroom in the dead of night. Alternatively, assets may be misappropriated through embezzlement or defalcation, that is, taken by employees who have been entrusted with their care. For instance, a cashier may pocket currency received by mail or a production worker may carry home a tool. Employees who embezzle often create deliberate errors in order to hide their thefts. Thus, the cashier who pockets currency may overstate the cash account.

*Breaches of Security.*   Unauthorized persons may gain access to the data files and reports of a firm. For instance, a "hacker" may break into a firm's computerized files via a distant terminal or an employee may peek at a salary report in an unlocked file drawer. Security breaches can be very damaging in certain cases, as when competitors gain access to a firm's confidential marketing plans.

*Acts of Violence and Natural Disasters.*   Certain violent acts cause damage to a firm's assets, including data. If sufficiently serious, they can interrupt business operations and even propel firms toward bankruptcy. Examples of such acts are sabotage of computer facilities and the willful destruction of customer files. Although violent acts are sometimes performed by outsiders such as terrorists, they are more often performed by disgruntled employees and ex-employees. Also, violent acts can arise from nonhuman sources, such as fires that engulf computer rooms or short circuits that disable printers.

## 11.10 DEGREE OF RISK EXPOSURE

To combat these risks effectively, the degree of risk exposure should be assessed. Exposure to risk is caused by factors such as:

- Frequency. The more frequent an occurrence, the greater the exposure to risk. A merchandising firm that makes numerous sales is highly exposed to errors in the transaction data. A contractor that bids on custom projects is exposed to calculating errors. A department store with numerous browsing shoppers has a significant exposure to merchandise losses from shoplifting.
- Vulnerability. The more vulnerable an asset, the greater the exposure to risk. Cash is highly vulnerable to theft, since it is easily hidden and fully convertible. A telephone may be vulnerable to unauthorized use for long distance calls, especially if it is left untended in a remote office.
- Size. The higher the monetary value of a potential loss, the greater the risk exposure. An accounts receivable file represents a high risk exposure, since it contains essential information concerning amounts owed and other matters that affect credit customers.

When two or more of the above factors act in unison, the exposure to risk is multiplied. Thus, an extremely high exposure occurs in the case of a firm that conducts numerous sales for cash, with each sale involving a sizable amount. As might be imagined, this situation requires more extensive controls than a situation in which the exposure to risk is slight.

## 11.11 CONDITIONS AFFECTING EXPOSURES TO RISK

The exposures to risk faced by a firm can be heightened by various internal conditions. Perhaps most serious are weaknesses in one or more of the control systems. Weaknesses may be caused by an inadequate selection of controls. Thus, thefts are abetted when the organizational structure is weakened by an inadequate segregation of duties. Control system weaknesses may also occur because of breakdowns. For instance, an accountant who is responsible for investigating variances from standards may not pursue the exception sufficiently to determine appropriate corrective measures. Other problem conditions involve collusion, lack of enforcement, and computer crime.

**(a) Collusion.** A frustrating condition of which a firm must be aware is internal collusion, the cooperation of two or more employees for a fraudulent purpose. For instance, an employee who has custody over inventory may remove inventory items from the store-room, while an employee who keeps inventory records deducts the removed items from the records; the consequence to the employer is lost inventory and covered tracks. Collusion may also involve an employee and a non-employee. This situation is known as external collusion. Either type of collusion is difficult to counteract, even with soundly designed control systems.

**(b) Lack of Enforcement.** Still another troubling condition is lack of enforcement. Thus, a firm may have adequate management policies and control procedures, but may overlook irregularities. For instance, an employee who has committed embezzlement may not be prosecuted on being detected, perhaps so that the firm can avoid embarrassment over its weak security measures. Such lack of action by a firm may encourage other potential wrongdoers. Management may even actively abet the problem. For example, a higher-level manager might improperly override an installed control procedure, either with the intent of personal gain or to enhance the firm's financial condition.

## 11.12 FRAUD AND CONTROL PROBLEMS RELATED TO COMPUTERS

Computer systems present special risk exposures and problems of control. After surveying these exposures and problems, we will examine the effects computer systems have on controls. Accountants are becoming increasingly concerned about these controls, since computer-based information systems are rapidly growing in number and complexity. Offenses involving computer-based systems have grown in quantity and seriousness in recent years. Computer fraud poses very high degrees of risk, since all three factors—frequency, vulnerability, and size—tend to be present. A computer-based system may process hundreds of transactions per hour, with each transaction being subject to error or to fraudulent activity. A computer and its stored data are often vulnerable to unauthorized access as well as to damage. To make matters worse, fraudulent activities—by either authorized or unauthorized persons—are very difficult to detect. Also, a loss from computer fraud tends to be several times larger than the average fraud loss when manual systems are involved. Huge quantities of data and lighting-fast processing speeds magnify the payoff from computer fraud. Individual losses from computer fraud can easily exceed millions. A famous case uncovered as early as 1973 illustrates the magnitude that computer fraud can assume. Equity Funding Corporation, an insurance holding company in Los Angeles, employed its computer system to create over 63,000 fictitious insurance policies with assumed values of hundreds of millions of dollars. In addition to reporting grossly inflated assets in its 1972 year-end financial statements, the higher-level managers of the firm embezzled millions of dollars by selling the policies to reinsurers.

## 11.13 TYPES OF COMPUTER CRIMES

Computer crimes take various forms. Two types already listed are unauthorized access of stored data and sabotage of computer facilities. Other types of computer crimes include the following:

- Theft of computer hardware and software. The latter, known as software piracy, is quite prevalent. It involves making copies of programs and software packages, usually from diskettes.
- Unauthorized use of computer facilities for personal use. This crime may be committed by a "hacker," who breaks into a computer system via a remote terminal or microcomputer, or by an employee who runs his or her own programs on the company computer.
- Fraudulent modification or use of data or programs. In most fraud cases the perpetrator intends to steal assets, such as cash or merchandise. For instance, a purchasing agent may enter unauthorized purchase transactions via a terminal and have merchandise sent to his home. A programmer employed by a bank may modify a withdrawal program in a manner that causes withdrawals against her personal account to be charged to an inactive account.

## 11.14 COMPUTER CRIMES ILLUSTRATED

The following sampling of reported computer crimes suggests the dimensions of the problem:

- A self-employed computer expert discovered the daily code that authorized funds to be transferred from a large bank to other banks. One day, five minutes before closing time, he called the wire room, gave the correct authorization code, and transferred $10 million into a bank account opened under his alias.
- A technician who helped design the computerized ticket system for a major league baseball club stayed around the office one day to show staff workers how to operate the system. Later, club officials discovered that he had also used that day to print 7000 tickets, which he illegally sold through ticket brokers.
- Automated teller machines (ATMs) installed by a large New York bank were the means of an ingenious fraud. Persons posing as bank employees would stop depositors in the middle of ATM transactions and direct them to other ATMs, explaining that the ATMs being used were inoperative. Then these persons would withdraw funds from the abandoned ATMs, which had been opened (but not closed) by the depositors.
- A number of unauthorized persons obtained the password into the files of the largest credit bureau in the country. From home computers they were able to view the credit reports of millions of credit card users.
- In a case similar to the preceding one the "414 gang" (a group of young computer "hackers") broke into the highly sensitive files of the Los Alamos National Laboratory.

The type of person who commits computer crimes seems to have a profile that is similar to that of a white-collar criminal. He or she tends to be intelligent and without a previous criminal record. This person is seldom prone to violence against others. Generally, however, he or she has a personal problem for which computer fraud appears to be the answer. Since the computer can be viewed as an inanimate object without human feelings, the perpetrator can treat any planned manipulation as a challenge rather than an illegal action.

## 11.15 OTHER COMPUTER-RELATED MISCHIEF AND CONTROL PROBLEMS

In addition to fraud, computer systems are subject to each type of risk exposure listed in the previous section. Unintentional errors in transactions and loss of assets, as well as breaches of security and acts of violence, can and do occur on a frequent basis. The following are a few of the many actual occurrences.

- A customer received a bill for $1 million instead of $100, because of an error in the invoicing program
- A disgruntled employee erased several reels of magnetic tape with the use of a small magnet
- A salesperson carried away in her briefcase a magnetic tape containing a publishing firm's list of customers
- A rookie computer operator accidentally wrote over the records in a customer master tape file and thus destroyed the account balances and related data
- A fire in a firm's tape library destroyed thousands of reels of magnetic tape
- A failure in an essential component of a computer caused the system to break down and the data to be lost

Furthermore, computer programs themselves are vulnerable to a variety of mischievous software techniques. These techniques, which often are programs but are never authorized by a firm's management, travel under such colorful names as viruses, worms, trapdoors, and Trojan horses. These examples show the havoc that such mischiefs can reap, as well as the extent of their spread:

- Bulgarian computer hackers have developed a computer virus known as the Dark Avenger. It attacks and destroys both computer programs and data. This particular virus is sophisticated, since it contains several perverse features. Dark Avenger operates on a delay cycle, infecting only once in every sixteen operating cycles. It modifies part of its structure after a certain number of executions, to frustrate virus detector software.

- Robert Tappan Morris, the son of the chief scientist of the U.S. National Security Agency, has been found guilty of creating the Internet Worm. This viruslike program disabled 600 computers that were connected to Internet (a nationwide VAX minicomputer network). He claimed that there was no malicious intent; rather, it was simply a misguided stunt gone wrong.

- A fired employee of the USPS & IRA Company planted a virus that destroyed 168,000 sales commission records.

## 11.16 REASONS UNDERLYING COMPUTER CONTROL PROBLEMS

Computer-based information systems manipulate and transcribe data with impeccable accuracy. In spite of this significant advantage, we have seen convincing evidence that computers do introduce severe problems of control. The major reasons for such problems can be traced to the following inherent characteristics of computer-based systems.

*Processing is Concentrated.*    In manual systems the processing is done by clerks in various departments, thereby providing for adequate segregation of duties. Employees can cross-check each other's work, thus detecting processing errors. In computer-based systems the processing is often concentrated within self-contained computer facilities. Certain organizational units are bypassed during processing operations. Consequently, less opportunity exists for detecting errors and fraudulent events such as unauthorized transactions, changes in programmed instructions, and thefts of assets.

*Audit Trails May be Undermined.*    Portions of the audit trail are more likely to be fragmented or eliminated in computer-based systems than in manual systems. Source documents may not be used, for instance, when sales orders are received via telephones and entered directly through terminals. Journals or other records may not be maintained when transactions or adjustments are posted directly to ledgers (master files). These shortcuts improve processing efficiency but cause partial losses of the audit trail. One consequence is that fraudulent acts are less likely to leave traces that can be detected.

*Human Judgment Is Bypassed.*    Computers perform programmed instructions blindly; that is, they exercise no judgment. Thus, fewer opportunities exist for persons to spot errors and questionable data or to observe processing steps. With no special programmed controls and reviews of processed results, transaction errors and irregularities in data can easily escape detection.

*Device-Oriented Data Storage.*   The stored data is invisible. Although this characteristic does not cause a serious problem, it is necessary for users to take specific steps to retrieve the data in readable form. The necessity for data retrieval increases opportunities for errors and often frustrates users. Another source of problems is that stored data (except for ROM memory) are erasable. Thus, valuable data, such as accounts receivable records, may be lost. Third, data are stored in compressed form. A single magnetic disk can hold as much data as several file cabinets. Thus, damage to a single device can cause the loss of a tremendous quantity of valuable data. Finally, stored data are relatively accessible. This condition is particularly acute in the cases of on-line computer systems and computer networks, since persons can access data from various points where terminals and on-line microcomputers are located. Thus, knowledgeable but unauthorized persons may more easily gain access to vital files.

*Vulnerable Computer Equipment.*   The computers are increasingly powerful but also complex and vulnerable. As a result of its processing power, a computer-based system can disseminate errors throughout files and reports more quickly. Because of its complexity, a computer system tends to be confusing to many employees, both at the clerical and the managerial levels. Such confusion can cause employees to make errors. It also may lead employees to resist improvements in computer systems, including improved control procedures. Complexity in computer hardware also causes a system to be vulnerable to breakdowns. If the breakdowns are not quickly repaired, serious interruptions to business operations may occur. Furthermore, computer hardware is often placed in fixed locations, thus rendering it relatively vulnerable to disasters such as fires, floods, and vandalism.

The control problems caused by computerization are summarized in Exhibit 11–1. The two columns on the left side of the figure contrast key characteristics of manual systems with those of computer-based systems. The two columns on the right list the added exposures to risks faced by computer-based systems and the types of controls and security measures needed to offset the risk exposures.

## 11.17 OTHER CONTROL CONSIDERATIONS

Building an effective and feasible internal control structure is not a simple task. It involves more than assembling all of the controls and security measures that come to mind. Audit, cost, and ethical issues need to be considered, since they too can impede the effectiveness of the controls.

**(a) Audit-Related Considerations.**   A typical AIS undergoes periodic audits. Normally, the internal control structure receives particularly close scrutiny during such audits. Thus, the internal control structure should be designed to be fully auditable. For instance, certain analyses and reconciliations can be generated on a routine basis for use by the auditors. Generally speaking, auditors should be consulted during the system design phase, so that all of the needed controls are considered beforehand. Adding controls after the system is designed usually tends to be more costly and difficult.

**(b) Cost-Benefit Considerations.**   Incorporating a control into an information system involves a cost. If every conceivable control were included within an internal control structure, the total cost would likely be exorbitant. Thus, a firm should conduct a cost-benefit analysis, in which the following key question is posed: Will the addition of a

| Element or Activity | Manual System Characteristics | Characteristics | Risk Exposures | Compensating Controls |
|---|---|---|---|---|
| | | **Computer-Based System** | | |
| Data collection | Data recorded on paper source documents | Data sometimes captured without use of source documents | Audit trail may be partially lost | Printed copies of source documents prepared by computer system |
| | Data reviewed for errors by clerks | Data often not subject to review by clerks | Errors, accidental or deliberate, may be entered for processing | Edit checks performed by computer system |
| Data processing | Processing steps performed by clerks who possess judgment | Processing steps performed by CPU "blindly" in accordance with program instructions | Errors may cause incorrect results of processing | Outputs reviewed by users of computer system; carefully developed computer processing programs |
| | Processing steps spread among various clerks in separate departments | Processing steps concentrated within computer CPU | Unauthorized manipulation of data and theft of assets can occur on larger scale | Restricted access to computer facilities; clear procedure for authorizing changes to programs |
| | Processing requires use of journals and ledgers | Processing does not require use of journals | Audit trail may be partially lost | Printed journals and other analyses |
| | Processing performed relatively slowly | Processing performed very rapidly | Effects of errors may spread rapidly throughout files | Editing of all data during input and processing steps |
| Data storage and retrieval | Data stored in file drawers throughout the various departments | Data compressed on magnetic media (e.g., tapes, disks) | Data may be accessed by unauthorized persons or stolen | Security measures at points of access and over data library |
| | Data stored on hard copies in human-readable form | Data stored in invisible, erasable, computer-readable form | Data are temporarily unusable by humans, and might possibly be lost | Data files printed periodically; backups of files; protection against sudden power losses |
| | Stored data accessible on a piecemeal basis at various locations | Stored data often readily accessible from various locations via terminals | Data may be accessed by unauthorized persons | Security measures at points of access |
| Information generation | Outputs generated laboriously and usually in small volumes | Outputs generated quickly and neatly, often in large volumes | Inaccuracies may be buried in impressive-looking outputs that users accept on faith | Reviews by users of outputs, including the checking of amounts |
| | Outputs usually in hard copy form | Outputs provided in various forms, including soft copy displays and voice responses | Information stored on magnetic media is subject to modification (only hard copy provides permanent record) | Backups of files; periodic printing of stored files onto hard copy records |
| Transmission of data and information | Usually transmitted via postal service and hand delivery | Often transmitted by communications lines | Data may be accessed or modified or destroyed by unauthorized persons | Security measures over transmission lines; coding of data; verification of transmitted data |
| Equipment | Relatively simple, inexpensive, and mobile | Relatively complex, expensive, and in fixed locations | Business operations may be intentionally or unintentionally interrupted; data or hardware may be destroyed; operations may be delayed through inefficiencies | Backup of data and power supply and equipment; preventive maintenance of equipment; restrictions on access to computer facilities; documentation of equipment usage and processing procedures |

**Exhibit 11–1. A table of control problems caused by computerization. Adapted from J. Wilkinson, *Accounting Information System: Essential Concepts and Applications,* 2nd ed. (NY: John Wiley & Sons)**

specific control provide expected benefits that exceed the costs of the control? If the answer is yes, and there is reasonable assurance that the control will achieve specified objectives, then the control is a desirable addition.

The broadest benefit provided by a control usually consists of reducing risk exposures, that is, of reducing the risks of failing to achieve one or more of the objectives pertaining to the internal control structure. Specific benefits may be either quantitative or qualitative in nature, such as reducing the losses due to thefts of assets; improving the reliability of information provided to management; and improving the reputation of the firm.

Costs of a control include one-time costs, recurring costs, and opportunity costs. One-time costs include the installation of security devices and training of clerks; recurring costs may be for supplies and salaries of new employees needed to implement the control. An example of an opportunity cost arises from the reduced efficiency in transaction processing caused by the added control; reduced efficiency translates into lost income.

When the costs of a control exceed its expected benefits, and the control is nevertheless installed, overcontrol exists. For instance, a control might be installed that detects certain errors that are missed by complementary controls. However, the costs deriving from these errors may not be as great as the costs of maintaining the added control.

A cost-benefit analysis includes the following steps:

1. Assessing the risks to which the firm is exposed, such as losses of vital records.

2. Measuring the extent of each risk exposure in dollar terms. For instance, if the exposure is the possible loss of an asset, the value would be the amount needed to replace the asset.

3. Multiplying the estimated effect of each risk exposure by the estimated frequency of occurrence over a reasonable period, such as a year. The resulting product is the potential loss that can be incurred by not reducing or avoiding a particular risk. Alternatively, it is the benefit to be gained by avoiding the risk or improving the reliability of information.

4. Determining the cost of installing and maintaining a control that is to counteract each risk exposure. Comparing the benefits against the costs for each control. On a broader level, this comparison should be employed for the group of controls pertaining to individual transaction processing systems and to the activities of the firm as a whole.

Cost-benefit analyses are difficult to apply. None of the factors are easy to measure. Also, in many situations several controls may be needed to mitigate a particular risk. However, new analytical techniques are being developed. For instance, a technique known as reliability analysis calculates reliability by measuring the error probabilities related to a process such as transaction processing.

**(c) Ethical Issues.**    Unfortunately, employees are often subject to negative influences in the workplace. Perhaps their firm gives them strong incentives to engage in questionable behavior. For instance, employees may be pressured to meet unrealistic short-term performance goals, with the consequence that they may "pad" the figures on reports. They may also be affected by appealing temptations. For example, an employee might be tempted to engage in the dishonest act of stealing assets when he or she is aware that controls are missing or penalties are likely to be trivial. Such undesirable behaviors can be reduced or eliminated by removing the incentives and temptations, that is, through an emphasis on ethics and controls.

A strong ethical climate is vital to a firm's well-being, since it contributes to the effectiveness of the component control systems. Thus, management, from the chief executive officer on down, should exhibit high ethical values. The firm should seek and hire employees who are not only competent but also possess integrity. To aid in attracting competent and trustworthy individuals, management should establish enlightened policies. Specific policies might encourage long-term goals, delineate fair human resource practices, and so on.

Ethical standards might also be stated in a written code of conduct and communicated widely to all employees. A sound code of conduct would set out acceptable practices with respect to employee behavior. It might address matters such as conflicts of interest, improper payments, anticompetitive actions, insider trading, and so on. Management should ensure that all employees are as aware of these standards as they are of their assigned job responsibilities. Furthermore, the managers of the firm should strictly follow the standards in their own daily behavior. The internal control structure of a firm is only as effective as the people who create it, operate it, and function within it. One of the primary concerns in designing the structure, therefore, is to provide controls that will influence in a positive manner the behavior of the employees and others who interact with the structure. Consequently, the management of a firm and the designers of the internal control structure must be concerned with the reactions of employees to controls. If controls are perceived by affected employees as being weak or unnecessary, the employees may circumvent the controls. Thus, it is highly desirable to inform employees of the purposes of the controls and to instruct them carefully in duties where controls are involved. Supervisors should watch for adverse reactions on the parts of employees, so that this understanding can be reinforced. For instance, it can be pointed out to accounting clerks that controls restricting their access to cash are desirable, in that temptation is thus removed from honest employees.

From the foregoing discussion we can conclude that ethical issues and internal controls are closely intertwined. Any internal control structure can be circumvented when the ethical climate is weak, while even a porous internal control structure will be enhanced if the ethical climate is strong.

## 11.18 FORCES FOR THE IMPROVEMENT OF CONTROLS

During earlier periods, many an AIS was deficient with respect to controls and security measures. Often, the system was intended primarily to provide the needed day-to-day documents and reports and to satisfy legal obligations. In recent decades, however, various forces have arisen to encourage the improvement of internal control systems. Perhaps the most influential forces have been managements, professional associations, and governmental bodies.

(a) **Needs of Management.** The managers of most firms have recognized their vital stake in adequate internal control systems. They have become aware of the huge losses and damages that can occur to the costly assets entrusted to their care. Newspapers and other media have publicized the increasing instances of "white-collar" crime, as well as overt thefts of merchandise and other portable assets. Managers have noted that the average loss from each crime has also been rising dramatically. On the other hand, they have grown concerned about the accuracy and reliability of the information they receive. Being primary users of information from their AISs, they appreciate the potential for making poor decisions owing to inaccurate and incomplete information. Furthermore, with the increasing dependence on computer systems, they have gained a realization of the seriousness of security breaches.

**(b) Ethical Concerns of Professional Associations.** Professional accounting associations such as the American Institute of Certified Public Accountants (AICPA) and the Institute of Internal Auditors (IIA) have established codes of ethics. These codes are self-imposed and self-enforced rules of conduct. Included are rules pertaining to matters such as independence, technical competence, and suitable practices during audits and consulting engagements involving information systems.

The codes of ethics have been expanded and clarified by various pronouncements, such as Statements on Auditing Standards issued by the AICPA and the standards promoted by the Institute of Internal Auditors. Particular attention has been given to internal controls, as the following pronouncements illustrate:

- A professional standard of field work for auditors specifies " . . . a proper study and evaluation of the existing internal controls as a basis for reliance thereon . . . " according to AICPA's Statement on Auditing Standards No. 1.
- The Statement on Auditing Standards (SAS) No. 3 requires that the computerized portions of an AIS be included in an auditor's study and evaluation of the internal control system.
- A study issued by the AICPA provides lists of control objectives and a step-by-step procedure for analyzing controls in computerized systems.
- A set of standards issued by the Institute of Internal Auditors pertains to the responsibilities of internal auditors for evaluating control and otherwise conducting the practice of internal auditing.
- SAS No. 48 amends SAS No. 3 with respect to the effects of computerized systems on accounting controls, as well as the need for specialized auditing skills when evaluating such systems.
- SAS No. 55 supersedes SAS No. 1 and calls for the need for obtaining a thorough understanding of the internal control structure (formerly called system) before undertaking audits of financial statements. It states that auditors should therefore carefully study the control environment, accounting system, and control procedures in the planning of audits.

# CHAPTER 12

# MORE ON CONTROL

## 12.1 INTRODUCTION

Those charged with documenting procedures must not only reckon with the fact that an effective internal control structure for a business firm consists of numerous and varied controls, but they must also allow for the variety in accounting information systems. There exists both manual and computer-based systems and they require both specific forms of controls as well as controls common to both. Specific controls, such as cross-checking of calculations, may be associated primarily with manual processing. Other specific controls, such as passwords, are applicable only to computer-based processing systems. Still other controls, such as supervision, are needed in either type of processing system.

Accountants and auditors, as well as those involved in documentation, need a thorough knowledge of controls in accounting information systems, since all are likely to come into contact with both manual systems and computer-based systems.

This chapter, building on the previous chapter, provides additional or alternative ways of classifying controls. Following is identification and discussion, within the frameworks of these classification plans, of various controls that broadly pertain to the environment within which the accounting function is performed. Those controls that pertain to manual processing are distinguished from those of computer-based processing, with additional references to controls that are used for both.

## 12.2 SYSTEMS WITHIN THE INTERNAL CONTROL STRUCTURE

The internal control structure may also be viewed as a collection of three component control systems or processes. Although known by various names, we will label these systems as:

- The management control system seeks to encourage compliance with management's policies and procedures with the help of administrative controls.
- The operational control system seeks to promote efficiency in operations with the help of administrative and technological controls.
- The accounting control system seeks to safeguard assets and ensure the accuracy and reliability of data and information with the help of accounting controls.

**(a) Management Control System.** The management control process focuses on managerial performance, rather than on technical operations. Its purpose is to encourage compliance with the firm's policies and procedures. Policies are the strategies or guidelines by which the overall goals and objectives are to be achieved; procedures are the

141

prescribed steps by which specified tasks are accomplished. If the policies and procedures are soundly developed, resources will be acquired and used effectively and efficiently. As the term implies, management control is exercised through the actions of managers. It follows the organizational structure of a firm, since the managers function through responsibility centers.

An example of the management control process can be drawn from the production function of a manufacturing firm. To fulfill his responsibility of controlling production-related resources, the production manager (planner) develops policies and production volume and cost targets (bench marks). These policies and targets are embodied in a production budget for the coming year, as well as in procedures concerning quality inspections and other matters. The budget is broken down by production departments (responsibility centers). Each department head is then evaluated by comparing the actual volumes and controllable costs against the budgeted amounts. When the variances in volumes and costs are significant, corrective action may be necessary. For instance, if the volume is below budget, the department head may be assigned more workers or he or she may be transferred.

All of the controls implemented by management control systems are known as administrative controls. These controls roughly parallel those encompassed by what are called the control environment. Specific administrative controls include:

- Preparing budgets
- Preparing documentation such as procedures manuals
- Establishing an organizational structure that provides adequate direction through supervision
- Providing adequate overseeing of accounting policies and practices via audit committees, internal auditors, and external auditors
- Carefully selecting, training, and supervising employees

**(b) Operational Control System.**   The process or system that promotes efficiency in performing tasks, the second listed objective of internal control, is called operational control. Examples of operational control systems within typical firms are the inventory control system, the credit control system, the production control system, and the cash control system.

Like the management control system, an operational control system incorporates administrative controls. In the context of operational control, however, the controls are related to specific transactions. For instance, a credit check is a central control within the credit control system; it is also a key control in a sales transactions processing system. Similarly, a bank reconciliation is an important part of the cash control system; it is also a key control in the cash receipts and cash disbursements transaction processing systems. Another example of an operational control system is the close electronic surveillance of the gambling area in Las Vegas casinos. The high degree of security in a casino ensures that the dealers do not take home any of the chips.

### 12.3 OTHER CONTROL CLASSIFICATIONS

Controls may also be classified by objectives, by risk aversion, by settings, and by system architectures. Such alternative perspectives on control enhance the perception of what control seeks to achieve within an organization.

**(a) Classification by Objectives.**   Two major control categories under the objectives classification plan are the administrative and accounting controls previously discussed. Administrative controls relate to the broad objectives of encouraging adherence to management's policies and procedures and promoting operational efficiency. Accounting controls have the broad objectives of safeguarding a firm's assets and ensuring accuracy of the accounting data and information.

Controls based on objectives can also be categorized as operations controls, financial reporting controls, and compliance controls. Operations controls are those needed in managing a firm's activities. Financial reporting controls are those that ensure reliable financial reports. Compliance controls are those that ensure compliance with laws and regulations.

**(b) Classification by Risk Aversion.**   Controls may also be classified according to the ways that they combat the risks to which a firm and its information are exposed. Preventive controls block adverse events, such as errors or losses, from occurring. An example is a manual of processing procedures. Preventive controls tend to be passive in nature. Detective controls discover the occurrence of adverse events. They are more active than preventive controls. An example of a detective control is key verification of data typed onto a magnetic medium by a clerk. Certain detective controls cause further processing to be halted, as when an input error in a transaction is detected. Corrective controls lead to the righting of effects caused by adverse events, usually by providing needed information. For example, information to the effect that the level of an inventory item is too low may trigger a suitable correction—a request that more inventory be ordered. In some instances, corrective controls may anticipate adverse events, as when insurance coverage is prescribed against the theft of assets. Corrective controls are generally more active than detective controls.

**(c) Classification by Settings.**   One of the most useful groupings of controls is in accordance with the elements described in Statement on Auditing Standard (SAS) No. 48, "The Effects of Computer Processing on the Examination of Financial Statements." This statement is further clarified by Statement on Auditing Standard No. 55. SAS No. 48, which separates controls into general controls and application controls. General controls are those controls that pertain to all activities involving a firm's accounting information system and resources (assets). These controls include those provided by the control environment, plus other control procedures. Application controls relate to specific accounting tasks or transactions; hence, they may also be called transaction controls. Application or transaction controls roughly correspond to the accounting system, as described in SAS No. 55. Another group of controls pertains both to the control environment and to transactions, and does not fit comfortably into either category. These controls, which we shall call security measures, are intended to provide adequate safeguards for access to and use of assets and data records.

These three categories of controls are intertwined, especially in computer-based information systems. An appropriate balance of controls in all three categories is needed for an internal control structure to function effectively.

**(d) Classification by System Architectures.**   As indicated, system architectures include manual systems, computer-based batch processing systems, computer-based on-line processing systems, computer-based database systems, and computer-based data communications systems.

## 12.4 ORGANIZATIONAL CONTROLS

A firm's organizational structure represents an underlying control because it specifies the work relationship of employees and units. The central control objective when designing the organizational structure is to establish organizational independence. When properly provided through a careful and logical segregation of assigned duties and responsibilities, organizational independence results in a complete separation of incompatible functions. It involves two or more employees or organizational units in each procedure, who can be assigned to check on the work of one another. Thus, errors made by one employee or unit will be detected by another. No single employee is able to commit a fraudulent act in the normal course of duties and then hide the deed. Fraud under such an arrangement can be perpetrated only by means of collusion. Thus, the chance for fraudulent activities is greatly reduced, since most persons who might consider fraud are afraid of being rejected if they propose the idea to a co-worker. Collusion by related persons, such as mother and daughter, can be prevented by employment rules that prohibit nepotism.

Although very important to organizational control, the segregation of duties is usually not sufficient. Most firms also depend on the diligence of independent reviewers as well as internal and external auditors. To be truly effective, these reviewers/auditors must stand apart from the procedures themselves. A typical large- or medium-sized firm has several types of reviewers. The higher-level managers, including the board of directors, represent reviewers who have broad perspectives and responsibilities. Lower-level managers, who receive and use the majority of outputs from the AIS, represent reviewers with narrower perspectives. Finally, internal and external auditors represent reviewers who are both expert and objective. A small firm having only a few employees may not be able to afford all these reviewers. However, since it cannot provide adequate segregation of duties, a small firm is in desperate need of review. The answer to this dilemma must be the manager or owner of the firm. He or she should carefully supervise the employees as they perform duties such as processing transactions, either manually or with the aid of computers. In addition, the manager or owner should perform key tasks such as opening mail, writing checks, and reconciling bank statements.

To understand how the concept of organizational independence enhances control, we should review its application in various types of systems by focusing on how the various functions to be performed must be segregated. We will begin with manual systems, since they present the most familiar situations. Then, we will apply the concept to differing computer-based environments.

## 12.5 MANUAL SYSTEMS CONTROL ORGANIZATION

Authorizing, record keeping, and custodial functions should be separated in manual systems. Thus, employees who handle assets, such as cash and inventory, should not authorize transactions involving those assets or keep the records concerning them. For instance, if an accountant is allowed to handle cash receipts and also keep the accounts receivable records, the accountant can easily conceal a theft. Making sales and cash receipts transactions independent is a logical division of duties. Numerous units are involved in the transactions involving sales, which means greater coordination, but control requires that independence be maintained. The sales order department originates and authorizes sales transactions, while the credit department authorizes the credit terms on which the sales are made; record keeping is performed by the billing and accounts receivable departments; and custodial duties are handled by the warehouse, shipping, and cash receipts departments.

The concept of organizational independence also prohibits the combining of duties in those cases where assets are endangered or adequate checks are not applied. For example: A clerk who is assigned the duty of handling a petty cash fund should not also handle cash received from customers, since the funds might become commingled and later lost. A buyer who prepares a purchase order should not also approve the prepared form; instead, the purchase order should be signed by the purchasing manager. An accountant who prepares a journal voucher should present the completed form to an accounting manager for signature. An accounting clerk who performs key steps in a procedure, such as posting cash receipts, should not perform a check on the procedure, such as preparing a bank reconciliation.

## 12.6 COMPUTER-BASED SYSTEMS ORGANIZATION

Organizational independence should also be maintained in computer-based systems, although adjustments are necessary. As with manual systems, the authorizing, custodial, and record keeping functions are to be separated. Our discussion of organizational controls in computer-based systems will first consider the needed segregation between user departments and the information systems function; then, we will look at needed segregation of responsibilities within the systems function itself.

**(a) Segregation of Responsibilities Between Departments.**   The information systems (IS) function has responsibilities relating to processing and controlling data (i.e., record keeping). Thus, it should be organizationally independent of all departments that use data and information, those departments that perform operational and custodial duties, and those persons who authorize transactions. All transactions and changes to master records and application programs should be initiated and authorized by user departments. For instance, sales transactions should be initiated by the sales order department, with the sales orders being processed within the information systems function. Changes to sales application programs should be initiated by sales management. Furthermore, errors in transactions should be corrected by user departments. All assets (except computer and data processing facilities) should reside under the control of designated operational departments; for example, the merchandise needed to fill customers' orders should be stored in a warehouse department.

Sometimes it may appear that complete separation of these functions is not possible. For example, when a computer-based system automatically approves an order from a customer on the basis of credit guidelines built into a sales application program, it may seem that the information systems function is authorizing the transaction. However, the authorizing function is in reality performed by the person, perhaps the credit manager, who established the guidelines and made them part of the software that processes orders from customers.

**(b) Segregation of Responsibilities Within the IS Function.**   To achieve organizational independence, it is necessary to subdivide several key responsibilities within the organizational structure of the IS function itself. The IS function has the overall purpose of providing information-related services to other departments within a firm. In those firms having computer-based systems, it has taken over the array of record preparation, record keeping, and processing activities traditionally performed by several accounting departments. That is, the presence of computers centralizes duties that should otherwise be segregated.

The major segregation of responsibilities is between systems development tasks, which create systems, and data processing tasks, which operate systems. The systems development function is concerned with analyzing, designing, programming, and documenting the various applications needed by user departments and the firm as a whole. Not only is it responsible for new computer-based applications, but it must also make changes in existing applications as needed. Furthermore, it can help users via an information center. The data processing function has responsibility for ensuring that transaction data are processed and controlled and the related files and other data sets are properly handled. These two major functions are separated—both organizationally and physically—for a very sound reason. If the same individuals had both detailed knowledge of programs and data and access to them, they could make unauthorized changes. Thus, systems analysts and programmers should not be allowed to operate the computer or to have access to "live" programs or data. Furthermore, computer operators and other data processing personnel should not have access to the documentation concerning programs or to various assets such as inventory and cash.

The IS function also includes other functions, such as technical services and database administration. Technical services has responsibilities with respect to computer-related areas such as data communications, systems programming, and decision modeling. Database administration is concerned with all aspects of the data resources. The database administrator must establish and define the schema of the database, control the use of the database via appropriate security measures, and control all changes in data and programs that use the database. The database administrator should have functional authority over the data library. However, he or she should not have direct access to "live" data or programs. A high degree of segregation is needed to reduce the risks of alteration to "live" data or programs. Without adequate segregation, a computer operator could make changes to his personnel records at will and escape detection, or a programmer could alter a computer program and not be caught.

**(c) Segregation of Responsibilities Within the Data Processing Function.**   This discussion illustrates how a suitable division of responsibilities provides needed segregation of duties in the course of a batch processing application:

- The data control unit serves as an interface between the various user departments and computer operations. It records input data (including batch totals) in a control log, follows the progress of data being processed, and distributes outputs to authorized users. As a part of its control responsibilities, it maintains the control totals pertaining to master files as well as transaction files, and reconciles these totals with updated totals shown on exception and summary reports. Finally, it monitors the correction of detected errors by the user departments. The data control unit must be independent of computer operations, since it helps to ensure that processing is performed correctly and that data are not lost or mishandled.
- The data preparation unit prepares and verifies data for entry into processing.

- Computer operations processes data to produce outputs. Its duties include loading data into input devices, mounting secondary storage devices such as magnetic tapes and magnetic disk packs, and performing operations as prescribed by run manuals and computer messages. One duty that computer operators should not perform is correcting errors detected during processing, since the corrections may introduce new, undetected errors. Computer operations should be physically as well as organizationally separated from the other units, so that persons such as the librarian and data control clerks do not have direct access to the computer.
- The data library unit maintains a storage room, called the library, where the data files and programs are kept. A librarian issues these files and programs to operators when needed for processing and keeps records of file and program usage. Thus the files and programs are better protected when not being used.

In contrast to the division of functions performed in the above batch-processing system, the division of duties is simplified while processing applications on-line. The various users enter the transactions via terminals. The transactions are checked by computer edit programs for accuracy and then are processed against on-line files. Outputs may be printed or displayed on printers, plotters, or terminals located in the departments of the recipients. The data control and data library tasks, as well as data processing, are performed by the computer system hardware and software.

## 12.7 MANAGEMENT PRACTICE CONTROLS

Some of the most severe risk exposures that a firm faces are related to possible deficiencies in management. For instance, an incompetent employee may be hired or an employee may be poorly trained, resulting in excessive errors. Or, a bank processing computer program could be changed by a programmer, so that bank funds are diverted into her personal accounts. A wide variety of administrative controls are needed to counteract such management-related risks. Most of the controls are those identified as being a part of the control environment discussed in Chapter 11. These controls are grouped under a category called "management practice controls." This broad category includes the organizational controls discussed above. In addition, management practice controls may be said to include asset accountability control, planning practices, personnel practices, audit practices, system change procedures, new system development procedures and documentation controls. (see Exhibit 12–1.)

## 12.8 ASSET ACCOUNTABILITY CONTROLS

A firm's assets are the productive resources that it possesses. These valued assets are subject to theft, pilferage, accidental loss, and damage. Another risk is that the assets will be valued incorrectly in the financial statements, perhaps as a result of miscounting the quantities. Thus, one of the objectives of internal control is to protect a firm's assets from these risks. The assets may be safeguarded and accounted for properly by doing the following:

- Permitting access to assets only in accordance with management's authorization
- Comparing the recorded accountability for assets with existing assets at reasonable intervals and taking appropriate action with respect to all differences

---

**Asset Accountability Controls**
Limited access
Periodic verification using:
   Subsidiary ledgers
   Logs and receipts
   Reconciliations/inventories
   Acknowledgement procedures
   Reviews and reassessments

**Planning Practices**
Budgets
   Develop standards
   Flexible budgeting
   Variance analysis
Schedules
New controls

**Personnel Practices**
Select competent and trustworthy employees
Train employees adequately
Provide clear job descriptions
Schedule periodic performance evaluations
Require employee bonding
Arrange employee job rotation
Require vacations
Maintain close supervision
Establish clear termination policies

**Audit Practices**
High-level audit committee
Periodic examinations by external auditors
Ongoing reviews by internal auditors

**System Change and New Development Procedures Documentation Control**
Administrative
System

---

**Exhibit 12–1.    Management practice controls.**

Restricting access to assets is achieved through various security measures, which are discussed in Chapter 13. Assuring that assets are properly valued in the accounting records involves the presence of asset accountability controls. Specific controls that aid in providing proper asset accountability include the use of subsidiary ledgers, logs, receipts, reconciliations/inventories, acknowledgment procedures, reviews, and reassessments. These controls are needed in manual as well as computer-based systems. However, because of the added computation power of computers, controls such as reconciliations can be performed more frequently in computer-based systems.

**(a) Accounting Subsidiary Ledgers.**    Subsidiary ledgers can be maintained for assets such as accounts receivable, inventory, plant assets, and investments. Amounts reflected in these ledgers are based on postings from transaction documents. The total of all balances in a particular subsidiary ledger should be equal to the balance in the corresponding control account in the general ledger. Since the postings are performed independently of each other, the use of a subsidiary ledger provides a cross-check on the correctness of the control account, and vice versa.

**(b) Reconciliations.** A reconciliation consists of comparing values that have been computed independently. Thus, a comparison of the balance in a control account with the total of balances in a corresponding subsidiary ledger is an example of a reconciliation. Reconciliations can also involve comparisons of physical levels of resources with the quantities or amounts reflected in accounting records. For instance, each item of a physical inventory should be counted periodically. These physical counts can then be reconciled with the quantities shown in the accounting records. If differences appear, they may signal the need to adjust the quantities in the accounting records to reflect the physical realities. Another important reconciliation, the bank reconciliation, compares the balance in the bank account with the cash balance in the general ledger.

In accordance with the principle of organizational independence, reconciliations should be prepared only by employees or managers not otherwise responsible for the processing of related transactions. For instance, no employee involved in the processing or handling of cash receipts or cash disbursements should prepare a bank reconciliation. Instead, it should be prepared by an accounting manager or internal auditor.

**(c) Logs and Registers.** Receipts, movements, and uses of assets can be monitored by means of logs and registers. For example, cash receipts are logged on remittance listings (i.e., registers). Later, the amounts recorded on deposit slips are reconciled to the amounts of these remittance listings, to ensure that all receipts are deposited intact. Files of data on magnetic tapes are noted on logs as they are moved from the data library into the computer room, and vice versa. When an employee uses a computer system from a terminal, the access can be recorded on a console log. Logs and registers thus help a firm to account for the status and use of its varied assets.

**(d) Reviews and Reassessments.** Reviews by outside parties provide independent verification of asset balances and, hence, accountability. For instance, an auditor may verify that the plant assets reflected in the accounts actually exist and are properly valued. A customer who receives her monthly statement will likely verify that the amount owed is correct. Reassessments are reevaluations of measured asset values. For example, accountants make periodic counts of the physical inventory and compare these counts to the inventory records. If necessary, the quantities and amounts in the records may be adjusted downward to reflect losses, breakage, and aging.

**(e) Acknowledgment Procedures.** In various transactions, employees are called on to acknowledge their accountability for assets. For instance, when merchandise arrives from suppliers, the clerks in the receiving department count the incoming goods, prepare a receiving report, and sign the report. In doing so they acknowledge accountability for the goods. When the merchandise later is moved to the storeroom, the storekeeper recounts the goods and signs for their receipt. Through such an acknowledgment, he or she accepts the transfer of accountability for the goods.

## 12.9 PLANNING PRACTICES

Another form of management practice controls is organizational planning. Sound planning practices involve activities such as preparing budgets, developing schedules, and approving new controls. Consider, for instance, budget preparation, which is closely allied with the management control system. A budget quantifies the financial objectives of a firm. That is, it establishes both the revenue levels that a firm expects to achieve and the cost levels within which it desires to constrain its operations. Through comparisons with actual revenues and costs, the budgeted amounts can help to detect inefficiencies, losses, and even fraudulent actions.

## 12.10 PERSONNEL PRACTICES

Sound personnel practices can significantly aid a firm in achieving efficient operations and maintaining data integrity. Trustworthy and competent employees should be sought, screened, and selected to fulfill all positions of responsibility. New employees should be provided with clear-cut job descriptions and be adequately trained concerning their responsibilities. In particular, employees should be made aware of important control requirements, such as the proper use and protection of confidential passwords. Each employee's performance should be evaluated periodically. Employees having access to cash and other negotiable assets should be bonded. A fidelity bond indemnifies a firm when it incurs losses of insured assets, due to events such as fraudulent activities of bonded employees. Operational employees such as computer operators should be rotated among jobs and shifts. Also, employees in key positions of trust should be required to take periodic vacations, to subject their activities to review by substitutes. In addition, they should be well supervised, so that they are encouraged to follow established policies and to avoid irregularities. For instance, the manager of data processing and/or shift supervisors should closely observe and review the actions of the computer operators. Finally, employees should be terminated only in accordance with reasonable and well publicized policies.

## 12.11 AUDIT PRACTICES

Sound evaluation practices are needed to provide independent checks on the performance of employees, the adequacy of the internal control structure, and the reliability of the accounting records. Evaluation of each employee's performance is normally performed by the employee's supervisor. However, checking computations or other work results is often performed by other employees, as we have discussed. Evaluating the internal control structure and accounting records can best be done by auditors.

Sound audit practices include oversight by audit committees, examinations by external auditors, and reviews by internal auditors. An effective audit committee usually reports directly to the board of directors. It takes an active role in overseeing the firm's accounting and financial reporting policies and practices. It is also a liaison between the board and the external and internal auditors. External auditors should perform periodic independent verifications or audits of the accounting records underlying the financial statements. Internal auditors should be assigned to a permanent organizational function within the firm. Their primary responsibility should consist of evaluating the adequacy and effectiveness of internal control structure policies and procedures. In addition, they may undertake specific verification procedures such as preparing bank reconciliations and distributing paychecks to employees.

## 12.12 SYSTEM CHANGE PROCEDURES

Changes in a computer-based information system most often pertain to application programs or the schema of the database. Both types of changes should follow clearly defined and sound procedures to prevent unauthorized manipulations and possibly well-meaning but injurious errors and mishaps. For instance, a programmer could incorporate into a program a feature that benefits him or her personally or that violates management policy.

If the change pertains to an application program, it should be initiated by a user-department manager, who explains the needed change in writing. The requested change should then be approved by the systems development manager (or by a committee of high-level managers if the modification is sufficiently large). After approval, the change or addition would be assigned to systems personnel, usually a maintenance programmer in the case of an application program. This programmer should use a working copy of the program, rather than the "live" version currently in use. The new or revised design is next tested jointly by systems personnel (including persons not involved in the design) and the user. Documentation should be thoroughly revised to reflect the change or addition. Finally, the documented change or addition and test results should be approved by the systems development manager and should be formally accepted by the initiating user.

## 12.13 NEW SYSTEMS DEVELOPMENT PROCEDURES

The design and development of new computer-based applications require controls similar to those needed for system changes. Each request for a new system development should be initiated by either a user-department manager or a higher-level manager, depending on its scope. Assume, for instance, that a new automated credit approval system is desired. The credit manager could initiate the request, which would then be authorized by a computer-system steering committee. The user-department and systems development personnel next would work together to clarify information needs, to define system requirements, and to develop necessary system design specifications. After implementing the designed system, they should jointly test all portions of the system, manual as well as computerized. Finally, the documentation concerning the design and test results would be reviewed and approved by the manager of the user department.

## 12.14 DOCUMENTATION CONTROLS

An AIS is a complex mixture of procedures, controls, forms, equipment, and users. If the instructions and guidelines for operating such a system are inadequate, the system is likely to function inefficiently. If and when breakdowns occur, the disruptions to operations are likely to be quite harmful. Consider the situation in which a system analyst has designed and installed a computer-based accounting application, but has neglected to prepare computer flowcharts that describe the programs. If a "bug" develops in one of the programs, the repairs to the program would be difficult and time-consuming.

Documentation, be it adminstrative or systems related, is potentially a very useful control tool for effective management. There needs to be some controls applied to documentation. More about such controls will be said subsequently in Chapter 13 dealing with the mechanics of documentation.

## 12.15 OPERATIONAL CONTROLS

A group of administrative controls may be described as operational controls. These controls pertain mainly to computer-based systems and may be subdivided into computer operating procedures, and computer hardware and software checks.

Computer operations are subject to a variety of problems and abuses. For example, operations might be poorly scheduled, with the result that needed computer printouts arrive so late in the purchasing department that out-of-stock parts are not ordered in time to keep the production line running. Sound and well-controlled computer operations are based on close supervision, careful planning, and organized procedures. Thus supervisory personnel, such as the manager of data processing and shift supervisors, should actively observe and review the actions of computer operators. Procedural manuals concerning all aspects of computer operations should be provided to computer operators, together with the appropriate console run books. Data processing schedules should be prepared as far in advance as feasible and revised as necessary. Preventive diagnostic programs should be employed to monitor the hardware and software functions, so that existing or potential problems may be detected. A variety of control reports should be prepared daily or weekly. Suggested reports include computer facilities utilization reports, employee productivity reports, and computer processing run-time reports. To be most effective, these reports should compare actual times against standard times. For instance, the last-named report can compare actual run times, as shown in the console log, against standard times shown in the data processing schedule.

Computer hardware and software checks represents another form of operational control specific to computer-based systems. Though modern computer hardware is generally reliable in operation, malfunctions can occur. Thus a variety of hardware and software checks are built into computer systems. They assure the reliability of computations, manipulations, and transfers of data within the systems. Typical built-in hardware checks include the parity check, echo check, read-after-write check, dual read check, and validity check. Other hardware checks or controls include duplicate circuitry, which allows the arithmetic unit within the CPU to perform calculations twice and to check the results, and scheduled preventive maintenance, which schedules periodic servicing of the computer system to reduce unexpected system failures and, hence, overall operating costs. Also, certain computers are provided with fault tolerance, the capability to continue functioning even in the face of a partial failure. Fault tolerance usually involves built-in software that constantly verifies the performance of each hardware element, plus the presence of redundant elements. If an element fails, the processing is shifted to an alternative element.

Software checks take a number of forms. Among the controls generally incorporated into operating systems are the label check and the read-write check. The label check automatically notifies the computer operator of the contents of internal labels on storage media such as magnetic tapes and magnetic disks. The read-write check automatically halts a program when reading or writing is inhibited (e.g., when a printer runs out of paper), or initiates an end-of-file routine when no further processing of a file medium is possible (e.g., when the end of a magnetic tape reel is reached). A control procedure involving checkpoints is useful for restarting computer programs when the computer system malfunctions during processing.

# CHAPTER 13

# DOCUMENTING ACCOUNTING INFORMATION SYSTEMS

## 13.1 INTRODUCTION

This chapter focuses on getting to know the workings of an accounting information system (AIS) and learning tools that can facilitate its documentation. We start by explaining what information is, describing a system, and defining an information system. Then, an information system and its components are described in detail, including the various types of information systems, such as management information systems (MIS) and accounting information systems. Then, the roles played by the accounting information system within an organization are explained.

Following the description of an accounting information system is a discussion of the tools, such as flowcharts and data flow diagrams, used to analyze and document an information system. Learning the use of such tools is crucial for those charged with the documentation of accounting procedures. Such documentation techniques allow accountants to understand, evaluate, design, and document accounting procedures.

## 13.2 WHAT IS INFORMATION?

Information is knowledge communicated or received concerning a particular fact or circumstance (a person, place, or thing). Information and data are not the same. Information is data that has been processed or organized into output that is meaningful to those receiving it. Data usually represent observations or measurements of events that can be of importance to potential users as a result of further processing. It must be seen as the input received by an information system for further processing, after which it converts into information. In contrast to data, information is the processed output that is organized, meaningful, and useful to the person who receives it.

The desirable characteristics of information include:

- Reliability
- Relevance
- Timeliness
- Completeness
- Understandability
- Verifiability

Turning data describing the activities of an organization into information requires the creation of a system, more specifically, an information system. Such a system can be defined as a combination of elements or parts forming a complex or unitary whole, an assemblage of a set of correlated members within an organization. In other words, a system is an entity of interrelated components or subsystems that interact to achieve goals. The system's concept requires that alternative courses of action be seen from the standpoint of the system as a whole, rather than from that of any of its components or subsystems. The concept of integration is crucial in order for a system to work optimally.

## 13.3 INFORMATION SYSTEMS

An information system is an organized means of collecting, entering, and processing data. It is also involved with storing, managing, controlling, and reporting information so that an organization can achieve its objectives and goals. The working of an information system may therefore be seen as consisting of four stages: input, processing, storage, and information output. The first three deal with data, while the fourth provides information. There are several ways to classify information systems (IS):

- Formal versus informal systems. A formal information system is one that has an explicit responsibility to produce information. Accounting, production, and marketing information systems are examples of formal information systems. An informal information system is one that arises out of a need that is not satisfied by a normal channel. The "grapevine" is an example of an informal information system.
- Manual versus computer-based systems. Manual information systems are those in which the processing tasks are performed by people. Computer-based systems are those in which processing is performed by a computer. The advantages of manual systems are that people are flexible and able to adapt to new situations, but at the same time people can be unreliable and slow. The advantages of computer-based systems are that they are fast and reliable, but they are less flexible and have high initial costs. Furthermore, at a time when technology is changing rapidly, there is a considerable risk of being stuck with a dinosaur of a system when the market provides a more advanced product.
- Mainframe or desktop systems. The systems may be either large or small in scale. Large-scale systems serve many users through a central mainframe computer, while small-scale systems use many personal computers distributed among users. However, thanks to improving technology, personal computers can now be linked to each other and to a common source.

## 13.4 ROLES PERFORMED BY INFORMATION SYSTEMS

Managers are shooting in the dark without information provided to them from throughout the organization. Individual managers cannot be everywhere, yet they are responsible for what goes on in every niche and corner of their organization. The proverbial buck stops with them. To make up for their inability to be everywhere at the same time, they must rely on information that reports to them the activities and the progress of their organization. By using such a substitute for their physical presence, they can provide surrogate leadership aimed at improving products or services by increasing quality, reducing costs, or adding features. An effective information system can better monitor quality control, improve inventory management, and facilitate the management process.

The ultimate goal of any business is to provide value to its customers. It can do so by enhancing any or all of the value activities that comprise a firm's value chain. In other words, the output of a firm, be it in the form of goods made or services provided, emerges after going through several stages. At each stage, the work in process assimilates greater value. At the end of the process, having gone through all the intervening steps, the finished product is potentially most valuable. A cake made for a customer becomes most valuable once it is mixed, baked, decorated, and delivered to the customer.

In the following paragraph, a typical organizational value chain is outlined and followed up with a list that supports activities needed for it to function in an optimum manner.

The value chain itself and the necessary support activities correspond to line and staff functions, in the more traditional vocabulary. Five primary value activities in the value chain:

- Inbound activities—deal with receiving, storing, and distribution of resources.
- Operation activities—transform raw materials or partially finished goods into final products.
- Outbound activities—involve distribution of products or services to customers.
- Marketing and sales activities—meant to ensure that customers learn about the value they will receive by buying given products or services. Customers will not rush to buy, if they have no knowledge of the value they will be purchasing.
- Service activities—deal with repair and maintenance for the products sold. They involve follow-up and learning from the customers how to improve the product or the service.

Four support activities in the value chain:

- Firm infrastructure—includes the information system
- Human resources
- Purchasing—procurement activities related to inbound activities
- Technology—includes research and development, new computers, and product design

Information systems can provide value in many ways. The more value an information system provides for a given cost to the users, the more likely it is to help the firm stay healthy or overcome its competition. The failure to keep up with rivals can lead a firm to go the way of carrier pigeons.

## 13.5 THE LIMITATIONS OF INFORMATION SYSTEMS

Not all information systems are equally good. They can provide better information on a more timely basis, but not all of them do. Among the reasons responsible for the failure of information systems is the tendency to see them only in technical terms. Information systems must also be understood in terms of the organizational contexts within which they operate. The degree of ambiguity fostered by an organization as well as its design cannot but impact the working of an information system. The success of an information system is also dependent on factors such as human relations, social psychology, and cognitive styles of those affected by and in turn affecting the information system.

The failure of information systems to deliver is fostered by the failure to think about the way individuals within an organization actually use the information provided to them by the system. It is also caused by simplistic explanations of the way decisions are made within an organization. The notion promoted in business courses that human beings driven by self-interest will make rational decisions is not always true: too many managers and employees often make seemingly irrational decisions. All too often, decisions get made first and justified later. Such irrational acts occur because the world of an organization is not well-planned, nor is the environment within which it works very predictable. Yet the assumptions driving an information system are often based on predictability in a well-defined world.

Information users and producers could learn a lot from the rules governing language-based communication. Language must contend with what is commonly known as denotation and connotation of words. Red is just a color, but it can cause one to think about, among other things, communism. Different individuals see a given event differently. Language-based communication also must grapple with establishment of communicational links between senders and receivers: speakers must use the language of the listeners in order to be understood. Listeners' comprehension is enhanced if the message they receive happens to be relevant to their concerns. Such linguistic and communicational truisms must be understood by those involved with the design and operations of information systems. Information can mean different things to different individuals, for example, budgets and variance analysis can have different meanings for different individuals within the same organization. Information systems will have greater impact if the output they disperse is understandable, meaningful, and relevant to those receiving it.

## 13.6 COMPONENTS OF AN INFORMATION SYSTEM

An information system is made of many distinct elements, described as follows:

*Objectives.*    A system cannot be everything to everyone, therefore it should be provided with objectives. Such objectives are related to key financial and operational variables that impact managerial decision-making.

*Inputs and Outputs.*    A system receives inputs from throughout the organization, which are then converted into outputs. The input may be timecards of the employee, while the output may be the payroll checks prepared for distribution to employees. Outputs are reports that the system provides after it has converted the data it received from throughout the firm. Output that is later reentered into the system as input is called feedback. Such feedback can take the form of variance analysis, which points out the difference between the budgeted and the actual costs of production, thus helping managers take care of problems preventing the realization of budgetary goals.

*Stored Data.*    In addition to providing periodic reports, a system also builds up its memory, which then allows managers to compare past performance with current activities. The stored data is akin to organizational archives or an individual's memory. When the new inputs are received, an updating of stored data takes place. Such updating is called file maintenance.

*Processor.*   This is the part of the system where inputs are converted to outputs using the technology provided by the system hardware and software. It is the computer, in other words. It is important to remember that a processor's effectiveness is dependent on software packages, without which a computer is no more than inert metal and wires.

*Instructions and Procedures.*   To ensure that the system works as designed, those con-nected with it must follow instructions and carry out the correct procedures. Otherwise, the integration of the system will fail, and the system will not run as a well-oiled machine.

*Boundaries and Constraints.*   These are the limits within which an information system must operate. Boundaries divide a system from its environment, while constraints are factors, such as capacity, that govern speed and efficiency of the system.

*Users.*   An information system exists in order to satisfy managers' needs for information. While simpler systems have a limited number of users, the same is not true of the more complex systems. Such complex systems must be capable of providing information reports that help a variety of users. Not all managers have the same needs, so an information system must be capable of satisfying their diverse needs. Furthermore, external users of organizational information have needs that are unlike the needs of the internal users.

*Security Measures.*   An information system is a valuable resource both for its technical machinery and for the information that is contained in its files. Steps must be taken to ensure the safety of the machines and what they contain.

*Subsystems.*   Increasingly, systems are becoming very complex. They can be asked to tackle, at the same time and often on an immediate, on-line basis, more than one set of users' needs. This has led to the emergence of systems that combine several subsystems. An example of an information system that contains more than one subsystem is the integrated accounting system, which may contain separate elements to deal with inven-tory, accounts receivable, and payroll subsystems.

*Information Interfaces.*   These are the shared boundaries at which information passes from one (sub)system to another.

## 13.7 TYPES OF INFORMATION SYSTEMS

Given the great diversity of information systems, there have evolved many ways to classify them:

- Management information systems collect and process data needed to produce information for planning and controlling activities of an organization at all levels of administration.
- Accounting information systems are subsets of the MIS that collect and process financial data. It is often the largest subsystem of the MIS, and, in many organiza-tions, may be the only formal information system.

- Decision support systems (DSS) help users to make decisions in unstructured environments where there is a high degree of uncertainty. DSS are typically used on an ad hoc, rather than ongoing, basis.

- Executive information systems (EIS) provide executives with the information they need to make strategic plans, to control and operate the company, to monitor business conditions in general, and to identify business problems and opportunities. An EIS accepts data from many sources and then combines and summarizes the data so that it can be displayed in an easy-to-understand way. The objective of an EIS is to filter and condense information.

- Expert systems (ES) support users with the expertise needed to solve specific problems in a well-defined area.

- Office automation systems (OAS) are systems designed to enhance the productivity of information workers through a variety of technologies including word processing, spreadsheets, databases, and so on.

- End-user systems (EUS) are information systems developed by users to enhance their personal productivity.

## 13.8 ACCOUNTING INFORMATION SYSTEMS

While it has all the attributes of an information system, an accounting information system usually is only one among several information systems within an organization. The concern with financial transactions and reporting differentiates an accounting information system from other systems within an organization. However, since very few transactions taking place within an organization lack monetary implications, an accounting information system can become a means to track almost all of the activities taking place within an organization. While more will be said about accounting information systems throughout the book, here we will describe the system in terms of its users' needs. The accounting information system serves two subsets of users, the internal and the external. We first talk about the needs of external users an accounting system must satisfy, then we will talk about the needs unique to internal users. The collection of information directed at external users is financial accounting, while that directed at internal users is managerial accounting. Following are two lists that summarize external and internal users and their needs.

External Information Users and Their Needs

- Customers need information about:
  a. account status
  b. discounts available
  c. date payments are due

- Suppliers need information about:
  a. items desired
  b. quantities desired

- Stockholders need information about:
  a. past performance to predict future performance
  b. routine information about dividends and stock transactions

- Employees need information about:
  a. wages, salaries, deductions, etc.
  b. fringe benefits

- Lenders need information about:
  - a. ability to meet present obligations
  - b. prospects for future success
- Governments need information about:
  - a. taxes due
  - b. wages paid
  - c. regulatory matters

Internal Information Users and Requirements

- Marketing management needs answers to questions about:
  - a. pricing of products
  - b. discounts, credit terms, and warranty policy
  - c. How much to spend on advertising
- Purchasing and inventory control management needs answers to questions about:
  - a. how much inventory to purchase
  - b. when to purchase inventory
  - c. which vendors to use
- Production management needs answers to questions about:
  - a. what to produce
  - b. when and how much to produce
- Human resource management needs answers to questions about:
  - a. how much to pay and deduct from each employee
  - b. skill experience of employees
  - c. turnover, efficiency, etc.
- Financial managers need answers to questions about:
  - a. amount and timing of cash inflows and outflows
  - b. planned capital expenditures (amount and timing)

The principal differences between internal and external information supplied are:

- Most of the information supplied to external users is mandatory or essential. In contrast, most of the information provided to internal users is discretionary. External reports must also follow a prescribed form and be in keeping with generally accepted accounting principles.
- Most managerial decisions require more detailed information than is needed for external reporting. Managers need the information as it develops. External reports are prepared long after the transactions being reported in an aggregate form actually occurred.

Even though there are calls for separate systems for the internal and external users, for now the needs of both are met by the same systems, even though it means the needs of internal users are given a lower priority.

## 13.9 THE SUBSYSTEMS OF AIS

Since an accounting information system can be quite complex, it is best to approach it in terms of its subsystems. We have chosen to divide it into the following elements to better illustrate the roles performed by the accounting information systems:

- Revenue cycle (sales)
- Resources management cycle

- Expenditure cycle
- Financial reporting cycle

Of the above elements, the last cycle is unique in that it integrates the financial conse-
quences of the transactions taking place throughout the firm. The activities that are
carried out by the firm to raise revenue, procure resources, convert them into products for
profitable sales, market them, and reimburse workers all become part of the information
flow received and processed by the financial reporting subsystem. Consequently, those
charged with financial reporting are also given the charge of directing the financial aspects
arising from a firm's other activities. Establishing credit policies, paying vendors, and
reimbursing employees are activities carried out by the accounting department, whose
principal task remains the production of financial statements.

## 13.10 DOCUMENTATION TOOLS AND TECHNIQUES

This section focuses on the tools and techniques used to analyze, develop, document, and
evaluate accounting information systems. Such techniques consists of both graphical
representations as well as narrative descriptions.
   Common systems documentation tools are:

- Flowcharts
- Data flow diagrams (DFDs)
- Decision tables
- Structured English
- HIPO (hierarchy plus input, process, and output) charts
- Structure charts
- Prototyping
- Narrative/manuals

Accountants as well as those in charge of writing procedures manuals need to understand
these tools because they add value by reducing the time needed to develop, evaluate, and
document systems. Data models provide a static view of the data used by a firm. By
contrast, flows of data through procedures present a dynamic view. Narrative descriptions
are often used to document procedural data flows. Many firms maintain system manuals
that are full of such descriptions. They are useful, however, narratives should be supple-
mented by visual representations. Two very suitable techniques for depicting data flows
are system flowcharts and data flow diagrams, but a variety of other techniques are in use
these days. They range from the very basic to the very detailed.
   Visual representations of the greatest interest to accountants are system flowcharts—
diagrams that pictorially portray the physical flows of data through sequential procedures. They highlight relationships among the elements within transaction processing
systems. That is, they provide answers to questions such as:

- What inputs are received and from where?
- What outputs are generated, and what is their form?
- What are the steps involved in the processing sequence?

- What files and accounting records are affected by the processing carried out?
- What type of accounting and organizational controls are employed?

## 13.11 VARIETIES OF VISUAL REPRESENTATIONS

System flowcharts and diagrams can be adapted to emphasize one or more aspects of a transaction processing system: a process flowchart emphasizes the procedural steps. It is therefore useful to systems analysts when they review present procedures for possible improvements. A document flowchart emphasizes the hard-copy inputs and outputs and their flows through organizational units. It is often used by auditors and accountants when analyzing a current system for weaknesses in controls and reports. A computer system flowchart focuses on the computer-based portions of transaction processing systems, including computer runs or steps and accesses of on-line files. The most important use of a computer system flowchart is to document a current procedure or a proposed new or improved procedure.

## 13.12 FLOWCHARTING SYMBOLS

The building blocks for a system flowchart are a set of symbols, most of which are generally accepted by accountants and analysts. Exhibit 13–1 displays the set of symbols to be used in this book. These symbols may be grouped as input-output symbols, processing symbols, storage symbols, flow symbols, and miscellaneous symbols. All of the symbols in the exhibit can be drawn with the assistance of a flowcharting template, which is readily available, or there are software packages that will do it for you. The following list provides an explanation of the symbols contained in the exhibit:

*Input-Output Symbols.* The top symbol in the leftmost column represents data on source documents or information on output documents or reports. The second and third symbols reflect the entry of data by keyboards or other on-line means and the display of information on terminal screens or other on-line devices. The term "on-line" refers to devices that are connected directly to a computer system. The last two symbols in the column, involving punched cards and punched paper tape, are seldom used in modern-day systems.

*Processing Symbols.* Symbols are available to indicate the processing of data by clerks (trapezoid), noncomputerized machines (square), and computers (rectangle). The decision symbol (diamond) is used to indicate when alternative processing paths exist. For instance, in a flowchart showing sales transaction processing, a decision symbol may be placed at the point just after a credit check. If an ordering customer's credit is found to be satisfactory, one path may lead to continued processing of the order. If the credit is not satisfactory, another path might lead to the writing of a rejection letter.

*Storage Symbols.* The top symbol (a triangle) is used to show documents and/or records being stored in an off-line storage device, such as a file cabinet or hold-basket. Remaining symbols are available to show data being stored on computerized media. The bottom symbol pertains to any on-line storage device, including a magnetic disk.

| Input/output | Processing | Storage | Flow |
|---|---|---|---|

Document or report
(input or output)

Processing operation
performed manually

File of stored
documents

*A* in symbol means filed alphabetically
*N* in symbol means filed numerically
*C* in symbol means filed chronologically

Terminal point of procedure

Manual input of data
into an online device
such as a terminal

Processing operation
performed by machine
other than a computer
(e.g., a bookkeeping machine)

Data stored on magnetic tape

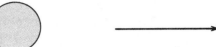

Flow of data or information

Online display on an
output device such as on
CRT terminal or plotter

Computer-performed
processing run or operation

Data stored on
magnetic disk

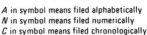

Communication link via
data communication line

Punched card
(input or output)

Decision
point

Data stored online
using a direct access
storage medium

Connector between two
points on a flowchart
on the same page
(i.e., on-page connector)

Data on
punched-paper tape

Connector between
two pages of a
flowchart (i.e.,
off-page connector)

Miscellaneous

Annotation

Adding-machine tape
for batch control

Records and master files,
including accounting
journals and ledgers

**Exhibit 13–1. A set of symbols for system flowcharting. Based on American National Standards Institute,** *Standard Flowchart Symbols and Their Use in Information Processing (X3.5)* **(New York: ANSI, 1971) and on other sources. Adapted from J. Wilkinson,** *Accounting Information System: Essential Concepts & Applications,* **2nd. Ed. (NY: John Wiley & Sons)**

***Data and Information Flow Symbols.***   The five symbols in the rightmost column provide direction throughout a flowchart chart. The oval terminal symbol marks a beginning or end point within the flowchart being examined, such as the receipt of an order from a customer. Often a beginning or end point is also a link to an adjoining procedure. The flow line shows the flow of data or information, usually in written form. The communication link symbol (the one that looks like a lightning bolt) represents the electronic flow of data from one physical location to another. Finally, two connector symbols are available to provide further linkages. The on-page connector (circle) is used within a single page of a flowchart, while the off-page connector (like homeplate) links two pages of a multipage flowchart.

***Miscellaneous Symbols.***   The annotation symbol (open-ended rectangle) can be connected to any symbol within a flowchart; its purpose is to provide space for a note concerning the procedure. For instance, it could indicate how often a particular processing step takes place, or who performs it. The remaining two symbols are useful in flowcharts portraying transaction processing through the accounting cycle. The parallelogram, for instance, is a specialized symbol that adds clarity to journalizing and ledger posting steps.

## 13.13 FLOWCHARTING A PROCESS

To help illustrate flowcharting, we will demonstrate it by using it to visually represent a common accounting procedure. The process begins with the following narrative describing the purchasing procedure for the Easybuy Company.

> A clerk in the accounting department periodically reviews the inventory records in order to determine which items need reordering. If the quantity on hand for a particular item has fallen below a preestablished reorder point, the clerk prepares a prenumbered purchase requisition in two copies. The original is sent to the purchasing department, where a buyer (1) decides on a suitable supplier by reference to a supplier file and (2) prepares a prenumbered purchase order in four copies. The original copy of the purchase order is signed by the purchasing manager/supervisor and mailed to the selected vendor. The second copy is returned to the inventory clerk in the accounting department, who pulls the matching requisition copy from a temporary file (where it has been filed chronologically), posts the ordered quantities to the inventory records, and files the purchase requisition and order together. The third copy is forwarded to the receiving department, where it is filed numerically to await the receipt of the ordered goods. The fourth copy is filed numerically, together with the original copy of the purchase requisition, in an open purchase order file. When the invoice from the supplier arrives, this last copy will be entered into the accounts payable procedure.

Several features of this procedure should be noted. It involves manual processing of transactions, which moves among three departments and generates documents having several copies. A system flowchart can present all these features in a clear manner by blending the features of a document flowchart with those of a process flowchart. Such a blended flowchart can be called a document system flowchart.

In order to plan a flowchart of this type, we begin by deciding that three organizational units are to be involved in the procedure. Then we section a sheet of paper into three columns, which we label "Accounting Department," "Purchasing Department," and "Receiving Department." Next, we select those symbols that pertain to manual processing, which will be combined in strict accordance with the sequence of the narrative.

For convenience, the work is subdivided into four key steps or functions. Each is discussed and then visually represented in Exhibits 13–2, 13–3, 13–4, and 13–5, respectively.

**(a) Preparation of the Purchase Requisition.**    The first step is the preparation of a purchase requisition. It consists of the following steps:

1. As the beginning segment shows, the flowcharted procedure begins in the accounting department with a terminal symbol.
2. This symbol is connected by a flowline to a clerical or manual processing symbol.
3. Inserted inside this second symbol is a notation that briefly states the actions taken by the inventory clerk.
4. To explain the basis on which the clerk prepares the document labeled "Purchase requisition," an annotation symbol is also attached to the manual processing symbol.
5. Another flowline connects an accounting record symbol, labeled "Inventory records," to the manual processing symbol. This connection from the inventory records to the manual processing symbol denotes that inventory data on file are used during the preparation of the purchase requisition.
6. A flowline from the manual processing to the document symbol indicates that a purchase requisition, in two copies, is an output from the processing step. Note that when multiple copies of a form are prepared, they are numbered and shown in an offset manner.
7. The final function of this flowchart segment is to show the disposition of the two copies of the purchase requisition. A flowline pointing to the right directs copy 1 to the purchasing department, whereas a downward flowline indicates that copy 2 is filed in a folder. The letter C in the file symbol means that copy 2 is arranged chronologically (by date) within the file. Note that it is not necessary to show a processing symbol that specifies a filing action between the document and the file.

**(b) Preparation of a Purchase Order.**    The second step in the process consists of preparing a purchase order. To do so, the following steps are undertaken:

1. The activity in the second segment also centers on a manual processing symbol. Two flowlines lead to this processing symbol, one from the first copy of the purchase requisition and the other from the supplier file.
2. Based on data from these two sources, a buyer in the purchasing department prepares a purchase order in four copies. Again, a flowline pointing from the processing symbol to the document symbol(s) designates the latter as being an output.
3. Another "output" flowing from the processing symbol is copy 1 of the purchase requisition. Since it entered the processing symbol, as noted by the flowline from the accounting department, it must also leave the processing symbol. As the segment shows, it is then deposited in the open purchase order file. An important rule of flowcharting is to show the final disposition of every copy.
4. The remainder of this flowchart segment depicts the disposition of the four purchase order copies.

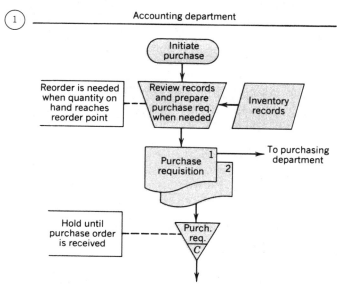

Exhibit 13–2. A diagram depicting the preparation of the purchase requisition. Adapted from J. Wilkinson, *Accounting Information System: Essential Concepts & Applications*, 2nd. Ed. (NY: John Wiley & Sons).

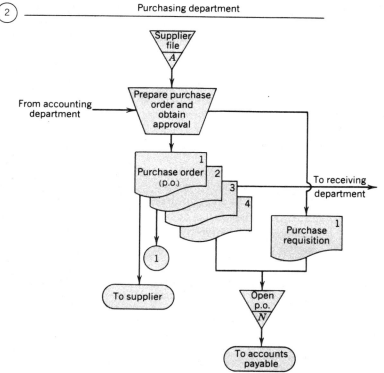

Exhibit 13–3. A diagram depicting the preparation of the purchase order. Adapted from J. Wilkinson, *Accounting Information System: Essential Concepts & Applications*, 2nd. Ed. (NY: John Wiley & Sons).

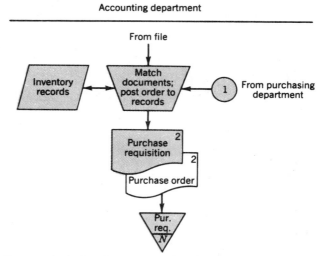

**Exhibit 13–4.** A diagram depicting the process of updating the inventory record. Adapted from J. Wilkinson, *Accounting Information System: Essential Concepts & Applications*, 2nd. Ed. (NY: John Wiley & Sons).

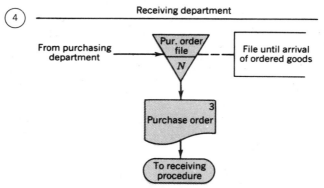

**Exhibit 13–5.** A diagram depicting the process of filing of receiving department's copy of the purchase order. Adapted from J. Wilkinson, *Accounting Information System: Essential Concepts & Applications*, 2nd. Ed. (NY: John Wiley & Sons).

    **a.** Copy 1 is mailed to the supplier. Since this mailing ends the treatment of copy 1 (as far as the flowchart is concerned), this is dictated by means of a terminal symbol. (Alternatively, we could have added a column on the flowchart labeled "Supplier" and shown the flow of copy 1 to that column. However, a column is necessary only if processing steps are to be shown within the column.)

    **b.** Copy 2 terminates with an on-page connector labeled "1." The next segment will continue the disposition of copy 2.

    **c.** Copy 3 is directed to the receiving department.

    **d.** Copy 4 is filed together with copy 1 of the purchase requisition.

  **5.** The terminal symbol below the file indicates that the filed copies will be used in the accounts payable procedure. Note that a column has not been allotted on this flow chart for the accounts payable department, since the department is not involved in the processing being portrayed.

6. One additional flowcharting convention is illustrated in this segment. When flowlines cross, a symbol called a "jumper" denotes the crossover.

**(c) Updating the Inventory Records.** The third stage in the process consists of updating the inventory records, which in turn depends on the following steps.

1. This flowchart segment, like the first segment, is located organizationally in the accounting department. Both copy 2 of the purchase requisition and copy 2 of the purchase order enter into the processing. The former is pulled from the file folder, while the latter arrives from the purchasing department. Note that the on-page connector in effect links to the on-page connector shown in the previous segment.

2. Processing is performed by the inventory clerk, who matches the documents, accesses the proper inventory records, posts the ordered quantities, and then replaces the posted inventory records within the inventory file. A bidirectional flowline (i.e., one with arrowheads on both ends) symbolically represents these accessing, posting, and replacing actions.

3. As the last step in this segment, the two documents leave the processing symbol and flow into a file. Note that when two or more documents move together, a single flowline is sufficient to represent both.

**(d) Filing Receiving Department's Copy of the Purchase Order.** The last stage of the process consists of disposal of the third copy of the purchase order. To do so, the following steps are taken:

1. In this brief segment, copy 3 of the purchase order is placed temporarily into a file maintained in the receiving department.

2. On the arrival of the ordered inventory goods, the copy is withdrawn (pulled) and entered into the receiving procedure.

3. Since the receiving procedure is shown on a different flowchart, a terminal symbol is employed to denote the interface with that procedure.

Exhibit 13–6 combines the four segments just described into a system flowchart of the purchases procedure. A variation of this flowchart, which omits the columns for organizational units, appears in Exhibit 13–7 which follows. Between them the various figures show the diversity of visual representations using flow diagrams.

## 13.14 GUIDELINES FOR PREPARING FLOWCHARTS

Given that different experts can draw differently a given process, it is imperative to have some guidelines. Good flowcharts result from sound practices consistently followed. Sound practices should be grounded on the following guidelines:

1. Carefully read the narrative description of the procedure to be flowcharted. Determine from the facts the usual or normal steps in the procedure, and focus on these steps when preparing the flowchart.

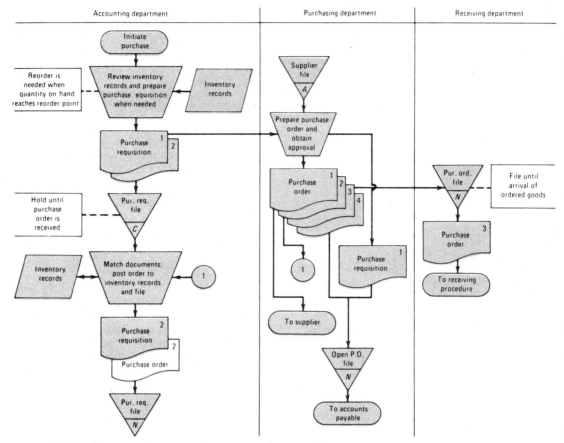

**Exhibit 13–6. A document flowchart of a manually performed purchases procedure.** Adapted from J. Wilkinson, *Accounting Information System: Essential Concepts & Applications*, 2nd. Ed. (NY: John Wiley & Sons).

**2.** Choose the size of paper to be used. Either use letter size (81/2 × 11 in.) or a larger size. Then gather materials such as pencils, erasers, and a flowcharting template.

**3.** Select the flowcharting symbols to be used. The variety of symbols used should be limited for clarity.

**4.** Prepare a rough flowchart sketch as a first draft. Attempting to draw a finished flowchart during the first effort usually results in a poorer final product.

**5.** Review your sketch to be sure that the following have been accomplished:

   **a.** The flows begin at the upper left-hand corner of the sheet and generally move from left to right and from top to bottom.

   **b.** All steps are clearly presented in a sequence, or a series of sequences. No obvious gaps in the procedure should be present.

   **c.** Symbols are used consistently throughout. Thus, the symbol for manual processing (an inverted trapezoid) should appear each time that a clerk performs a step in the procedure.

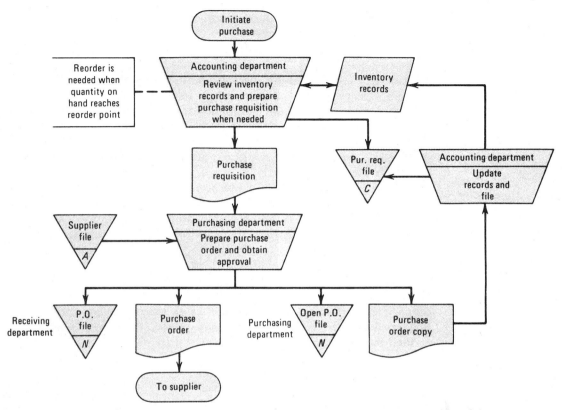

**Exhibit 13–7. A variation of the document flowchart shown in Exhibit 13–6. Adapted from J. Wilkinson, *Accounting Information System: Essential Concepts & Applications*, 2nd. Ed. (NY: John Wiley & Sons).**

**d.** The dispositions of all documents and reports are shown. In fact, the final "resting place" of every copy of every prepared document should be specified. Typical dispositions include placing documents in files, sending documents to outside parties such as customers, forwarding documents to connecting procedures (such as a general ledger procedure), distributing reports to managers and even destroying documents.

**e.** The "sandwich" rule is consistently applied. This rule states that a processing symbol should be sandwiched between an input symbol and an output symbol.

**f.** When a document crosses an organizational line within the flowchart, the document is normally pictured again in the new organizational unit. However, the repetition is not necessary in some instances if the organizational units are adjacent.

**g.** All symbols contain brief but specific labels written inside the symbols. For instance, "Sales invoice" might appear inside a document symbol. Do not simply write "Document" inside a document symbol, since the shape of the symbol indicates its nature. When lengthy labels are needed, draw the symbols sufficiently large to contain the labels completely. That is, the size of a symbol may vary without affecting its meaning.

**h.** Multiple copies of documents are drawn as an overlapping group and are numbered in the upper right-hand corners; these numbers remain with the copies during their flows through the procedure.

**i.** Added comments are included within annotation symbols and are attached to appropriate symbols, such as the processing symbols to which the comments relate.

**j.** Ample connections (cross-references) are provided. The symbols used in forming the connections depend on the situation. Thus, if two sheets are needed to contain the flowchart, the flows between pages are formed by off-page connector symbols. In those cases where the procedure being flowcharted links to an adjoining procedure, the connection can be formed by a terminal symbol.

**k.** Exceptional occurrences, such as backorders, are clearly noted. They may appear as: (i) comments within annotation symbols, (ii) separate flow charts, with references to the main flowchart, or (iii) decision branches.

**l.** Special presentation techniques are adopted when their use increases both the content and clarity of the procedure. An apt illustration of this rule is the portrayal of batch control totals in computer-based batch processing systems. Batch control totals are generally computed from key data in each batch of transactions prior to processing runs. Then, during each processing run the totals are recomputed and compared to the precomputed totals. These run-to-run comparisons may be performed at the direction of the computer processing programs, and the results may be shown on printed exception and summary reports. If the results show differences in the totals, the differences must be located before processing can continue. This batch control procedure may be diagrammed with the help of dashed lines that depict the run-to-run comparisons with the precomputed totals.

**6.** Complete the flowchart in final form. A finished flowchart should be neatly drawn and uncrowded. Normally it also should contain a title, the date, and the name(s) of the preparers.

## 13.15 DATA FLOW DIAGRAMS

Unlike the flowcharts, data flow diagrams (DFDs) emphasize the logical view of data rather than the physical view. That is, they focus on what happens, rather than on the mechanics of how it happens. While both are visual representations of procedures used within an organization, differences exists between them: DFDs emphasize the flow of data, while flowcharts emphasize the flow of documentation. DFDs concentrate on *what happens*, but flowcharts focus on *how* the processing of data into information occurs. DFDs use only four basic symbols, but flowcharts use many more. DFDs leave out the timing or sequence of data processing steps; in contrast, flowcharts show both the timing and the sequence of steps that occur during the process.

A system flowchart clearly indicates how data are being processed, for example, manually or with the aid of computers. It also specifies certain physical aspects. For instance, a flowchart may portray the storage media as being file cabinets or magnetic disks, and the input method as being documents on hard copy or data entry screens. In contrast, a data flow diagram emphasizes the specific data and what is being done to it. The diagram reveals the content of data flows, the processes involving data, the stores of data, and the sources and destinations of data. Because of this simplified focus, only four symbols are needed.

The two techniques also differ with respect to their roles. A system flowchart is mainly used to document the physical elements of an AIS, either the system that is currently in use or a newly designed system. However, a data flow diagram is better suited for analyzing a processing system. That is, it allows systems analysts and accountants to visualize the essential flows and processes without being concerned about the physical design features.

Only four symbols are needed to draw data flow diagrams. They include a rectangle, which represents an entity that is a source or destination of data that resides outside the system being diagrammed. Examples of sources (and also destinations) are customers, suppliers, banks, and managers. Another symbol is a circle (also called a bubble) representing a process that transforms data inflows into data outflows. An example is the process of handling incoming cash receipts. The third symbol, a pair of parallel lines, represents a data store, that is, a place where data can be kept. An example is a file of cash receipts transaction data. The last of the four symbols, a curved line having an arrowhead, represents a data flow. It may connect together any of the above symbols, but at least one end of the flow will generally be connected to a process, shown as a bubble.

## 13.16 ILLUSTRATING DATA FLOW DIAGRAMS

A "data flow diagram" is in reality a hierarchical set of diagrams. Each diagram in the set is a decomposition (in effect, an "explosion") of the preceding diagram. In other words, each succeeding diagram provides a greater degree of detail concerning a process. Data flow diagrams within a set may delve deeper and deeper until extremely detailed views appear.

The example of a sales process will consist of only two levels, as depicted in Exhibits 13–8 and 13–9. The diagram in the former is called a context diagram. It is the top level of a set of data flow diagrams, since the process is encompassed within a single circle. The four entities shown in the diagram are outside the scope of the process being documented. Two of the entities (the customer and credit department) serve as both sources and destinations with respect to data involved in the sales process. One of the entities (the management) is only a destination (i.e., a recipient of data and information) from the process. Yet another (the warehouse) is only a source. Notice that the type of data involved in each flow is written along the flowline. Also note that no data stores appear on this diagram, since they are incorporated within the process itself.

The second data flow diagram, Exhibit 13–9 shows certain details of the sales process. It contains four subprocesses, numbered 1.0, 2.0, 3.0, and 4.0. Because of the zeros following the decimal points, this level just below the context diagram is called the level-zero data flow diagram. Included with the subprocesses are four data stores (customer, inventory, shipping, and credit data) and a variety of data flows.

Additional data flow diagrams can be prepared to show more details concerning each process. For instance, a level 1 diagram of the "obtain sales order data" would include sub-subprocesses shown by bubbles coded 1.1, 1.2, 1.3, and so on; level 2 diagrams would show sub-sub-subprocesses detailing each of these sub-subprocesses. An illustration showing the context diagram, level 0 and level 1 of the registration process in a college is shown in Exhibit 13–10.

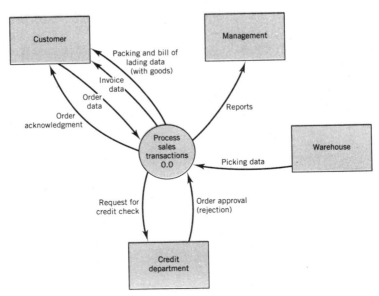

**Exhibit 13–8.    A context diagram depicting the process of a sales transaction. Adapted from J. Wilkinson, *Accounting Information System: Essential Concepts & Applications*, 2nd. Ed. (NY: John Wiley & Sons).**

To summarize, the data flow diagrams are a logical representation of data flows that focuses on a process. A variation of this type of representation, called a physical data flow diagram, specifies where or by whom the processes are performed. It also may indicate how the processing is performed.

### 13.17 GUIDELINES FOR PREPARING DATA FLOW DIAGRAMS

Effective data flow diagrams also require established guidelines. The exhibits used to illustrate data flow diagrams follow several guidelines that should be helpful when you prepare data flow diagrams:

1. Begin with a context diagram that shows the interactions of the selected process with the outside entities. Include all interactions as separate data flowlines, even though there may be more than one flow from the process to an outside entity, or vice versa. Include in the context diagram only those outside entities that are directly involved in data flows to or from the process.

2. Then break the context diagram into a data flow diagram at the zero level. Verify that the data flows are balanced between the context diagram and the level-zero diagram. That is, make sure that all of the data flows in the context diagram also appear in the level-zero diagram.

3. "Explode" the subprocesses in the level-zero diagram into successively more detailed sub-subprocesses in level-1 diagrams, level-2 diagrams, level-3 diagrams, and so on. Continue the balancing of data flows between successive levels, for example, between the level-zero diagram and its set of "exploded" diagrams at level 1. To check that the diagrams are balanced, first verify that all of the outside

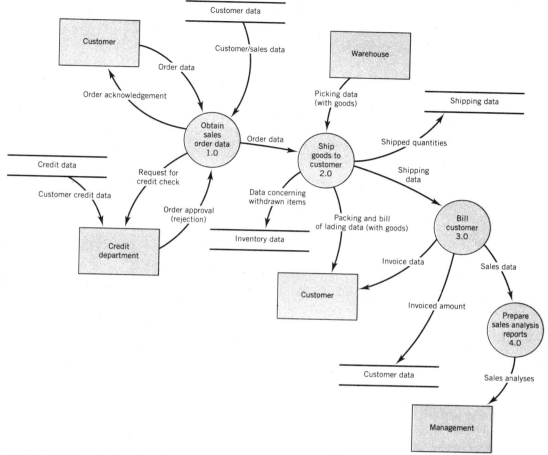

**Exhibit 13–9.    A data flow diagram (level 1) of the four major subactivities in processing sales transactions. Adapted from J. Wilkinson,** *Accounting Information System: Essential Concepts & Applications,* **2nd. Ed. (NY: John Wiley & Sons).**

entities have been carried to the lower level, e.g., from level zero to level 1. Then count the data flows between each outside entity and the subprocesses in the diagrams.

4. Do not incorporate too many details into any single diagram. For instance, limit each detailed diagram to only a few bubbles, probably no more than six or seven.

5. Code the sub-subprocesses carefully in each detailed diagram, so that they can be identified easily with their "parent" subprocess. For instance, if three subprocesses are coded as 1.0, 2.0, and 3.0, the sub-subprocesses for 1.0 would be 1.1, 1.2, 1.3, and so on; the sub-subprocesses for 2.0 would be 2.1, 2.2, 2.3, and so on; and the sub-subprocesses for 3.0 would be 3.1, 3.2, 3.3, and so on. In turn, those for 1.1 would begin as 1.11, 1.12, and 1.13.

6. In instances where multiple entities function in the same manner, use a single encompassing label to represent all. For instance, the reference to "customer" in the example means all customers.

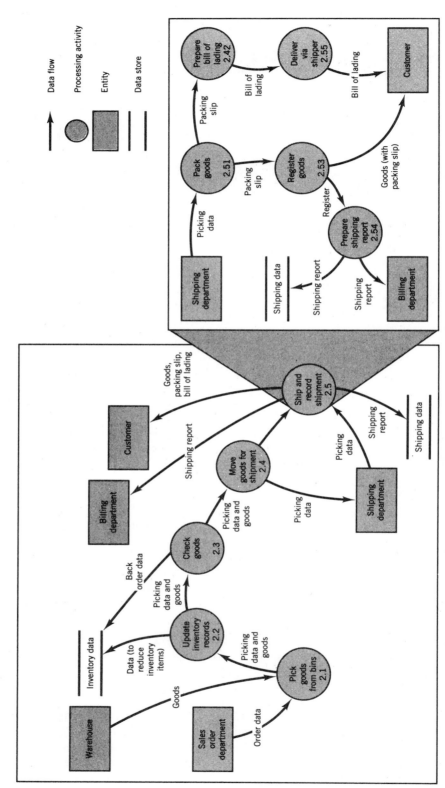

**Exhibit 13–10. Two Levels of data flow diagrams. Adapted from a transparency for use with J. Wilkinson, *Accounting Information System: Essential Concepts and Applications*, 2nd edition (NY, John Wiley & Sons, 1993).**

7. Where multiple entities function differently, use separate boxes. For instance, assume that a process involves both credit and cash customers, and that the processing steps relating to credit customers are different from those for cash customers. Separate boxes, labeled "credit customer" and "cash customer," should be employed.

8. Do not allow data flowlines to cross over each other. To avoid crossovers, repeat an entity box or data store symbol as necessary in a single data flow diagram.

9. Show only normal processing sequences in a single data flow diagram. That is, avoid exceptional situations or show them as a separate set of data flow diagrams.

10. Show process bubbles that progress generally from left to right and from top to bottom in a single data flow diagram. The first bubble (1.0) appears near the upper left and the last bubble (4.0) appears near the lower right.

The guidelines above are further illustrated in the exhibits. They carry the analysis to more than one level.

## 13.18 ADDITIONAL TECHNIQUES

This section takes a brief look at some of the newer documentation tools. They have emerged to help some of the difficulties encountered while working with flowcharts.

Decision tables are a tabular alternative to a program flowchart, and can help further clarify a given process. There are four parts to a decision table:

- The condition stub contains a row for each logical condition for which the input data will be tested.
- The condition entry consists of a set of vertical columns that specify a pattern of yes, no, or not relevant for the set of tests in the condition stub.
- The action stub lists the actions to be taken.
- The action entry links which actions should be taken with the pattern specified in the condition entry.

A fairly basic example of a decision table is given in Exhibit 13–11. In Exhibit 13–12 an alternative to a decision table is exemplified.

Sometimes called pseudocode, structured English is a narrative alternative to a program flowchart. It can be used by systems designers to communicate with programmers,

| Purchase Discounts Allowed | 1 | 2 | 3 |
|---|---|---|---|
| Purchase > $10,000 | Y | N | N |
| Purchase $5,000 to $10,000 | | Y | N |
| Purchase < $5,000 | | | Y |
| Take 5% discount | 1 | | |
| Take 3% discount | | 1 | |
| Pay full amount | | | 1 |

**Exhibit 13–11. A decision table showing purchase discount procedures. Adapted from J. Burch & G. Grudnitski, *Information Systems*, 5th. ed. (NY: John Wiley & Sons, 1989).**

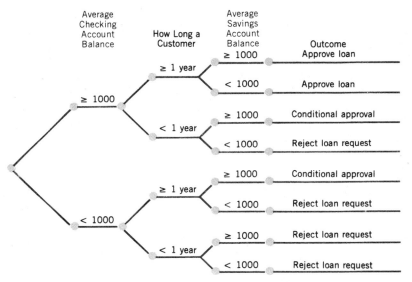

**Exhibit 13–12.   A decision tree describing the loan approval process. Adapted from J. Burch & G. Grudnitski, *Information Systems*, 5th. ed (NY: John Wiley & Sons, 1989).**

since it is easy to understand. It reduces the time required to code a program. A fairly basic example of structured English is given in Exhibit 13–13.

HIPO charts are yet another tool. HIPO is an acronym for **Hierarchy** plus **Input**, **Process**, and **Output**. It shows the modular design of a system and its components. HIPO charts are another alternative to program flowcharts and there are two advantages to such use of HIPO charts. First, they are easy to create and understand because they focus more on the logic rather than the mechanics of what is to be accomplished. They also focus on the process and how its various components are related to each other. Doing this allows one to see the trees as well as the forest. An illustration is in Exhibit 13–14.

Structure charts display the structure of systems application. Exhibit 13–15 is a good example of a structure chart describing the payment of accounts payable. Prototyping is another technique that seeks to represent perceived users' requirements. If the first set of requirements are not acceptable, then another set is drawn up. The process continues until an acceptable prototype emerges. Exhibit 13–16 describes prototyping.

## 13.19 NARRATIVE DOCUMENTATION OF INFORMATION SYSTEMS

In addition to flowcharts and data flow diagrams, an information system needs further documentation. Such documentation consists of manuals and other means of describing the information system and its operations. It also should include those aspects of a firm that have an impact on the system, such as policy statements, organizational charts, and job descriptions. Ordinarily, the documentation of a manual accounting information system should be the same as accounting procedures documentation. However, since very few accounting systems in use by organizations of any substance are totally manual anymore, a major segment of documentation is devoted to describing the technical aspects of the system. Here, the documentation refers to technical aspects of the accounting system and not to the documentation of accounting procedures, the subject of the

```
ORDER ENTRY:
*Process customer and item file;
 FOR ALL orders
     Access CUSTOMER record;
     IF CUSTOMER NUMBER is valid
         Access ORDER form;
     ELSE
         Issue "invalid customer code" message;
         STOP ORDER ENTRY;
     ENDIF;
     FOR ALL items ordered
         Access ITEM record;
         IF ITEM NUMBER is valid
             IF QUANTITY ON HAND GE QUANTITY ORDERED
                 Enter ORDER;
                 Decrease QUANTITY ON HAND by
                     QUANTITY ORDERED;
             ELSE
                 Access BACKORDER form;
                 Write BACKORDER;
             ENDIF;
         ELSE
             Issue "invalid item code" message;
         ENDIF;
     ENDFOR;
     Access SHIPPING DOCUMENT form;
     Enter shipping data;
     Print documents;
 ENDFOR;
EXIT ORDER ENTRY.
```

**Exhibit 13–13. An example of order entry procedure using structural English. Adapted from J. Burch & G. Grudnitski, *Information Systems*, 5th. ed. (NY: John Wiley & Sons, 1989).**

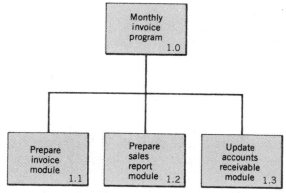

**Exhibit 13–14. A HIPO visual table of contents. Adapted from J. Burch & G. Grudnitski, *Information Systems*, 5th. ed. (NY: John Wiley & Sons, 1989).**

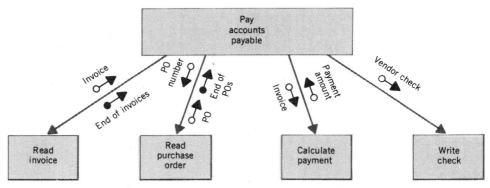

**Exhibit 13–15.   A structure chart. Adapted from J. Burch & G. Grudnitski, *Information Systems*, 5th. ed. (NY: John Wiley & Sons, 1989).**

book. It is important to recognize, however, that similar considerations govern both kinds of documentation.

System documentation can also be an important cog in the internal control structure pertaining to an information system. Auditors are able to examine the internal control system more quickly and thoroughly. Consequently, a general control objective is to prepare complete and clear system documentation and to maintain it in an up-to-date condition. It helps employees to understand and interpret policies and procedures. Data processing clerks are more likely to perform their appointed tasks correctly and consistently if they know what they are. Systems analysts and programmers can redesign transaction processing systems more easily and reliably, especially when the original designers have left the firm.

Below is a description of the typical documentations needed by manual and computerized systems. Both have different needs, while having a great deal in common.

### 13.20  MANUAL SYSTEMS

Documentation for manual systems should include all of the usual components associated with accounting systems: source documents, journals, ledgers, reports, document outputs, charts of accounts, audit trail details, procedural steps, record layouts, data dictionaries, and control procedures. Numerous examples should be provided, such as typical accounting entries and filled-in source documents. Procedures should be documented by narrative descriptions, system flowcharts, and data flow diagrams.

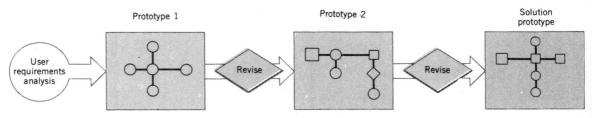

**Exhibit 13–16.   A flowchart showing prototype iteration until acceptable prototype is achieved. Adapted from J. Burch & G. Grudnitski, *Information Systems*, 5th. ed. (NY: John Wiley & Sons, 1989).**

Equally necessary are those elements related to the effective operation of the system. Clear policy statements encourage employees to adhere to management's policies. An organization chart and job descriptions inform employees of their roles and responsibilities with respect to data processing.

Since a manual system does not have to reckon with the mechanics of computer-based processing, its documentation is simpler, and indeed is no different from basic accounting manuals containing procedures and policies in effect. But once computers are introduced, additional documentation specific to the computers and the constraints they impose is required.

## 13.21 COMPUTER-BASED SYSTEMS

All of the documentation appropriate to manual systems is likewise needed in computer-based systems. Even the most automated system contains some manual processing steps and involves interactions with human users and clerks. However, additional documentation is needed because complex hardware and nonvisible programs are used.

Computer-related documentation concerns the computer system itself and the persons who interface with it. In the former category are the overall system standards, system application documentation, program documentation, and data documentation. In the latter category are operating documentation and user documentation. Although the contents of each type of documentation will differ as circumstances dictate, the following descriptions represent a reasonable coverage. Exhibit 13–17 itemizes the array of documentation needed in a computer-based system.

System standards consist of policy statements pertaining to systems development and other system-related matters. For instance, a systems development standard might describe suitable methods and procedures for analyzing, designing, and implementing information system modules. One firm's standard might specify, for example, the use of techniques such as data flow diagrams and entity-relationship diagrams.

System application documentation includes the purpose of the application and descriptive materials such as computer system flowcharts, input-output descriptions, error procedures, and the components needed for manual as well as computer-based systems. This type of documentation is of primary interest to systems analysts, systems users, and auditors.

Program documentation includes program flowcharts or other logic diagrams, source program listings, printouts of inputs and outputs, record layouts or data structures, and information pertaining to operations, testing, changes, and errors. Program documentation is usually organized around individual programs and packaged into run manuals. In the case of applications programs, program documentation may be combined with system application documentation. It is of primary interest to programmers; however, database administrators have concerns with respect to data manipulation language verbs that alter data. Auditors may also need to review program documentation to detect unauthorized changes to programs, as reflected in program listings and printouts of outputs.

Data documentation includes the descriptions of data elements stored within the firm's database, including their relationships. This type of documentation, usually incorporated within a data dictionary, is of particular importance to database administrators and auditors. It is also of interest to application programmers, but only insofar as it relates to the data elements required by the programs that they develop or change.

**System Standards Documentation**
  Systems development policy statements
  Program testing policy statements
  Computer operations policy statements
  Security and disaster policy statements

**System Application Documentation**
  Computer system flowcharts
  Data flow diagrams
  Narrative descriptions of procedures
  Input/output descriptions, including filled-in source documents
  Formats of journals, ledgers, reports, and other outputs
  Details concerning audit trails
  Charts of accounts
  File descriptions, including record layouts and data dictionaries
  Error messages and formats
  Error-correction procedures
  Control procedures

**Program Documentation**
  Program flowcharts, decision tables, data structure diagrams
  Source program listings
  Inputs, formats, and sample filled-in forms
  Printouts of reports, listings, and other outputs
  Operating instructions
  Test data and testing procedures
  Program change procedures
  Error listings

**Data Documentation**
  Descriptions of data elements, including names, field sizes, and so on
  Relationships of specific data elements to other data elements

**Operating Documentation**
  Performance instructions for executing computer programs
  Required input/output files for specific programs
  Setup procedures for specific programs
  List of programmed halts, including related messages and required operator actions,
    for specific programs
  Recovery and restart procedures for specific programs
  Estimated run times of specific programs
  Distribution of reports generated by specific programs

**User Documentation**
  Procedures for entering data on source documents
  Checks of input data for accuracy and completeness
  Formats and uses of reports
  Possible error messages and correction procedures

**Exhibit 13–17.    A table of types of documentation needed in computer-based systems.
Adapted from J. Wilkinson, *Accounting Information System: Essential Concepts &
Applications*, 2nd. ed. (NY: John Wiley & Sons).**

Operating documentation includes all of the performance instructions needed to
execute computer programs, plus instructions for distributing the outputs. Operating documentation, generally organized into console run books, is of primary interest
to computer operators, since they need very explicit directions. Note, however, that operating documentation does not contain program flowcharts and listings, since

| Element or Activity | Manual System Characteristics | Computer-Based System | | |
|---|---|---|---|---|
| | | Characteristics | Risk Exposures | Compensating Controls |
| Data collection | Data recorded on paper source documents | Data sometimes captured without use of source documents | Audit trail may be partially lost | Printed copies of source documents prepared by computer system |
| | Data reviewed for errors by clerks | Data often not subject to review by clerks | Errors, accidental or deliberate, may be entered for processing | Edit checks performed by computer system |
| Data processing | Processing steps performed by clerks who possess judgment | Processing steps performed by CPU "blindly" in accordance with program instructions | Errors may cause incorrect results of processing | Outputs reviewed by users of computer system; carefully developed computer processing programs |
| | Processing steps spread among various clerks in separate departments | Processing steps concentrated within computer CPU | Unauthorized manipulation of data and theft of assets can occur on larger scale | Restricted access to computer facilities; clear procedure for authorizing changes to programs |
| | Processing requires use of journals and ledgers | Processing does not require use of journals | Audit trail may be partially lost | Printed journals and other analyses |
| | Processing performed relatively slowly | Processing performed very rapidly | Effects of errors may spread rapidly throughout files | Editing of all data during input and processing steps |
| Data storage and retrieval | Data stored in file drawers throughout the various departments | Data compressed on magnetic media (e.g., tapes, disks) | Data may be accessed by unauthorized persons or stolen | Security measures at points of access and over data library |
| | Data stored on hard copies in human-readable form | Data stored in invisible, erasable, computer-readable form | Data are temporarily unusable by humans, and might possibly be lost | Data files printed periodically; backups of files; protection against sudden power losses |
| | Stored data accessible on a piecemeal basis at various locations | Stored data often readily accessible from various locations via terminals | Data may be accessed by unauthorized persons | Security measures at points of access |
| Information generation | Outputs generated laboriously and usually in small volumes | Outputs generated quickly and neatly, often in large volumes | Inaccuracies may be buried in impressive-looking outputs that users accept on faith | Reviews by users of outputs, including the checking of amounts |
| | Outputs usually in hard copy form | Outputs provided in various forms, including soft copy displays and voice responses | Information stored on magnetic media is subject to modification (only hard copy provides permanent record) | Backups of files; periodic printing of stored files onto hard copy records |
| Transmission of data and information | Usually transmitted via postal service and hand delivery | Often transmitted by communications lines | Data may be accessed or modified or destroyed by unauthorized persons | Security measures over transmission lines; coding of data; verification of transmitted data |
| Equipment | Relatively simple, inexpensive, and mobile | Relatively complex, expensive, and in fixed locations | Business operations may be intentionally or unintentionally interrupted; data or hardware may be destroyed; operations may be delayed through inefficiencies | Backup of data and power supply and equipment; preventive maintenance of equipment; restrictions on access to computer facilities; documentation of equipment usage and processing procedures |

**Exhibit 13–18.** A table of control problems caused by computerization. Adapted from J. Wilkinson, *Accounting Information System: Essential Concepts & Applications*, 2nd. Ed. (NY: John Wiley & Sons).

| Purpose of Security Measures | Physical Noncomputer Resources in Both Manual and Computer-Based Systems | Hardware Facilities of Computer-Based Systems | Data in Computer-Based Systems |
|---|---|---|---|
| 1. Protect from theft or access by unauthorized persons | Security guards<br>Fenced-in areas<br>Burglar alarms<br>Television monitors<br>Safes and vaults<br>Locked cash registers<br>Lockboxes<br>Close supervision<br>Insurance coverage<br>Logs and registers | Security guards<br>Television monitors<br>Locked doors<br>Locked terminals<br>Inaccessible terminals<br>Employee badges<br>Passwords<br>Segregated test terminals | Locked doors, terminals, stacks of blank forms<br>Off-line data library<br>On-line data and program storage partitions<br>Encoded data<br>Paper shredders<br>Passwords<br>Limited terminal functions<br>Automatic lockouts<br>Callback procedures |
| 2. Protect from natural environment or disasters | Sprinkler systems<br>Fireproof vaults | Air-conditioning<br>Humidity controls<br>Fireproof vaults<br>Halon gas spheres<br>Auxiliary power supplies<br>Insurance coverage<br>Prudent locations<br>Disaster contingency plans | |
| 3. Protect from breakdowns and business interruptions | Preventive maintenance<br>Backup equipment<br>Insurance coverage | Preventive maintenance<br>Backup hardware systems<br>Graceful degradation<br>Insurance coverage | |
| 4. Detect attempted access or change | | | Access logs<br>Control logs<br>System and program change logs |
| 5. Protect from loss or alteration | | | Read-only memory<br>Tape file protection rings<br>External file labels<br>Internal file labels<br>Library log<br>Transaction logs<br>Batch control logs<br>Lockouts |
| 6. Reconstruct lost files | | | Backup procedures<br>Reconstruction procedures<br>Rollback and recovery procedures |

Exhibit 13–19.  A table of selected security measures for physical resources, computer facilities, and data. Adapted from J. Wilkinson, *Accounting Information System: Essential Concepts & Applications*, 2nd. Ed. (NY: John Wiley & Sons).

operators should not be informed of the detailed logic of the programs that process data.

User documentation includes instructions for entering data on source documents, information relating to the formats and uses of reports, and procedures for checking for and correcting errors in data. User documentation is of primary interest to user-department clerks and managers.

## 13.22 ADDITIONAL DOCUMENTATION NEEDED

The system documentation must also address problems caused by the introduction of computers within the accounting systems. Such problems could occur in the context of collections, processing, storage, and retrieval of data as well as the generation and transmission of information. Alternatively, problems could occur because of the equipment itself.

The existence of documentation allows these problems to be addressed more effectively. The fact is, no system is immune to such problems. What can go wrong eventually will go wrong, but dealing with the unpleasant or unexpected is easier if documentation anticipating the problems exists.

The following two Exhibits, 13–18 and 13–19, suggest the documentation needed if a system is to be ready for the unforeseen. The first is a listing of control problems caused by computerization and the second lists selected security measures needed to protect the system resources from an assortment of dangers.

## 13.23 CONCLUSION

This chapter examined the nature of an accounting information system and studied the tools that help to analyze and document the system. To better appreciate and communicate the working of an accounting information system, and in fact to better design one, it is helpful to ask a series of questions pertaining to how the system works and what it supplies, namely information. Such questions can be: Why is information needed? Where does that information come from? Who decides what information is relevant for a particular purpose? How is such a decision made? Who is going to use the information, and who is going to develop it? What steps are required in order to obtain the relevant information and make it available to users? What resources are consumed in obtaining the information and making it available? Is the value of the information worth the cost of producing it? How can the organization ensure that the information is available on a timely basis, and is accurate and reliable? The posing of such questions and the documenting of the answers received is facilitated by the discussions in this chapter.

# CHAPTER 14

# THE WORKING OF AIS

## 14.1 INTRODUCTION

In this chapter the focus is on the actual working of an accounting information system, in other words, the bricks and mortar that makes up an accounting information system. The chapter starts by providing an overview of the accounting system and follows it by describing the ways in which data resulting from various transactions throughout the organization flow into the accounting system and are processed. Given the large number of transactions that need to be tracked and then processed, a variety of tools are used to facilitate the processing of such transactions. One approach classifies the transactions into different cycles. Another tool uses codes to help classify data. Both are described in the chapter.

## 14.2 THE BRICKS AND MORTAR OF AIS

Business organizations must routinely process a vast quantity of data in an orderly and efficient manner to accomplish the following:

- Provide financial statements on a timely basis
- Pay bills in a timely manner
- Ensure that resources in the right quantity are available to do business with customers
- Ensure that the issue of invoices to customers and checks to employees are conducted in an orderly and purposeful manner
- Provide reports required by the government or regulatory agencies properly and in a timely manner

The above objectives can be accomplished by orderly and efficient processing of account data by the accounting system.

The basic accounting system is the manual system with one journal and one ledger. This basic system is generally used by very small businesses. However, as a business grows, its number of business transactions will also grow, and the company looks for ways to speed up the accounting process by developing its accounting system.

An accounting system can be defined as a set of records (journals, ledgers, work sheets, trial balances, and reports) plus the procedures and equipment regularly used to process business transactions. To be effective, accounting systems should:

- Provide for the efficient, cost-effective processing of data
- Be accurate

- Allow for control to prevent theft or fraud and for managerial efficiency
- Provide room for growth within the system to keep up with the growth of the business

Accounting systems consist of a set of record keeping tools: they are journals, ledgers, work sheets, trial balances, and statements. This discussion begins with the most basic forms of such tools.

A ledger is the collection of all the accounts of a company. The accounts are used to record the changes occurring in the financial values of a firm's assets, liabilities, owner's equity, revenues, and expenses. It is best visualized in the form of the letter "T," and such a visualization is the source of the word "T-account," used to describe the various elements of an accounting ledger. Accounts can have debit or credit balances. Assets, expenses, and dividends have debit balances, while the liabilities, revenue, and owner's equity accounts will have credit balances. The debit accounts are increased by means of debits, and decreased by means of credits. The credit accounts, on the other hand, are increased by means of credits, and decreased by means of debits.

The revenue and expense accounts, also called nominal accounts, are closed out at the end of each accounting period. Such closing of those accounts is needed to measure the income earned for the accounting period. The balances, if carried over to the next period, will distort the firm's performance expressed in terms of income, the difference between revenue and expenses for a period.

The ledger accounts only list the increases and decreases in the account balances. They do not reveal how those changes came about. To track those changes, a journal is needed. A journal is a chronological record of business transactions expressed in monetary terms. It lists all the internal and external transactions that occur in the course of business in the form of a journal entry. After transactions have been recorded in the journal, the information they contain gets transferred to the various ledger accounts. This step, called "posting," is undertaken periodically.

A journal entry is a representation of a transaction, expressed in terms of the monetary changes the transaction will bring to a company's assets, liabilities, revenue, or owner's equity. A given entry consists of a debit and a credit—which translates as changes in various accounts. If a transaction is going to represent the payment of an amount owed, say $1,000 by a customer to a business, it will be represented by means of a debit and a credit to the two accounts that will be affected. Cash paid to the firm will increase its cash balance, hence cash account will be debited. At the same time accounts receivable will decrease, since a customer is paying the amount owed. The amounts owed to a firm are listed in the account called accounts receivable.

In an accounting universe governed by the logic of double-entry bookkeeping, there cannot be a debit without a credit. There is no such thing as an immaculate debit or credit in the accounting world. Consequently, at any point in time the balances in debit accounts will equal the balances in credit accounts, unless some mistakes were made in record keeping.

In addition to regular journal entries, there exists adjusting entries. Accountants make adjusting entries as part of accounting transaction processing in order to document information on an activity that has occurred but has not yet been documented. Such absence of documentation is most likely due to two causes: it is more convenience to wait until the end of the accounting period to record the activity or the source documents concerning the particular activity have not yet reached those in charge of documenting it. One way to explain adjusting entries is by a reference to the way people mark their birthdays. Even though we grow older every second, we do not put aside everything in order to mark advancing age in very small fractions of time; instead, we conveniently defer the passing of time until a "birthday," at which time we suddenly grow, in a legal sense, a year older. Birthdays are like adjusting entries. The same holds for economic activities involving assets and liabilities that can live with periodic adjustments in their values as listed in accounting records, such as marking depreciation within accounting books.

There are also entries that pertain to the end of the accounting period. Such closing entries are needed to close out the nominal accounts in order to prepare the financial statements.

## 14.3 THE IMPACT OF COMPLEXITY ON AIS

One journal and one ledger is adequate if a business is small and needs to process only a very limited amount of information. To efficiently process a large quantity of information, an accounting system must adapt itself to the type and quantity of information needed. The large number of transactions that occur in a typical business organization requires the use of subsidiary ledgers and special journals.

**(a) Subsidiary Ledgers.** A subsidiary ledger is a grouping of related accounts, such as receivables or payables, showing the details of the balance of a general ledger control account. Subsidiary ledgers are segregated from the general ledger to simplify record keeping and eliminate a lot of details from the general ledger. The subsidiary ledger tracks accounts receivable, and lists the amounts owed by each customer. In effect, this system lets each customer have its own space. This subsidiary ledger can be used to bill customers. Similarly, the subsidiary ledger for payables can be used to pay vendors and suppliers.

While relieving the general ledger of the details, it is still a good idea to keep a summary account within the general ledger, representing receivables and payables for control purposes. In this "control account" the balance should equal the sum of all balances in the subsidiary accounts.

The general ledger has backup subsidiary ledgers for accounts in addition to the accounts receivable account. Some examples of accounts that frequently have backup subsidiary ledgers are those for office equipment, delivery equipment, and store fixtures. The number of subsidiary ledgers maintained by a company varies according to the company's information requirements. Control accounts and subsidiary ledgers are generally set up when a company has many transactions in a given account and detailed information is needed about these transactions on a continuing basis.

**(b) Special Journals.** As the transactions of a company increase, the first step in altering the manual accounting system is usually to use special journals along with the original general journal. Each special journal records one particular type of transaction, such as sales on account, cash receipts, purchases on account, or cash disbursements.

The following advantages are obtained from the use of special journals:

- Time is saved in journalizing. Only one line is used for each transaction; usually a full description is not necessary. The amount of writing is reduced because it is not necessary to repeat the account titles printed at the top of the special column or columns.

- Time is saved in posting. Many amounts are posted as column totals rather than individually.
- Detail is eliminated from the general ledger. Column totals are posted to the general ledger, and the detail is left in the special journals.
- Division of labor is promoted. Several persons can work simultaneously on the accounting records. This specialization and division of labor pinpoints responsibility and allows for more rapid location of errors.
- Management analysis is aided. The journals themselves can be useful to management in analyzing classes of transactions, such as credit sales, because all similar transactions are in one place.

Special journals, then, are designed to systematize the original recording of major recurring types of transactions. The number and format of the special journals actually used in a company depend primarily on the nature of the company's business transactions. The special journals usually used are the sales, cash receipts, purchases, and cash disbursements journals:

- The sales journal is used to record all sales of merchandise on account (on credit).
- The cash receipts journal is used to record all inflows of cash into the business.
- The purchases journal is used to record all purchases of merchandise on account (on credit). Merchandise refers to items of inventory that are available for sale to customers.
- The cash disbursements journal is used to record all payments (or outflows) of cash by the business.

The general journal is not eliminated by the use of special journals; it is used to record all transactions that cannot be entered in one of the special journals. All five of these journals are books of original entry. If a transaction is recorded in a journal, it will be posted and is part of the accounting records. Therefore, if a transaction is recorded in a special journal, it should not be recorded in the general journal because this would record the transaction twice. Since the journals are posted to ledger accounts, the posting reference column in the ledger should indicate the source of the posting and it should be posted in both the control and the subsidiary ledgers.

## 14.4 TRANSACTION PROCESSING CYCLES

Even though subsidiary ledgers and special journals help process transactions more efficiently and speedily, their usefulness is limited. Additional sophistication is realized by using a cycle approach to transaction processing. A transaction processing cycle combines one or more types of transactions having related features or similar objectives. It groups together recurrent, cyclically occurring activities. Any such groupings are arbitrary and subject to overlapping. Nonetheless, the groupings are useful, and the following groups will be used in this book: revenue cycle, expenditure cycle, resource management cycle, and financial reporting cycle. Following this chapter, each group of transactions is examined in greater detail.

The revenue cycle facilitates the exchange of products or services with customers for cash. The transactions involved span the activities from the point the customer order is received to the point at which the payment is received.

The expenditure cycle seeks to facilitate the payments to vendors for the goods and services provided to the firm. This cycle covers the span between the time the need for an order to vendor(s) is felt and the time the payment is disbursed for the acquired services and products.

The resources management cycle consists of all activities pertaining to the operation of the business. This could include obtaining the capital to invest in the business; acquiring and maintaining the plant and other fixed assets used by the business; selling and marketing; and, lastly, human resource management. This particular cycle can be seen as encompassing a series of self-contained cycles. Given their relative independence, the sub-cycle can be grouped in a variety of ways. For instance, the cash receipts from customers and cash disbursements could be removed from those cycles and could be included with the funds resource cycle. However, this would truncate those cycles. Thus, we will define the funds cycle to include only the funds acquired from bank loans, bond issues, stock issues, contributions from owners and others, and disposal of plant assets.

The resources management cycles can be defined in other ways. Alternate descriptions are the financial cycle (those transactions relating to funds received from owners and facilities acquired with those funds), the administrative cycle (those transactions relating to cash receipts and disbursements, as well as to the acquisition of facilities), and the inventory cycle (those transactions relating to acquiring, maintaining, and selling inventory).

At the center of the aforementioned cycles is the general ledger and financial reporting cycle. This cycle is unique in that the processing of individual transactions is not its sole, or even its most important, function. Its primary inflows arise as outputs of a variety of transaction processing systems. In addition, the cycle processes the relatively few non-routine transactions that arise during each accounting period and the adjustment type transactions that occur at the end of each period.

**(a) Transaction Processing Cycles for Varied Organizations.** The transaction processing cycles described above pertain only to merchandising firms. Many other types of organizations exist, however. Although all organizations employ most of the cycles just discussed, some significant differences appear. Thus, certain types of organizations can dispense with certain key cycles. For instance, the typical governmental agency does not make sales and therefore does not need a revenue cycle. More often, though, organizations require added cycles that are unique to them. A manufacturing firm requires a product conversion cycle. A bank requires demand deposit and installment loan cycles. Health care, insurance, and transportation firms also require special cycles. All of these special cycles display certain peculiarities not found elsewhere. Nevertheless, they exhibit many features that are very similar to those of the basic transaction processing cycles.

## 14.5 COMPUTER TRANSACTION PROCESSING APPROACHES

Transaction processing typically involves three stages: data entry, file processing, and output preparation. Processing may be performed either manually or with the use of computers. Although we will contrast these alternative means of processing, our main focus is on the latter. The two basic computer-based transaction processing approaches are batch processing and on-line processing. One or the other of these approaches must be selected by a firm for each of its transaction processing applications. For instance, a firm might decide to employ the batch processing approach for its payroll application and the on-line processing approach for its sales application. The choice of one approach over the other depends on which attributes—such as efficiency, timeliness, economy, accuracy—are most critical for a particular application.

## 14.6 BATCH PROCESSING APPROACH

Batch processing involves the periodic processing of data from groups of like transactions. Transactions are collected and stored temporarily until a sufficiently large number is accumulated or until a designated time arrives. Then the batch of transactions is processed to post the transaction data to one or more files. This approach is most frequently employed to process routine transactions that occur in relatively large volumes. In addition to payroll transactions, batch processing can feasibly be applied to applications such as general ledger accounting, cash disbursements, cash receipts, and student registration transactions. Batch processing is also used to extract information from a file. Thus, the quantity of items on order could be extracted from an inventory file and listed on an inventory status report. Often, posting and extracting/reporting are combined.

Since time is required to collect transactions into batches, batch processing is marked by delays. Periods of time—days, weeks, months—elapse between successive processings. Hence, this approach is sometimes referred to as delayed processing, and the periods of time between successive processings are called processing cycles.

Most batch processing applications involve sequential file access and off-line data entry. Sequential file access consists of processing transactions one after the other, according to the order in which they have been sorted. Off-line data entry consists of entering batches of data onto computer-readable media by off-line input devices. For instance, a data-entry clerk may key transaction data from a batch of source documents onto magnetic tape via a key-to-disk system. A batch processing application can alternatively involve direct file access and on-line data entry.

**(a) A Sequential Batch Processing Application.**   Exhibit 14–1 diagrams the steps in a typical sequential batch processing application. Major steps in the application consist of off-line data entry, sequential file access processing, and printed output preparation. For purposes of discussion, we will assume that the application pertains to sales transactions for a merchandising firm.

Data entry begins when the source documents (i.e., sales invoices) are gathered into a batch and one or more batch totals are computed. For instance, the sales amounts shown on the invoices might be totaled. This batch total will be used as a key accounting control over the accuracy of the processing step. Next, the sales transaction data are converted (transcribed), by an off-line device such as a key-to-tape encoder, to magnetic tape. The data elements transcribed by a clerk from each sales invoice source document, such as customer number and name, are arranged on a magnetic tape to form a record. To increase processing efficiency, the records are grouped into blocks within the created file of sales transactions. Each block of transaction records is then entered into the computer processor, using a magnetic tape drive. On being entered, the transaction data are validated by an edit procedure within the sales application program. Any program function performed within a computer and affecting a batch of transactions is called a run.

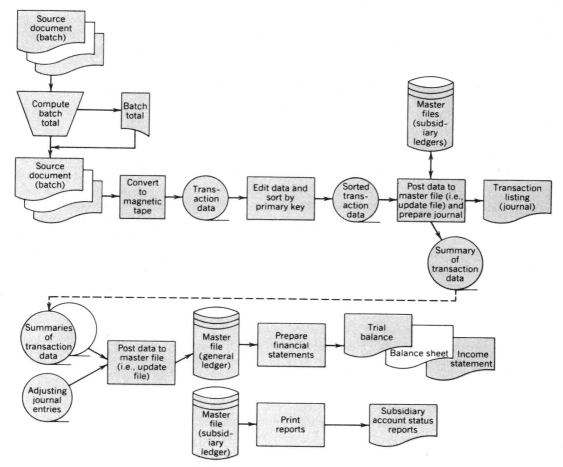

**Exhibit 14–1.   A flowchart depicting computer-based batch processing of transactions. Adapted from J. Wilkinson, *Accounting Information System: Essential Concepts & Applications*, 2nd. ed. (NY: John Wiley & Sons).**

This editing procedure is generally the first run performed in a computer-based application, since it is important to check the input data for correctness and completeness.

File processing in this application involves accessing a file in which the records are arranged sequentially. Thus, it begins with the sales transactions being sorted into the same order as the records in the file that is to be posted. This sorting step is performed by reference to a sort key, the data element by which the records are to be rearranged. For sales transactions, the sort key is likely to be the customer number, since records in the accounts receivable master file are generally arranged according to customer numbers. In a computer-based system the accounts receivable master file is, in effect, the subsidiary ledger.

A critical step in transaction processing is to post data from the batch of transactions to one or more master files. This step is called updating. With respect to sales transactions the two master files of concern are the accounts receivable file and the general ledger. These master files are shown as being stored on magnetic disks (designated by the cylindrical symbols). Since the transaction records have been sorted according to customer numbers, the accounts receivable master file is updated first. In updating the file, the processing program starts at the beginning accounts receivable record and reads every record in the file, changing (updating) each record affected by a transaction. Summary data from the updated transactions are copied onto a magnetic tape. Periodically, the daily totals from the summary tapes are combined with those from other types of transactions and with adjusting journal entries. Then these summary data are posted to the affected accounts within the general ledger.

Various reports—outputs—are generated from the application. A listing of the sales transactions is printed during the file updating run. This transaction listing is, in effect, the sales journal. The total amount that it shows as being added to the accounts receivable account balances should be compared with the predetermined batch total to verify the accuracy of the updating. In addition, the account balances from the general ledger are periodically arranged to produce trial balances, balance sheets, and income statements. Also, status reports of the subsidiary ledger, such as an accounts receivable aging report, are generated. These outputs are provided as needed by users, whether weekly, monthly, or at the end of each accounting period.

## 14.7 CONTRASTS WITH MANUAL BATCH PROCESSING

Accounting systems before the advent of computer traditionally employed the batch processing approach. As a result, the computerized batch processing and its manual counterpart have much in common. Exhibit 14–2 portrays manual batch processing in a flowchart that parallels the computer-based batch processing shown in in Exhibit 14–1. By comparing these two flowcharts, it is easy to see both similarities and differences between manual batch processing and computer-based batch processing. Key similarities are as follows:

- Both involve the same essential steps, such as computing batch totals, sorting transactions, posting to subsidiary and general ledgers, and preparing financial outputs.
- Both generate the same outputs—trial balances, financial statements, status reports—that are similar in terms of content and printed formats.
- Both rely on batch totals to provide an important control over the accuracy and completeness of processing.
- Both employ processing cycles of constant lengths.

Major differences include:

- Computer-based batch processing steps are performed more rapidly and accurately than those in manual batch processing applications.
- In computer-based batch processing applications the transaction data and master files are stored on magnetic media and thus not readable by humans; in manual batch processing applications the data and files are stored on hard copy records.
- In computer-based batch processing applications, the journal is produced as a byproduct of the posting step, rather than during a prior step in the processing sequence. In fact, the primary value of a journal in a computer-based system is to provide a visible person-readable link in the audit trail.

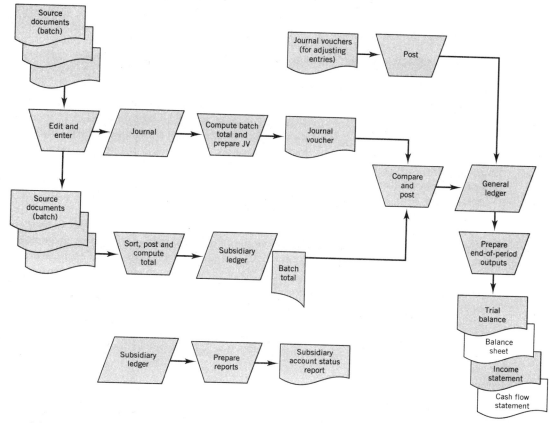

**Exhibit 14–2. A flowchart depicting batch processing of transactions by a manual system. Adapted from J. Wilkinson, *Accounting Information System: Essential Concepts & Applications*, 2nd. ed. (NY: John Wiley & Sons).**

- In computer-based batch processing applications many of the processing steps are combined and performed by the system itself; in a manual system all of the steps are performed by human clerks, who are located in separate organizational units. Thus a computer-based system performs the validation (edit) checks customarily performed by human clerks.

Given the similarities between manual and computerized batch transactions processing, it's possible to argue that accountants using computers were simply using faster tools to do the same things they did before. This is not the most optimum use of technology. But, as shown in our chapters on reengineering, advances are being made in the world of accounting, where information technology is starting to have an impact on the nature of functions performed by accountants.

## 14.8 ADVANTAGES AND DISADVANTAGES OF BATCH PROCESSING

Computer-based batch processing provides several advantages. First, it is relatively simple and requires less complex and expensive hardware and software than the on-line processing approach. Secondly, because it normally involves sequential processing, the batch processing approach is relatively efficient, especially when large batches are accumulated. Third, this approach facilitates the use of controls such as batch totals and thorough audit trails. Fourth, it was similar to the way things were before the advent of computers, which made the change easier.

The main disadvantage of the batch processing approach is that the records in the master file become out of date very quickly. Consequently, most reports from a batch processing application are tied to the processing cycles. Another drawback is that errors detected in transaction data cannot easily be corrected at the time of entry; they must be corrected and reentered, either in a separate "run" or during the next processing cycle. A third disadvantage is that the processing step of sorting must be performed before a master file can be updated. Finally, all records in the master file must be read during the updating step, thus increasing the processing time when relatively few records are affected.

**(a) Variations in Batch Processing Applications.**    Sequential batch processing applications may be modified with respect to data entry, data storage and posting, and report preparation.

*Transaction Data May Be Entered into the Computer System by On-Line Devices Such as Terminals.*    In this variation, the data would be validated (edited) as entered. Then it would be accumulated temporarily on an on-line storage medium, such as magnetic disks, until time for processing as a batch.

*Direct Access May Be Used for File Processing.*    Instead of processing a batch of transactions sequentially against the entire master file, each transaction could be posted directly to the particular record in the master file that it affects.

*Data Could Be Captured on Scannable Documents.*    Credit card transaction slips could contain information that could then be read with OCR devices.

*Outputs Could Be Spooled.*    Instead of printing the reports directly onto hard copy, the information could be stored temporarily on a magnetic tape or disk. At a later time the reports could be printed off-line, thus making more efficient use of the computer processor.

## 14.9 COMPUTER-BASED ON-LINE TRANSACTION PROCESSING

On-line processing consists of processing each transaction as soon as it is captured and entered. A flowchart of such processing is shown in Exhibit 14–3. Data from each transaction are entered via an on-line device and posted immediately and directly to the affected record in one or more master files. Thus, the stored data in the master files, which are continuously on-line, are kept up-to-date. Also, the application software or programs that direct the transaction processing are continuously on-line and are available for use. Often, this type of processing is called immediate or direct processing, since it so obviously exhibits these characteristics. An on-line processing system is also called an interactive processing system, since it involves direct interactions between persons and a computer-based system.

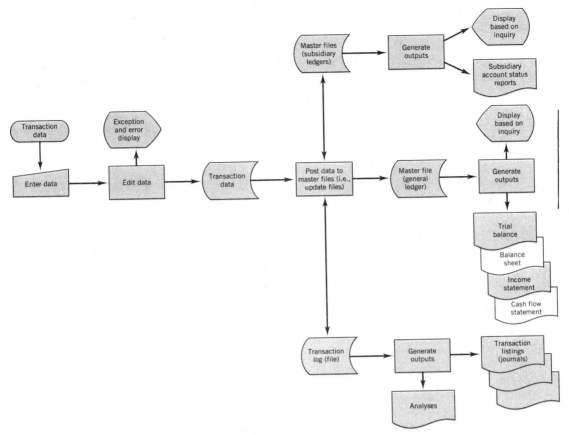

**Exhibit 14–3. A flowchart depicting computer-based on-line processing of transactions. Adapted from J. Wilkinson,** *Accounting Information System: Essential Concepts & Applications,* **2nd. ed. (NY: John Wiley & Sons).**

As shown in Exhibit 14–3, steps needed for an on-line approach consist of data entry and edit, file processing, file inquiry, and report generation. For purposes of illustration, we again assume that the application involves sales transaction processing. To simplify the application, we will further assume that sales orders are billed as soon as they are received.

The need for data entry arises when a sales order clerk receives an order from a customer. The sales transaction data may be recorded on a source document, such as a sales order. Alternatively, the data may be acquired without benefit of a source document. For instance, the sales clerk may receive sales orders by phone. To begin the interactive data entry procedure, the sales order clerk accesses the system via an on-line device, such as a terminal. The specific steps taken depend on the method of assistance provided by the system software. If the dialogue prompt method is employed, the clerk would key the code SO (for sales order transaction) when the system prompts: "What is the transaction type code?" If the menu method is used, the clerk would key numbers that lead to the option for sales order data entry. Under either method the response by the clerk causes the sales order application program to be activated, that is, loaded into the primary storage unit and initialized. The data entry portion of this program aids the clerk in entering the sales data (perhaps by means of a preformatted screen) and edits the data. If errors or omissions are found by the edit routine, the edit routine notifies the clerk by some type of message displayed on the screen. After all errors and omissions are corrected, the data are accepted and readied for processing.

File processing consists mainly of updating the affected files immediately and directly. In the sales application, it also involves calculating the amount of the sale from the sales order. First, the proper customer's account in the accounts receivable master file (subsidiary ledger) is posted to reflect the amount of the sale. Secondly, the accounts receivable and sales accounts in the general ledger are also posted to reflect the debit and credit effects of the transaction. Finally, the transaction data are added to a transaction log, generally just following the most recently posted transaction. Note that in a typical on-line processing system, various types of transactions may be entered in the order received. Thus, a sales transaction may be preceded by a purchase or cash receipts transaction. Inquiries of up-to-date information in files may be made at any time. One may inquire about a particular sale or a customer's account in the accounts receivable file. Responses to the inquiries would be displayed on terminal screens.

The same reports and other outputs provided by a batch processing procedure can be generated by an on-line processing system. The sales application program might generate status reports concerning accounts receivable, sales summaries, and also sales invoices. A general ledger program would periodically generate trial balances, balance sheets, income statements, and cash flow statements. Other programs might produce the transaction listings. If desirable, these listings could be analyzed by type of transaction and in other ways.

## 14.10 CONTRASTS WITH COMPUTER-BASED BATCH PROCESSING

On-line processing is similar to computer-based batch processing in that both approaches generate the same basic outputs and employ the same data records. The two approaches exhibit significant differences, however. Batch processing generally involves sequential processing of transactions, whereas on-line processing always employs direct processing. In the batch processing approach, the transaction data either may be prepared off-line or may be entered via an on-line device as received. In the on-line processing approach the data are always entered via an on-line device. Batch processing focuses a separate program within a particular application on all the transactions comprising a batch. In contrast, on-line processing employs a single application program to perform all the processes for each transaction before beginning the processing of the next transaction. Although batch processing applications may on occasion employ magnetic tape to store master files, on-line processing applications use magnetic tape to store the backup files and records. Finally, batch processing applications are relatively unintegrated, whereas individual on-line processing applications tend to be highly integrated with related applications.

## 14.11 ADVANTAGES AND DISADVANTAGES OF ON-LINE PROCESSING

The major advantage of on-line processing is better service to users. Since records are updated immediately and are directly accessible, up-to-date information is available in a timely manner. Also, data are accurate and complete, since transactions are checked thoroughly on capture and detected errors are corrected. Since processing is performed in a direct manner, no sorting or transcribing is necessary. Finally, because on-line processing applications are integrated, fewer duplicates of files are required, and the informational needs of users can be satisfied more fully.

A variety of applications are greatly affected by the foregoing advantages. Many firms have found the sales order application to be an apt example, especially if it is integrated with an on-line inventory system. An integrated system of this type allows a sales clerk to determine if adequate inventory is available to fill a sales order when received. Other suitable candidate applications for on-line processing include airline reservations, job order manufacturing, hospital patient care, and materials management.

The main disadvantage of the on-line processing method is its complexity relative to batch processing. Not only must the operating system handle varied users and diverse functions but the data entry routines in the application programs usually are expected to provide more user guidance and perform more elaborate checks. Also, the needed hardware, such as terminals and direct-access storage devices, is more expensive than the hardware needed in batch processing. Furthermore, transaction data processed by the on-line processing method cannot be supported by controls such as batch totals. In addition, reliable and thorough audit trails are more difficult to maintain. Finally, the entry of data is slower, since the system must wait for the clerks to key in the data and to respond to error messages.

But the disadvantages notwithstanding, on-line information technology offers revolutionary potential for changes that are more than superficial. A Chicago-based cosmetics maker, Helene Curtis, recently changed its strategy of being a low-cost manufacturer to being an on-line supplier with excellent service to retailer customers. To foster this new strategy, the firm developed a transaction processing system for on-line order entry and finished goods inventory. It also added a 300,000-square-foot automated warehouse. When the new warehouse is fully tied into the on-line processing systems, the firm's shipping costs are expected to decrease by 25 percent.

**(a) Variations in On-line Processing Applications.**   The variations in on-line processing applications are relatively limited. The data must be entered on-line and direct access storage devices must be employed. The options essentially concern (1) whether to use source documents as the basis for data entry, (2) whether to enter transactions individually or in batches, (3) which type of on-line input device to use, (4) which method of interactive assistance to provide to users, and (5) which mode of output to provide (e.g., soft-copy display, hard-copy report, voice response).

A major alternative is to convert the on-line processing system into a real-time system, a support system that provides information in a sufficiently timely manner to control a process. Real-time systems may control physical processes or transaction-oriented business processes. For instance, the physical process of cold-rolling steel sheets is often controlled by a real-time system through the use of process control computers. As another example, a transaction-oriented real-time system is activated each time a consumer purchases merchandise with a credit card and the clerk makes an on-the-spot credit check.

## 14.12 FILE PROCESSING AND MANAGEMENT

A firm's data have traditionally been stored in a variety of files. A file of data may be defined as a collection of data elements that have been organized into records. All applications involving transaction processing require one or more files. Therefore, accountants should be keenly aware of the characteristics and types of files, as well as the nature of file storage and the several functions of file management. In addition, accountants should recognize the alternative file structures that may be used for organizing and accessing data.

## 14.13  NATURE OF FILE STORAGE

Before delving into the details of file management, let's clarify the differences between data storage in manual and computerized systems and between logical storage and physical storage. In a manual information system, the file data are generally stored in folders, tubs, in/out baskets, and file drawers throughout the departments of a firm. Often, the same data are duplicated in multiple copies of a form and stored in several places. Certain data may be abbreviated or even omitted on forms, since such shortcuts are understandable to experienced clerks. In general, an air of informality permeates the storage and management of files. In computer-based information systems, the data comprising files are stored on magnetic or optical media. File storage and management require formalized procedures that must be strictly followed. The identity and location of each item of data stored in a computerized system must be precisely specified. Each step in file storage must be carefully planned. But the rewards of such planning comes from the ease of access and the compactness of storage: a small disk replaces volumes of bulky ledgers.

Data used in processing transactions are stored in logical records within files. In turn, the logical records and files are stored on physical media (e.g., magnetic disks) in the form of physical records and files. Logical files include data files—such as master, transaction, history, reference, and non-data (such as program, text, and index files) files. In logical terms, data files consist of records, data elements, and characters. Each record is organized into a series of fields, either fixed or variable in length. Data elements, whose values are placed within the fields, reflect attributes of entities or events. Data elements also may serve as primary or secondary keys within the records. Three major functions of file management are file storage, file maintenance, and retrieval of stored records. Files may be arranged sequentially or may be random. Records within files may be accessed either by following a specified sequence or by means of a direct method. Three widely used methods of physical file organization are sequential files, indexed sequential files, and random files. Indexed sequential files may be updated sequentially but their records may be accessed directly. Random files cannot be updated sequentially but their records may be accessed very quickly. Each file organization method is selected for particular circumstances by reference to such measures as activity ratio, volatility, and required response time.

In summary, a random file (on a direct-access storage medium) is most suitable when up-to-dateness and fast response times are important, the activity ratio is low, the file volatility is high, and the file is relatively limited in size. A sequential file (on magnetic tape) is most suited to the opposite set of conditions. An indexed sequential file (on a direct-access storage medium) is a good choice when both processing efficiency and rapid accessibility are important.

## 14.14  CLASSIFICATION AND CODING OF DATA

Coding is critical to the efficient storage and processing of data. Thus, an important step when creating a database is to devise codes suited to the various types of data needed by a firm in its transaction and information processing activities. This section examines the relationships of coding to classification, the major coding systems, and the attributes of sound codes.

**(a) Classification Versus Coding.** Classification is the act of grouping into classes. In the context of an information system, classification refers to the grouping of data and information. For instance, transaction data may be grouped according to the accounts to which they pertain, such as the sales account. Sales data may be classified or grouped according to the sales offices making the sales, the product lines sold, and the dates of the sales.

Classification plans or schemes are designed with certain objectives in mind. Consider, for example, a chart of accounts, the basic financial classification plan of every organization. The chart of accounts has the primary purpose of satisfying key internal and external information needs. If it is well designed, a chart of accounts enables useful financial statements and other reports to be prepared. Not only should such financial statements and reports aid managers in controlling operations, but they should assist external parties in making investment decisions.

Coding is the assignment of symbols, such as letters and numbers, in accordance with a classification plan. A coding system provides unique identities for specific events and entities. For instance, the coding system of a firm may assign the letter code "CS" to identify a credit sales transaction; the number code "711" to identify the sales invoice issued in a particular credit sales transaction; the number code "1346" to identify the customer John Henry Johnson of Akron, Ohio; and the alphanumeric code "XQ7" to specify a particular item that it sells. As we can see, codes are much briefer than the events or entities for which they stand. A coding system can therefore ease the entry of data, enhance processing efficiency and accuracy, speed the retrieval of data from files, and aid the preparation of reports.

## 14.15 CODING SYSTEMS

A variety of coding systems have been devised. Familiar examples are bar codes, such as those used to identify merchandise items; color codes, such as those used to distinguish copies of multicopy forms; and cipher codes, such as those used to protect confidential messages. However, the codes most widely useful to business firms are based on alphabetic, numeric, and alphanumeric characters.

Four coding systems that use these characters are mnemonic, sequence, block, and group coding systems.

**(a) Mnemonic Coding System.** A code of this type provides visible clues concerning the objects it represents. For instance, AZ is the code for Arizona and WSW-P175R 14 represents a white sidewall radial tire of a specific size. Thus, a mnemonic code is relatively easy to remember. However, its applications are more limited than those of the other three coding systems.

**(b) Sequence Coding System.** The simplest type of coding system is the sequence code, which assigns numbers or letters in consecutive order. Sequence codes are applied primarily to source documents such as checks and sales invoices. A sequence code can facilitate document searches, such as a search for a particular sales invoice. Furthermore, a sequence code can help prevent the loss of data, since gaps in the sequence signal missing documents. Sequence codes are inflexible, however. New entities or events can be added only at the end of the sequence. Moreover, a sequence code generally is devoid of logical significance. For instance, a specific sequence code assigned to a customer does not identify the sales territory within which he or she resides, nor the customer class to which he or she belongs.

**(c)  Block Coding System.**   A third coding system partially overcomes these drawbacks. A block coding system assigns a series of numbers within a sequence to entities or events having common features. Consequently, a block code designates the classification of an individual entity or event. Block codes have varied applicability within a firm. Customer numbers, for instance, may be blocked by sales territory. To illustrate, customers in the southern sales territory may be assigned numbers ranging from 1 to 4999, whereas customers in the northern sales territory may be assigned numbers from 5000 to 9999. In other applications, products may be blocked by product line, employees by department. An important use of block codes is to designate blocks of numbers for the major account groupings within charts of accounts. While a block is reserved for individual codes, it is not necessary to assign every possible number within a block. In fact, one advantage of a block code is that unassigned numbers are usually available to be assigned to new objects (e.g., products) as they are added to the firm's scope of activity.

**(d)  Group Coding System.**   The group code is a refinement of the block code, in that it provides added meaning to the users. That is, a group coding system reveals two or more dimensions or facets pertaining to an object. Each facet is assigned a specific location, called a field, within the code format. Code segments (i.e., subcodes) appear within the respective fields, thus identifying the facets pertaining to a particular object.

An example of a group code for an entity should clarify this description. Raw materials stored for use by a metal products manufacturer may be coded by means of a six-digit code, where the first digit may represent the type of material, the next two represent the storage location, while the last three define the size of the material. A particular raw material item could therefore be coded 573201 or 5–73–201 for greater clarity, where 5 stands for the type of material (steel rod), 73 represents seventh row and third bin, while 201 represents a 20 by 1 square inch material.

Group codes are extremely versatile. They may be expressed as hierarchical structures. For instance, the left-most field in the code of a plant asset may designate the major or broad classification, the fields to the right designating increasingly detailed classifications. They may include block, sequence, or other types of codes. For instance, the employee code may be assigned sequentially as each employee is hired, with the department code being added as a suffix. Although this combinational feature of group codes tends to make them relatively lengthy and cumbersome, it also packs such codes with much useful information. For instance, a code that captures relevant features of a transaction can provide the basis for key analyses and reports. Thus, a variety of sales analyses can be prepared when a group code of the following format is employed to record sales transactions:

ABBBBCCDDDEEE or A-BBBB-CC-DDD-EEE
In the above code, the first letter, A, represents the class of customer to whom a sale is made.
The next four letters, BBBB, stand for the customer's account number.
The following two letters, CC, represent the sales territory where the sale is made.
The following three letters, DDD, represent the code of the salesperson making the sale.

The last three, EEE, represent the type of product (product line) sold. From an accounting perspective, coded charts of accounts are of immense help, since transaction processing will be very cumbersome without the coded ledger accounts. In the appendix following Chapter 15, a coded chart of accounts for a mid-sized company is provided.

## 14.16 CODING AND COMPUTERS

Computer-based systems also dictate certain attributes. For example, they require that codes be fixed in length. In addition, they encourage the use of numeric codes, since numeric sorting can be performed more easily. However, the presence of computers should not discourage the application of sound design principles. Thus, alphabetic codes provide twenty-six choices for each character position of a code. Their use can therefore aid conciseness.

Certain codes are peculiar to computer-based systems. Two such codes, the user code and the cipher, relate to security and access to the system. Two other computer-oriented codes are the transaction code and the action code. Transaction codes identify types of transactions to the computer system. Such codes designate the specific application programs to be used in processing the various transactions entered into the system. By specifying a particular transaction code, a user informs the computer system of which accounts to debit and credit and which data elements to update in which files. In effect, a transaction code is a concise replacement for the familiar journal entry format used in manual processing systems.

Action codes identify specific operations pertaining to file maintenance or data retrieval. For instance, they may specify the addition of new master records, the deletion of current master records, or the display of stored records. Action codes are employed primarily with on-line processing systems. Often, they are entered with the aid of menu screens.

## 14.17 ATTRIBUTES OF CODES

A sound coding system furthers the primary purpose of a classification plan, which is to satisfy certain informational needs of users. In addition, each of its codes:

- Uniquely identifies objects, such as customers and sales.
- Is concise and simple, so it is easy to remember and apply and economical to maintain.
- Allows for expected growth, so that it will not need to be changed in the foreseeable future. For instance, a growing firm with nine product lines should allow two digits (assuming a numeric code) for the product line coding system.
- Is standardized throughout all functions and levels within a firm, so that reporting systems can be fully integrated.

## 14.18 CONCLUSION

The transaction processing systems of an organization comprise activities that focus on particular economic events. They may be grouped to form transaction cycles such as the revenue cycle, expenditure cycle, resources management cycle, and general ledger and financial reporting cycle.

Transaction processing typically involves three stages: data entry, file processing, and output preparation. Processing can be performed either manually or with the use of computers. The two basic computer-based transaction processing approaches are batch processing and on-line processing. Batch processing generally consists of accumulating groups of like transactions, computing batch totals, entering the data for processing on magnetic media, sorting the data by appropriate sort keys, posting the transaction data to the sequentially organized records in master files, and generating printed outputs.

Computerized batch processing is similar in most respects to manual processing of transaction data, although the benefits of computers are realized and the journal is produced as a byproduct of the posting step. Whether it is performed with the aid of computers or manually, batch processing offers two advantages: relatively efficient processing and the important control known as batch totals. Its major drawback is that the records in the master files become out of date between processing cycles.

On-line processing consists of entering single transactions, via on-line input devices; posting the transactions directly to the records in the master files; and generating outputs as needed. This immediate and direct type of processing offers the important advantage of up-to-date records that are readily accessible to users. It also allows data to be edited and corrected as soon as it is entered and eliminates the need for sorting transaction records. The major drawbacks of on-line processing are the added complexity and cost of software and hardware, as well as the lack of the batch total control.

Sound classification and coding techniques can improve the efficiency of data storage and processing. Four coding systems that use alphanumeric characters are the mnemonic, sequence, block, and group coding systems. A sound coding system satisfies the information needs of users, employs unique identifiers, allows for expected growth, and is concise and simple. Two codes peculiar to computer systems are transaction and action codes.

# CHAPTER 15

# GENERAL LEDGER AND FINANCIAL REPORTING CYCLE

## 15.1 INTRODUCTION

This chapter will describe the objectives and functions of the general ledger and financial reporting cycle and its relationship with other business transaction cycles. In addition, the data sources, inputs, files, data sets, data flows and processing, accounting controls, and outputs pertaining to the general ledger system will all be identified and discussed. The chapter will also consider the nature of financial statements and the characteristics of the responsibility accounting system, and provide additional exposure to account coding and charts of accounts.

In Chapter 14 the transaction cycles used to facilitate the transactions having related features or similar objectives were discussed. Exhibit 15–1 summarizes the transaction cycles pertaining to revenues, expenditures, resources management, product conversion, and the general ledger and financial reporting. We begin our coverage of transaction cycles with the general ledger and financial reporting cycle, since it provides an overview of the objectives associated with the accounting information system and leads into the remaining transaction processing cycles and systems. As Exhibit 15–2 shows, transactions from component transaction processing systems flow into the general ledger. The information appears in the form of a data flow diagram, in order to show more clearly the logical data flows among entities (e.g., customers, suppliers, banks), data stores (e.g., customer records, inventory records), and processes (e.g., order receiving process, cash receipt process). Beyond updating the general ledger, the transaction flow permits the system to prepare financial reports.

## 15.2 OBJECTIVES AND FUNCTIONS

The general ledger and financial reporting cycle—to be called the general ledger system—represents the very core of an accounting information system. After discussing the objectives associated with this cycle, we go on to cover other topics associated with the cycle such as data sources and inputs, data flows and processing, the database itself, accounting controls, charts of accounts, and reports relating to this particular cycle. Both manual and computer-based systems are discussed.

As stated earlier, the overall objectives of the general ledger system are to gather together all of the transactions and then to provide information for the array of reports concerning an entity (e.g., a business firm, a governmental agency). The following attributes will allow a general ledger system to function in an optimum manner:

| Cycle | Typical Included Events |
|-------|-------------------------|
| Revenue | Sales of products or services |
| | Cash receipts from credit sales of products or services |
| Expenditure | Purchases of materials or services |
| | Cash disbursements in payment of acquired materials or services |
| Resources management | Acquisition, maintenance, and disposal (or disbursement) of funds, facilities, and human (e.g., employee) services |
| Product conversion | Conversion of raw materials into finished goods through the use of labor and overhead |
| General ledger and financial reporting | Compilation of accounting transactions from the remaining transaction cycles |
| | Generation of financial reports |

**Exhibit 15–1. A table of common transaction cycles. Adapted from J. Wilkinson,** *Accounting Information System: Essential Concepts & Applications*, **2nd. ed. (NY: John Wiley & Sons).**

- Records all accounting transactions promptly and accurately
- Posts transactions to the proper accounts
- Maintains an equality of debit and credit balances among the accounts
- Allows for needed adjusting journal entries
- Generates reliable and timely financial reports pertaining to each accounting period

In order to achieve these objectives, a general ledger system performs several functions. The manner in which these functions are performed depends in part on the extent and type of computerization employed. However, all general ledger systems must perform the following:

- Collect transaction data. Transactions arise from a variety of sources, such as sales and purchases. The more numerous types of transactions are grouped by component processing systems. These component systems then interface with and feed their summarized data to the general ledger system. Other daily transactions are recorded individually, generally on specially designed forms and journal vouchers. In addition, varied adjusting entries are posted at the end of each accounting period.
- Process transaction inflows. Collected transaction data undergo several processing steps before coming to rest in the general ledger. First, they are checked to see that debit amounts equal credit amounts, that eligible account names are used, and so on. Also, individual transactions may be verified to see that they are in conformity with generally accepted accounting principles. Then the transactions are posted to the general ledger accounts. Proof listings of the posted transactions may be prepared.

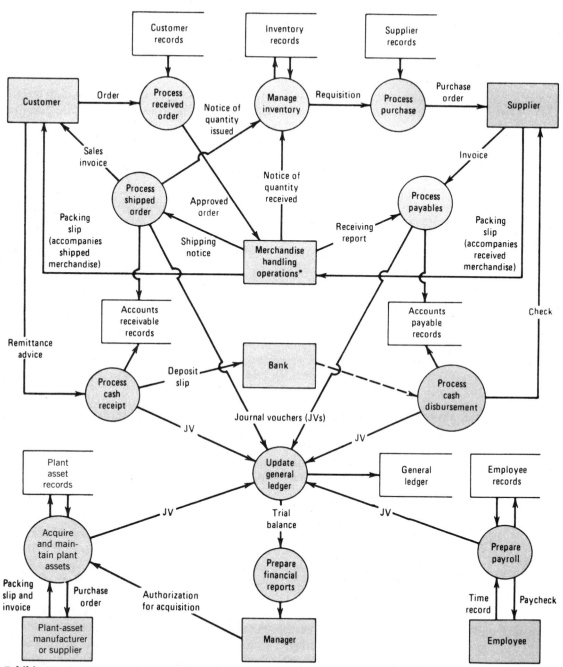

**Exhibit 15–2.** An overview data flow diagram of basic transaction processing systems. Adapted from J. Wilkinson, *Accounting Information System: Essential Concepts & Applications*, 2nd. ed. (NY: John Wiley & Sons).

- Store transaction data. The general ledger, as well as varied subsidiary ledgers, reflects account balances. Thus, they represent master files within a firm's database. If the balances in the accounts of the general ledger "master file" are to be kept current, they must be updated through the posting of transaction data.

- Assist in maintaining controls. The general ledger system is an integral part of the accounting system, it incorporates a number of controls. Indeed the general ledger itself represents a form of control, since the credit account balances must constantly equal the total of the debit account balances. Other controls range from checks or edits of the transaction data as they are entered into the system as well as reviews of the financial reports by managers and other recipients.

- Classify and code transaction data and accounts. Underlying the maintenance of the general ledger system are adequate classification and coding systems. Classification is necessary in assigning the effects of transactions to various accounts within the general ledger. Coding is desirable for identifying accounts, transactions, files, and other elements that have an impact on the general ledger.

- Generate financial reports. The most familiar financial reports generated by the general ledger system are the income statement and the balance sheet. However, a wide variety of other beneficial reports may be prepared. Some of these reports aid in the verification of the general ledger accounts while other reports, such as operating budgets, aid managers in planning and control responsibilities.

## 15.3 DATA SOURCES AND INPUTS

As should be clear by now, the general ledger system receives inputs from a wide variety of sources. Exhibit 15–3 illustrates this variety by showing the source documents of the various component or feeder transaction processing systems being entered into special journals. Summary totals from these entries are posted to the general ledger, as well as to any subsidiary ledgers that are maintained. Other data entered into the general ledger usually are recorded in a general journal or on journal vouchers. Regardless of the source, transaction data should include elements such as the date, accounts affected and amounts, and an authorization code or initials.

The types of transactions that affect the general ledger of a firm can be summarized as follows:

- Routine external transactions that arise during an accounting period from exchanges with independent parties who are in the environment of the firm. An example is a credit sale.

- Routine internal transactions that arise during an accounting period from internal activities of the firm. An example is the transfer of raw materials inventory to work-in-process inventory within a manufacturing firm.

- Non-routine transactions that arise, usually on an occasional basis and externally, during an accounting period from non-routine activities. An example is the donation of public land to the firm or even the purchase of a business in exchange for company stock. Company X will buy company Y and pay for it through its stocks.

- Adjusting entries that usually occur on a recurring basis at the end of accounting periods although they may be nonrecurring and take place at any time. Four types of adjusting entries are accruals, deferrals, revaluations, and corrections.

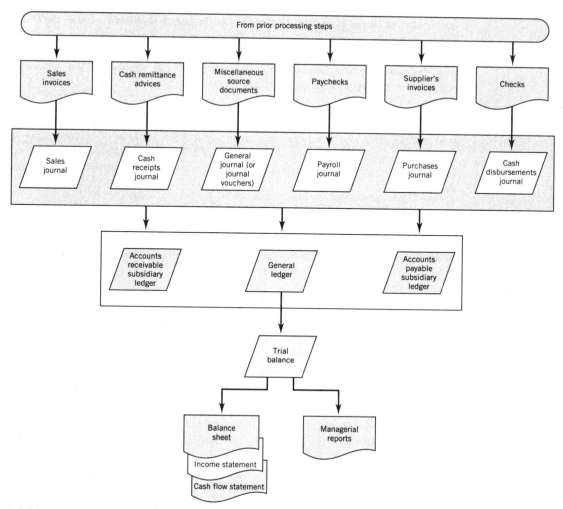

**Exhibit 15–3. A diagram of sources of inputs and their flows through the general ledger system. Adapted from J. Wilkinson, *Accounting Information System: Essential Concepts & Applications*, 2nd. ed. (NY: John Wiley & Sons).**

a. Accruals are recurring entries that arise from the passage of time and that reflect accumulated amounts as of the end of accounting periods. For instance, a firm may provide services that are not billed or paid by the end of an accounting period. The adjusting entry would show a debit to accrued fees receivable and a credit to service fees revenue.

b. Deferrals are recurring entries that also arise from the passage of time, but they represent the amount not yet due or earned. For instance, a firm may receive revenue for which services are still unearned at the end of the accounting period. At the time of the receipt, the entry would show debit to cash and a credit to the liability account named unearned service fees. The adjusting entry at the end of the accounting period, to reflect the portion earned, would show a debit to unearned service fees and a credit to service fees revenue.

c. Revaluations are nonrecurring entries that arise when the value of a physical asset is found not to correspond to the values reflected in the accounting records, or when an accounting measurement method is changed. An example of the former is the adjustment of the merchandise inventory account to reflect a shrinkage in the physical inventory. An example of the latter is a change in inventory cost flow assumptions from the FIFO to the LIFO method.

d. Corrections are entries that reverse the effects of errors, thus restoring affected accounts to proper balances.

- Reversing entries that are entered at the beginning of accounting periods in order to reverse the effects of adjusting entries made at the end of the previous periods. An example is the entry on January 1, 1993, that reverses the accrual of payroll expense made on December 31, 1992.

- Closing entries that transfer the amounts in all temporary accounts to the appropriate owners' equity account, leaving the temporary accounts with zero balances.

The general ledger can be manual or computer-based. Although computer-based processing is gaining in popularity, it is best to include a discussion of a manual system.

The primary source document used as an input to the general ledger system is the general ledger journal voucher, which has largely replaced the general journal sheet. A journal voucher is a document that lists the details of a single transaction. It is typically prepared for each non-routine, adjusting, reversing, and correcting transaction. A journal voucher is also often prepared to summarize the results of a batch of routine transactions that have been manually entered into special journals. For instance, after entering checks written today into the check register, a clerk may prepare a journal voucher that reflects the total of the cash disbursements.

Although the journal voucher remains the primary input to the general ledger in computer-based systems, its form will likely vary from that employed in manual systems. Three variations worth noting are the individual journal entry form, the batch entry form, and the on-line screen.

The journal voucher for entering individual transactions most likely contains columns for entering account descriptions and numbers, as well as columns for debit and credit amounts. After the data pertaining to an individual transaction have been manually written onto this form, the data are keyed by data entry clerks onto transaction files stored on magnetic tape or disk.

The batch entry forms are similar to journal voucher forms. They have replaced the special journal in computer-based systems, and are used to record and process the miscellaneous transactions. Data are entered manually onto the batch entry form and then are keyed onto tape or disk transaction files. Often the entered data are then listed on a hard-copy printout, to provide a permanent record of the journal.

A data entry screen can be used to record transactions when the general ledger system employs on-line files. Transactions are entered interactively with the aid of a CRT screen that is preformatted to exhibit a journal voucher. Either individual transactions or batches of transactions may be entered in this manner. It may be preceded by a hard-copy authorization form, which has been signed by a data entry clerk and supervisor. Alternatively, the authorization may be indicated by an on-screen code, as provided by a supervisor. The screen can have provision for several debits and credits; also, running totals are computed and shown on the screen by the application program. There are several advantages to on-line data entry:

- A preformatted screen that aids the proper entry of data
- Automatic data entry (e.g., date) and calculations (e.g., total debit amount and credit amount) by the application software
- Extensive use of codes to reduce input strokes
- Main and detailed level menus that allow easy access to the preformatted screen

On-line data entry can also provide other assistance. Thus, on-line help is available, as well as built-in editing features such as validity checks.

## 15.4 DATABASE USED FOR GENERAL LEDGER SYSTEM

The database for the general ledger and financial reporting system contains a variety of master, transaction, and history files and data sets. In addition to financial data concerning past events and current status, the database contains budgeted data that relate to planned future operations and status. Although their exact contents and composition will vary from firm to firm, the following set of files is representative:

*General ledger master file.*   Each record of the general ledger master file contains data concerning one general ledger account. In effect, the general ledger master file is the computerized version of the general ledger maintained in hard-copy form in a manual system. Taken together, the records in the general ledger master file constitute the complete chart of accounts for the firm and the current status of all account balances. Thus, the primary key is the account number.

*Current journal voucher file.*   This transaction file contains all of the significant details concerning each transaction that has been posted to the general ledger during the current month. In addition to individual non-routine transactions, it contains summary vouchers pertaining to daily routine transactions and adjusting entries. Included for each record would be the journal voucher number, date of the transaction, accounts debited and credited and the corresponding amounts, and a description of the transaction. Alternatively, in on-line processing systems, a transaction code may replace the chart of account numbers for amounts debited and credited. In effect, this file is a summary of all of the journals for the current month. Its primary key is the journal voucher number, which also provides the audit trail for transaction data. In the case of batch entries, the audit trail is provided by the batch number. In a database system, the journal voucher records can be linked to the appropriate accounts in the general ledger master file. For a database having a relational structure, the common linking columns in the transaction and master tables would contain the transaction codes.

*General ledger history file.*   The general ledger history file contains the actual balances of general ledger accounts for each month for several past years. It is used to provide financial trend information.

*Responsibility center master file.*   The responsibility center master file contains the actual revenues and costs for the various divisions, departments, work centers, and other responsibility centers within a firm. It is used in the preparation of responsibility reports for managers.

*Budget master file.*    The budget master file contains the budgeted amounts of assets, liabilities, revenues, and expenses allocated to the various responsibility centers of the firm. The budgeted values may be broken down on a monthly basis for the next year, whereas the budget period may extend for five or more years into the future. Together with the responsibility center master file, the budget file provides the basis for the preparation of responsibility reports.

*Financial reports format file.*    This file contains the information necessary for generating the formats of the various financial reports. Included are factors such as the report headings, column headings, side labels (such as descriptions of accounts and subtotal and total lines), spacing and totaling instructions, and the like.

In addition to the above-mentioned files, a firm will need a journal voucher history file for past months and years. It will also need detailed transaction files that support the current and historical journal voucher files. Certain firms also need various reference files, such as a cost allocation file that contains factors for allocating incurred costs (e.g., administrative costs) to responsibility centers and other segments within the firm. Systems in which transaction codes are employed will need reference files that link the specific accounts to debit and credit with each transaction code analyses, financial statements, and managerial reports. Exhibit 15–4 portrays the variety of outputs that may be generated from the files. Although several output preparation programs may be necessary to produce the displayed outputs, the exhibit summarizes the relationship between files and outputs, and previews the specific types of reports to be discussed later.

### 15.5 DATA FLOWS AND PROCESSING

In the traditional manual system, transaction data flow into journals and are posted to subsidiary ledgers; then, the total amounts are posted to the general ledger. Data from source documents involving high volume transactions, such as sales and cash disbursements, are entered into special journals. Columns in the journals are summed to generate batch totals. These totals are entered on journal vouchers. After the source documents are posted to the appropriate subsidiary ledgers, the posted totals are compared with the totals on the journal vouchers. Then the total amounts from journals are posted to the general ledger. Various financial statements and reports are periodically prepared from the general ledger and subsidiary ledgers.

In computer-based systems, the transaction data are entered from the source document forms described earlier. While there are systems that input on-line to the database, resulting in immediate file maintenance, most systems are on-line and batch processing. Periodically, the individual transactions are processed to update the master files (known as subsidiary ledgers), and their summarized amounts are processed to update the general ledger. If the transaction data are stored on magnetic tapes, they are sorted by account numbers and then processed sequentially against the general ledger accounts. If the transaction data are stored on magnetic disks, they will likely be processed directly to the general ledger accounts.

Many general ledger processing systems are a combination of on-line processing and batch processing. One likely combination is portrayed by Exhibit 15–5 and Exhibit 15–6.

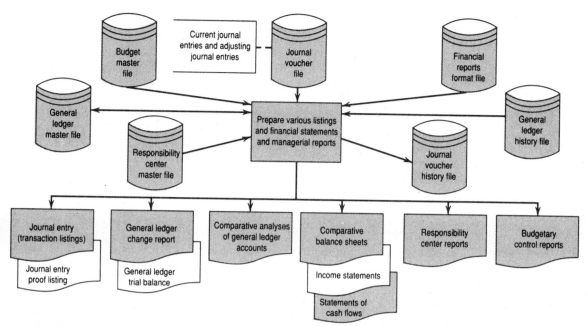

**Exhibit 15—4.   A system flowchart showing period-end preparation of outputs related to the general ledger. Adapted from J. Wilkinson, *Accounting Information System: Essential Concepts & Applications*, 2nd. ed. (NY: John Wiley & Sons).**

Exhibit 15–5 shows the on-line processing of transaction data during an accounting period. The sales, cash receipts, purchases, cash disbursements, payroll, plant asset, and application procedures process volumes of transaction data daily. In the course of their processing, the application programs accumulate and summarize the transaction amounts. Since the general ledger is on-line, these application programs also can post the summary amounts to the respective general ledger accounts. In fact, the general ledger postings can be done concurrently with, and as a byproduct of, the updating of the various subsidiary ledgers (master files). As nonrecurring special transactions arise, they are entered by accountants via CRT terminals from manually prepared journal vouchers. Each day the general ledger system generates a journal entry listing and a trial balance. These outputs enable the accountants to verify the correctness and accuracy of the postings. For instance, they can trace the summary amounts on the journal entry listing back to predetermined batch totals for each of the types of transactions. The journal entry listing may also serve as an integral part of the audit trail. Other outputs (not shown) can also be generated. Thus, prenumbered journal vouchers that reflect the summary amounts can be printed, thereby strengthening the audit trial. In addition, an error listing of all transactions in error can be printed, so that users can be monitored in correcting detected errors or omissions.

Exhibit 15–6 shows a batch computer system for processing adjusting journal entries at the end of each accounting period. The system begins with the entry of the journal entry data onto a transaction file. Standard adjusting journal entries, which are stored in an on-line file, are transferred to the journal entry data file by a special end-of-period program. Nonrecurring adjusting journal entries are entered via a CRT terminal, using manually prepared journal voucher coding forms prepared by accountants. The entered adjusting

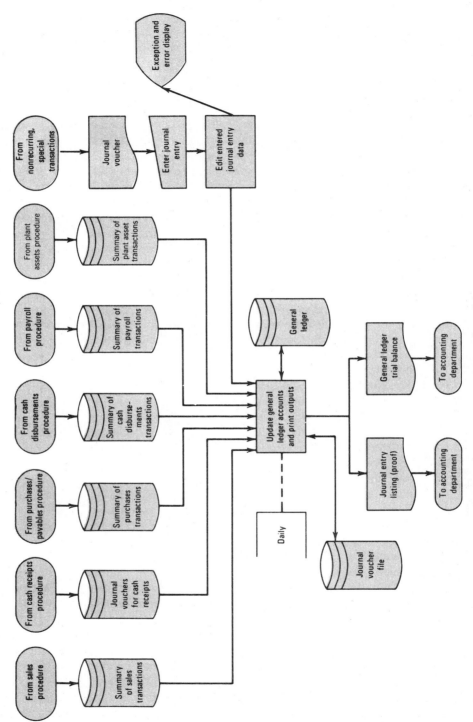

**Exhibit 15–5.** A computer-based general ledger processing procedure. Adapted from J. Wilkinson, *Accounting Information System: Essential Concepts & Applications*, 2nd. ed. (NY: John Wiley & Sons).

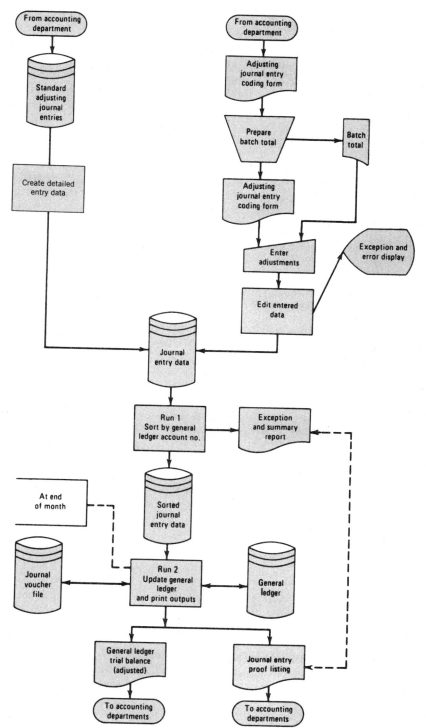

**Exhibit 15–6.** A month-end general ledger processing procedure. Adapted from J. Wilkinson, *Accounting Information System: Essential Concepts & Applications*, 2nd. ed. (NY: John Wiley & Sons).

entries are edited for mistakes. After being accepted by the system, the adjusting journal entries are merged with the standard recurring adjusting entries. Then the entries are split, with each portion being tagged by its identifying journal entry number. In computer processing run 1, the split entries are sorted into the general ledger account number sequence. In run 2 the general ledger accounts are updated, and two printed outputs are generated. One output is a journal entry proof listing, whose totals are compared with the run-to-run totals on the exception and summary report from run 1. The other output is an adjusted trial balance. After accountants review the outputs for correctness, an end-of-period program produces and posts closing entries and prepares a postclosing trial balance and financial statements. Individual firms may use some variation of the processing illustrated in Exhibit 15–6.

## 15.6  ACCOUNTING CONTROLS

The general ledger system is expected to provide reliable financial reports for a variety of users. To do so, it must independently check on the component transaction processing systems, carefully monitor the array of non-routine transactions that it accepts, and accurately record and post data from all transactions. A general ledger system faces a number of risks that could keep it from providing reliable reports.

The first step in providing reliable reports is to determine the risks to which the general ledger is exposed. Among the risks that exist in the maintenance of the general ledger are the following:

- Journal entries may be improperly prepared
- Journal entries may be left unposted
- Total debit balances and total credit balances in the accounts in the general ledger may become out of balance
- Control account balances in the general ledger may become out of balance with the totals of balances in subsidiary ledgers
- Unauthorized persons may gain access to data in the general ledger
- The audit trail that links the general ledger with source documents may become obscured
- Data pertaining to the general ledger and source documents may be lost or destroyed

These risks lead to exposures such as inaccurate financial statements and related reports. For instance, assets and liabilities may be valued incorrectly; they could be either overstated or understated. They also can lead to "leaks" of important financial data to competitors and added costs in locating or reconstructing needed transaction data. Adequate general and transaction controls are necessary to counteract such risk exposures.

**(a)  General Controls.**   Suitable general controls include the following:

- Organizationally, the function of posting journal vouchers to the general ledger should be separated from the functions of preparing and approving journal vouchers and from the function of preparing the trial balances of the general ledger.

- Documentation should consist of a fully descriptive chart of accounts plus a manual of general ledger process in effect.
- Operating practices should be clearly established, including period-end schedules and the preparation of control reports.
- Security should be maintained (in the case of on-line systems) by such techniques as (a) requiring that clerks enter passwords before accessing the general ledger file, (b) employing special terminals for the sole entry of journal voucher data, (c) generating audit reports (access logs) that monitor entries, and (d) dumping the general ledger onto magnetic tape backups.

**(b) Transaction Controls.** The following control procedures relating directly to general ledger accounts and processing are generally adequate:

- Prenumbered journal vouchers are prepared in the appropriate departments. For example, a journal voucher that reflects the declaration of a dividend may be prepared in the treasurer's office. These prepared journal vouchers are then approved by responsible managers.
- The data in journal vouchers, such as the account numbers, are checked for accuracy, in order to provide assurance that errors and omissions are detected before processing. In manual systems, general ledger clerks perform the checks. In computer-based systems, the checks are mainly performed by computer edit programs. Exhibit 15–7 lists several suitable programmed checks that are employed to detect errors and omissions in transaction data being entered for processing. Any errors detected by the programs are either listed in exception reports or are signaled on CRT data entry screens, depending on the type of processing system being used.
- Detected errors in journal entries are corrected before the data are posted to the general ledger.
- Approved journal vouchers are posted by specially designated persons who are not involved in their preparation or approval:
  a. In manual systems, the journal vouchers are posted by general ledger clerks directly to the general ledger sheets.
  b. In batch computer-based systems, the journal vouchers are keyed by data entry clerks onto a magnetic medium; then, the batches of entries are sorted by general ledger account numbers and posted during computer runs to the accounts. Exhibit 15–8 lists various checks that should be made during the posting or updating run, to provide assurance that processing errors are detected.
  c. In on-line computer-based systems, the journal vouchers are entered directly into the system, with the aid of preformatted screens on CRT terminals; then, the entries are posted by the computer system to the accounts.
- The equality of debits and credits for each posted journal entry is verified.
- The totals of amounts posted from batched journal entries to general ledger accounts are compared with precomputed control totals.
- Adequate cross-references are included to provide a clear audit trail. For instance, the journal voucher number and general ledger account numbers are shown in printed listings of the general journal, and the source page number of the general journal or journal voucher number is shown in each posting to the general ledger.

| Type of Edit Check | Typical Transaction Data Items Checked | Assurance Provided |
|---|---|---|
| 1. Validity check | General ledger account numbers; transaction codes | The entered numbers and codes are checked against lists of valid numbers and codes stored within the computer system. |
| 2. Field check | Transaction amounts | The amount fields in the input records are checked to see if they contain the proper modes of characters for amounts (i.e., numeric characters). If other modes are found, such as alphabetic characters or blanks, an error is indicated. |
| 3. Limit check | Transaction amounts | The entered amounts are checked against preestablished limits that represent reasonable maximums to be expected. (A separate limit is set for each account.) |
| 4. Zero-balance checks | Transaction amounts | The entered debit amounts are netted against the entered credit amounts, and the resulting net amount is compared to zero. (A non-zero net amount indicates that the debits and credits are unequal.) |
| 5. Completeness check | All entered data items | The entered transaction is checked to see that all required data items have been entered. |
| 6. Echo check | General ledger account numbers and titles | After the account numbers pertaining to the transaction are entered at a terminal, the system retrieves and "echoes" back the account title; the person who has made the entry can visually verify from reading the titles that the correct account numbers were entered. |

**Exhibit 15–7. A table of programmed checks which edit and validate journal entry data.** *Note:* **When entered data do not match or otherwise meet the expected conditions or limits, alert messages are displayed by the edit program on the CRT screen in the case of an on-line data-entry system. Adapted from J. Wilkinson,** *Accounting Information System: Essential Concepts & Applications,* **2nd. ed. (NY: John Wiley & Sons).**

Also, transaction numbers on source documents may be carried to transaction listings and into proof reports such as listings of account activity.

- Journal vouchers are filed by number, and periodically the file is checked to make certain that the sequence of numbers is complete.
- Standardized adjusting journal entries (including accruals and reversing entries) are maintained on preprinted sheets (or on magnetic media), to enhance the posting step at the end of each accounting period. Standardized journal entries are those that are prepared to reflect period-end adjustments of a recurring nature.
- Trial balances of general ledger accounts are prepared periodically, and differences between total debits and credits are fully investigated.
- General ledger control account balances are reconciled periodically to totals of balances in subsidiary ledger accounts.
- Special period-end reports are printed for review by accountants and managers that help correct mistakes.

| Type of Edit Check | Typical Transaction Data Items Checked | Assurance Provided |
|---|---|---|
| 1. Internal label check | | The internal header label of the file to be updated (posted to) is checked before processing begins to ascertain that the correct general ledger master file has been accessed for processing. |
| 2. Sequence check | General ledger account numbers | The transaction file, which has been sorted so that the amounts in the various journal vouchers are ordered according to account numbers, is checked to see that no transaction item is out of sequence. |
| 3. Redundancy matching check | General ledger account numbers and titles | Each transaction debit and credit is checked to see that its account number matches the account number in the general ledger record it is to update. Then the account titles in the transaction record and master file record are also matched, thus providing double protection against updating the wrong file record. |
| 4. Relationship check | Transaction amounts and account numbers | The balance of each general ledger account is checked, after the transaction has been posted, to see that the balance has a normal relationship to the account. If the balance in an account balance that normally exhibits a debit balance (e.g., accounts receivable) appears as a credit, the abnormality will be flagged by the check. |
| 5. Posting check | Transaction amounts | The after-posting balance in each updated account is compared to the before-posting balance, to see that the difference equals the transaction amount. |
| 6. Batch control total checks | Transaction amounts | The amounts posted are totaled and compared to the precomputed amount total of the batch being processed; also, the total number of transaction items processed is compared to the precomputed count of the transactions. |

**Exhibit 15–8.** A table of programmed edit checks which validate batched data during posting (updating) runs. Adapted from J. Wilkinson, *Accounting Information System: Essential Concepts & Applications*, 2nd. ed. (NY: John Wiley & Sons).

- Periodic review of journal entries and financial reports are performed by accountants and managers, while internal auditors may check the procedures being used.

## 15.7 FINANCIAL AND MANAGERIAL REPORTS

The financial reports generated by the general ledger system may be classified as general ledger analyses, financial statements, and managerial reports. A wide variety of examples are found in each category.

**(a) General Ledger Analyses.**    Most general ledger analyses are prepared as control devices. That is, they are intended to aid accountants in verifying the accuracy of postings. Two examples are the general journal listing and the general ledger change report. Other analyses related to general ledger processing may contain the following:

- Listings of all transactions posted during an accounting period, arranged by account numbers
- Allocations of expenses to the respective cost centers
- Comparisons of account balances for the current period with those for the same period last year, and comparisons of year-to-date account balances for this year with those for last year

**(b) Financial Statements.**    The most visible financial statements are the balance sheet, income statement, and cash flow statement. These statements, which are based directly on information in the general ledger, are provided to various parties outside the firm. Since a major reason for getting these reports prepared and disseminated is due to legal obligations, the accounting information system must incorporate rules and procedures that ensure that financial statements and derivative reports will conform with generally accepted accounting principles. Furthermore, since the array of financial reports include income tax returns and other tax reports, the AIS must also comply with federal, state, and local tax regulations.

Often the financial statements are accompanied by additional information that is useful to the recipients. For instance, they may be accompanied by comparative statements for previous years, by budgetary amounts, and by detailed schedules. Variances may also be computed and reported.

**(c) Managerial Reports.**    Numerous reports based on the general ledger may be prepared for the use of the firm's managers. Most of these managerial reports are generated from the same data used to prepare the financial statements. However, the financial statements for managers tend to be much more detailed than those provided to outside parties. For instance, analyses based on individual general ledger accounts are often prepared. Examples are analyses of sales, broken down by products or sales territories or markets; analyses of cash, broken down by type of receipts and expenditures; analyses of receivables, broken down by customers and ages of amounts due.

For managerial control and performance evaluation, comparative financial statements are also useful to managers. To be genuinely useful, though, the comparative information must be detailed and performance-oriented. Consequently, financial reports have been incorporated into reporting systems that focus on managerial responsibilities and segment profits. They enable managers to exercise more precise control over cost centers, profit centers, investment centers, products, sales territories, and so on. These responsibility-oriented reports are derived largely from the same information used in preparing financial statements.

A responsibility accounting system compares actual financial performance results with planned budgetary amounts for the various responsibility centers of a firm. In its report the focus is on controllable versus noncontrollable costs. One cannot hold managers responsible for costs they do not control. Controllable costs are usually those pertaining to direct materials, direct labor, supplies, and the like. Controllable costs are those that the manager heading the cost center can affect. Every cost is controllable at some level within the organization.

## 15.8 CODING OF ACCOUNTS

A sound coding system is critical to the success of a responsibility accounting system. Costs must be gathered and coded by responsibility centers (as must revenues for those centers having profit responsibilities). Exhibit 15–9 illustrates responsibility center codes based on a hierarchical group coding scheme. For instance, all costs coded 1333 will be compiled for the report to be received by the foreman of production unit 3. If the code 761 is the general ledger account code for direct labor, the code 761–1333 attached to a cost of $200 would identify the cost as being direct labor incurred by cost center 1333. All costs having the same code for a month would be totaled and listed as one of the items on the foreman's report.

## 15.9 CHARTS OF ACCOUNTS

We should not leave the general ledger system without discussing the chart of accounts, the AIS component around which the general ledger is organized. A chart of accounts is a coded listing of the asset, liability, owners' equity, revenue, and expense accounts pertaining to a firm. Taken together, these accounts facilitate the preparation of needed financial statements and reports. As an aid in its use, the chart of accounts should also describe accurately (but concisely) the contents of each account. For instance, it might identify the specific transactions that will affect a particular account; for instance, all

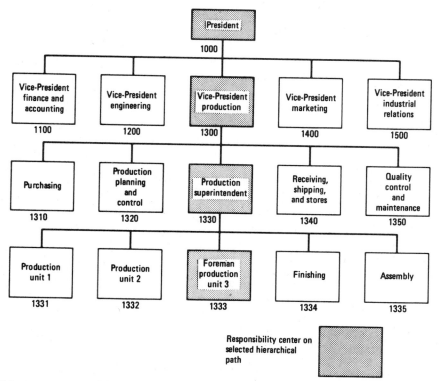

**Exhibit 15–9.  A portion of an organization chart used to illustrate responsibility reporting. Adapted from J. Wilkinson, *Accounting Information System: Essential Concepts & Applications*, 2nd. ed. (NY: John Wiley & Sons).**

purchases of merchandise and purchase returns will affect the purchases account (assuming that the periodic inventory method is to be employed).

The chart of accounts is closely related to the general ledger and subsidiary ledgers as well as to the financial statements. Each record in the general ledger is identified by one of the accounts listed in the chart of accounts. Details of the transactions affecting the accounts are reflected in the records of the general ledger. Certain accounts in the chart of accounts are control accounts, since the corresponding records in the general ledger contain balances that equal the total of balances in subsidiary ledgers. An example is the accounts receivable control account in the general ledger that controls the balances in the accounts receivable subsidiary ledger. It must be emphasized that the financial outputs dictate the composition of the chart of accounts as well as the accounting records. This will explain why managers find the systems set up to meet external financial reporting needs inadequate for internal managerial decision making.

Having described the place of the chart of accounts within the system, let us examine more closely its account classifications and coding systems.

(a) **Account Classifications.**  As noted previously, financial statements and reports are needed by a wide variety of external, as well as internal, users. Thus the listing of accounts should include all accounts required by law and by tax regulations and other legal obligations. For instance, public utilities are required to have specific accounts prescribed by the Federal Power Commission and by state utility regulatory bodies.

The accounts are also affected by the activities and customs of the industries within which firms operate. Thus, firms that manufacture goods need three inventory accounts and one supplies account (at least), whereas firms that provide services need only supplies accounts. Service firms, however, must focus on the performance and pricing of services. Not-for-profit organizations differ from both manufacturing and service firms in that they are more concerned with costs than with revenues.

With respect to the internal users such as managers, the accounts should be sufficiently detailed to aid in planning and control functions. For example, the revenues of an automobile dealer could be separated into accounts for new car sales, used car sales, parts sales, and repairs services. To provide sufficient details for managers, the number of accounts maintained within a chart of accounts is generally greater than the items appearing in the financial statements.

Although a chart of accounts pertains directly to asset and equity accounts (including revenues and expenses), it is closely tied to a firm's organizational structure and segments. As we have noted, an expense account within the chart of accounts can be linked to responsibility centers. We will consider this expanded view of the chart of accounts in the following discussion of account coding.

(b) **Account Coding.**  The coding system used to form the broad framework within a chart of accounts is usually of the block type. For instance, current assets may be assigned numbers from 100 to 199, noncurrent assets from 200 to 299, liabilities may be 300 to 399, while revenues could be 500 to 599, product expenses may be 600 to 699, 700 to 899 could be operating expenses, while 900 to 999 may be the nonoperating expenses. Each block may contain subordinate blocks. Consider the operating expenses, which are blocked from 700 through 899. Within this broad block could be subordinate blocks representing the functional categories of expenses such as administrative travel. The particular accounts and the details they represent are based on the various needs of internal and external users.

A group coding system can provide added meaning to individual account codes. The left-most digit can represent a major category (e.g., asset), the middle digit an intermediate classification (e.g., plant assets), and right-most digit a minor classification (e.g., the type of plant asset). For example, the account code 112 may refer to the type of plant asset known as buildings. This example of a code is similar in construction to the hierarchical group code employed with responsibility centers. Also, it is interesting to observe the close relationship between block and group codes.

Group codes are often employed as the coding structure for general ledger accounts, since they allow several facets to be incorporated. For instance, direct labor expense in a production department might be coded 761–1333. Other coding fields might also be added, relating to subsidiary ledgers, locations, and products. Group codes that capture these facets of transactions aid the preparation of useful analyses.

An illustrative group code having fields pertaining to general ledger accounts, subsidiary ledger accounts, and responsibility centers would appear as follows:

AAA-BBBB-CC

A sales transaction coded according to this format could show a code 121–563–00 to identify the debit and the code 820–1738–08 to identify the credit. The 121 and 820 refer to the general ledger accounts entitled accounts receivable and sales. The 5634 refers to the customer account against which the credit sale is charged, and the 1738 refers to the salesperson whose account is credited in the commissions payable ledger. The 08 refers to the sales region (responsibility center) where the sale is made. Finally, the 00 indicates that accounts receivable are general in nature, and not applicable to a particular responsibility center.

The Appendix represents a coded chart of account for a mid-sized firm. Through it one can gain further understanding of how ledger accounts are coded.

## 15.10 CONCLUSION

General ledger systems are involved in collecting, classifying, coding, processing, storing transaction data, maintaining accounting controls, and generating financial reports. The general ledger system is also called the general ledger and financial reporting cycle, which is at the center of such transaction-oriented cycles as the revenue, expenditure, and resources management.

The general ledger system receives routine recurring and nonrecurring transactions, non-routine transactions, adjusting entries, reversing entries, and closing entries. The inputs are normally on general ledger journal vouchers, which in computer-based systems may be transferred to batch entry forms or to data entry screens. The entered data, in journal entry form, are then posted to the general ledger to update the accounts during the accounting period. At the end of each accounting period standard and nonstandard adjusting journal entries are posted. The exact methods of processing vary, depending on whether the processing is performed by manual or computer-based systems. The database includes files such as the general ledger master file, general ledger history file, responsibility center master file, budget master file, financial reports format file, current journal voucher file, and journal voucher history file.

A variety of general and transaction controls are needed to offset a number of risks to which the general ledger is exposed. These controls include adequate documentation, organizational segregation, data security, carefully designed and prepared journal vouchers, and periodic trial balances. Financial reports typically generated by the general ledger system include analyses of general ledger accounts, financial statements, and managerial reports such as sales analyses and responsibility reports. Many of the managerial reports involve comparisons between budget and actual amounts.

The chart of accounts is a coded listing of all of the classified accounts of a firm. Charts of accounts vary widely among firms, since they must satisfy the needs of both external and internal users.

## APPENDIX: TOTAL QUALITY APPLIANCE CORPORATION—CLASSIFICATION OF ACCOUNTS

### *BALANCE SHEET ACCOUNTS*

**Current Assets**
**10100** Petty Cash
**10200** Cash in Bank
**11100** Accounts Receivable–Customers
**11100A** Allowance for Uncollectible Accounts
**11200** Accounts Receivable–Officers and Employees
**11300** Sales/Representative's Drawing Accounts
**11400** Notes Receivable
**11400D** Notes Receivable Discounted
**11500** Accrued Interest Receivable
**12000** Merchandise Inventories
**12100** Kitchen Appliances
**12200** Laundry Appliances
**12300** Other Appliances
**13100** Prepaid Rent–Outside Warehousing
**13200** Prepaid Insurance
**14000** Supplies Inventories
**14100** Office Supplies
**14200** Warehouse and Shop Supplies
**14300** Advertising Supplies

**Property, Plant, and Equipment**
**15000** Land
**16100** Building
**16100A** Allowance for Depreciation–Building
**16200** Warehouse Equipment
**16200A** Allowance for Depreciation–Warehouse Equipment
**16300** Delivery Trucks
**16300A** Allowance for Depreciation–Delivery Trucks
**16400** Service Shop Equipment
**16400A** Allowance for Depreciation–Service Shop Equipment
**16500** Service Trucks
**16500A** Allowance for Depreciation–Service Trucks
**16600** Office Furniture and Fixtures
**16600A** Allowance for Depreciation–Office Furniture and Fixtures

**Current Liabilities**
**17100** Vouchers Payable–Appliance Suppliers
**17200** Vouchers Payable–Other
**17300** Accrued Payroll
**17500** Notes Payable
**17600** Bank Loans
**17700** Accrued Interest Payable
**17800** Dividends Payable
**18100** Accrued Payroll Taxes
**18200** Withholding Tax Deducted
**18300** Federal Income Tax Accrued
**18400** Accrued Real Estate Taxes
**18500** Accrued Personal Property Taxes
**18600** Accrued Sales Tax–Illinois
**18700** Accrued Sales Tax–Other

**Reserves**
**18900** Reserve for Warranty

**Long-Term Liabilities**
**19000** Mortgage Payable

**Net Worth**
**19500** Capital Stock
**19600** Retained Earnings
**19900** Profit and Loss

## INCOME STATEMENT ACCOUNTS

**Sales and Purchases**
**21000** Sales and Service Income
**21100** Sales of Kitchen Appliances and Parts
**21200** Sales of Laundry Appliances and Parts
**21300** Sales of Other Appliances
**21400** Service Income–Inside
**21500** Service Income–Outside
**21600** Service Contracts
**21700** Labor Used on Service Contracts (debit)
**22000** Sales Returns and Allowances
**22100** Kitchen Appliances and Parts
**22200** Laundry Appliances and Parts
**22300** Other Appliances
**23000** Purchases
**23100** Kitchen Appliances and Parts

**23200** Laundry Appliances and Parts
**23300** Other Appliances
**24000** Freight In
**24100** Kitchen Appliances and Parts
**24200** Laundry Appliances and Parts
**24300** Other Appliances
**25000** Purchase Returns and Allowances
**25100** Kitchen Appliances and Parts
**25200** Laundry Appliances and Parts
**25300** Other Appliances

**Administrative Expenses**
**30000** Administrative Expenses Control
**31100** Administrative Salaries
**31200** Office Salaries
**31900** Overtime Premium
**32100** F.I.C.A. Tax Expense
**32200** Unemployment Insurance Expense
**33100** Stationery and Office Supplies
**34100** Repairs to Office Furniture and Equipment
**36100** Postage
**36200** Telephone and Telegraph
**36300** Association Dues
**36400** Professional Services
**36500** Donations
**37000** Travel
**38000** Unclassified
**39100** Rentals of Office Equipment
**39200** Depreciation–Office Furniture and Equipment
**39300** Insurance–Office Furniture and Equipment
**39400** Taxes–Office Furniture and Equipment

**Selling Expenses**
**40000** Selling Expenses Control
**41100** Sales Supervisory Salaries
**41200** Inside Salesmen
**41300** Outside Salesmen
**41400** Commissions
**42100** F.I.C.A. Tax Expense
**42200** Unemployment Insurance Expense
**43100** Stationery and Office Supplies
**45100** Newspaper and Magazine Advertising
**45200** Dealer Aids

**45300** Convention Expense
**45400** Advertising Allowances to Dealers
**46100** Association Dues
**46200** Professional Services
**46300** Freight Out
**47100** Automobile Expense
**47200** Other Travel and Entertainment
**48000** Unclassified Expense
**49000** Warranty Expense

### Receiving, Warehousing, and Shipping Expenses
**50000** Receiving, Warehousing, and Shipping Expenses Control
**51100** Supervisory Salaries
**51200** Warehouse/Employees
**51300** Truck Drivers and Helpers
**51900** Overtime Premium
**52100** F.I.C.A. Tax Expense
**52200** Unemployment Insurance Expense
**53100** Packing Supplies
**54100** Repairs to Warehouse Equipment
**54200** Repairs to Trucks
**54300** Gasoline and Oil
**58000** Unclassified Expenses
**59100** Rent–Outside Warehousing
**59200** Depreciation–Warehouse Equipment
**59300** Depreciation–Trucks
**59400** Insurance–Warehouse Equipment
**59500** Insurance–Inventories
**59600** Insurance–Trucks
**59700** Taxes–Warehouse Equipment
**59800** Taxes–Inventories
**59900** Taxes and Licenses–Trucks

### Occupancy Expenses
**60000** Occupancy Expenses Control
**61100** Supervisory Salaries
**61200** Janitors and Security Staff
**61900** Overtime Premium
**62100** F.I.C.A. Tax Expense
**62200** Unemployment Insurance Expense
**64100** Janitor Supplies
**64200** Repairs to Building
**66100** Heat, Light, and Air Conditioning
**69100** Depreciation–Building

**69200** Insurance–Building
**69300** Taxes–Building

**Service Department–Inside Expenses**
**70000** Service Department–Inside Expenses Control
**71100** Supervisory Salaries
**71200** Repair Workers
**71900** Overtime Premium
**72100** F.I.C.A. Tax Expense
**72200** Unemployment Insurance Expense
**73000** Operating Supplies
**76100** Power
**79100** Depreciation–Service Shop Equipment
**79200** Insurance–Service Shop Equipment
**79300** Taxes–Service Shop Equipment
**79900** Labor Used on Service Contracts (Credit)

**Service Department–Outside Expenses**
**80000** Service Department–Outside Expenses Control
**81100** Service Employees
**82100** F.I.C.A. Tax Expense
**82200** Unemployment Insurance
**83000** Operating Supplies
**84100** Repairs to Service Trucks
**84200** Gasoline and Oil–Service Trucks
**88000** Unclassified
**89100** Depreciation–Service Trucks
**89200** Insurance–Service Trucks
**89300** Taxes and Licenses–Service Trucks
**89900** Labor Used on Service Contracts (Credit)

**Other Income and Expense**
**91100** Purchases Discounts
**91200** Interest Income
**95100** Sales Discounts
**95200** Interest Expense
**95300** Warranty Expense (Labor and Parts)
**95400** Bad Debts

# CHAPTER 16

# THE REVENUE CYCLE

## 16.1 INTRODUCTION

This chapter will provide an understanding of the working of the revenue cycle of a firm, as well as the documentation used to carry out the functions associated with it. Furthermore, since realizing revenue involves the entire organization in various roles, this chapter will outline the involvement of various departments with respect to the functions performed during the revenue cycle and the information needs of those involved. Key activities and data processing operations included in the revenue cycle are also detailed. The flowcharting of data and information flows in typical sales order processing systems will be highlighted, along with information that prepares managers and accountants to evaluate and recommend control policies and procedures for a sales order processing system. In addition, this chapter discusses data files used for the cycle, typical accounting entries used, and coding used to facilitate processing.

The revenue cycle of an organization seeks to facilitate transactions that exchange products or services for cash. Most organizations depend on the revenue they raise through such transactions for their survival. The functions related to revenue generation, either through providing services or products, comprise the revenue cycle of an organization.

Specifically, the revenue cycle for a business includes: soliciting customer orders, executing sales transactions, and delivering products or services to the customers. The revenue cycle also involves the following data processing operations: sales order processing, billing, updating inventory records, and maintaining customer account records. Additionally, it is one of the business cycles that interfaces with the general ledger and financial reporting system. Information about the revenues earned is a major component of the financial reports periodically issued by a firm.

In other words, the revenue cycle consists of both transaction-related and data-related functions. But the efforts needed to earn revenue cannot be limited to elements specifically related to the revenue cycle. Indeed, the entire organization has to be involved if the business is to continue, and every function affects the continued ability of a firm to earn revenue. While showing how the involvement of the entire organization results in revenue may be difficult, we will describe the involvement of some of the departments in the revenue cycle and the information needs of those involved.

## 16.2 ORGANIZATIONAL CONTEXT

One department that plays a major role in the revenue cycle is the marketing staff. Marketing managers are usually charged with functions such as establishing policies with respect to base prices, discounts, credit terms, return policies, and warranties. In addition, they participate in staff decisions such as those pertaining to advertising and new product development. They also have to review and evaluate the performance of marketing personnel. Their time is taken up, to a large extent, with allocating key resources, such as people and funds, to achieve marketing goals both strategic and operational.

These functions are facilitated by the information available to them. To perform their duties, marketing managers usually need information pertaining to the following: economic trends, competitors' prices, national and international sociopolitical developments, sales forecasts, market research data, product costs, and costs of returns, credit, and warranties.

Sales personnel are also heavily involved in the revenue cycle. Their duties include: establishing standards and quotas (with the help of others such as marketing managers), selecting effective distribution channels, and reviewing and evaluating performance of sales personnel. Sales managers need information pertaining to: sales forecasts, sales analyses (breakdown of sales by category), profitability analyses (showing marginal contribution to profit), operational data on individual salespeople, including information about sales calls made, the ratio of calls made to sales concluded, and inventory availability.

The advertising and promotion department is also heavily involved in the revenue cycle of a business. Their duties include: creating advertising campaigns, establishing dealer incentives, and managing other promotional activities. Key information needed by the advertising and promotion department to perform those duties are: sales forecasts, profitability analyses, surveys of customer attitudes, advertising costs, and measures of advertising effectiveness.

The department of product planning also affects the revenue function. Its duties include: planning characteristics of product lines, styling and packaging, and introducing new products and reviewing performance of existing products. And the information needed to perform those duties includes: sales analyses, profitability analyses, product costs, and customer attitudes.

Finally, customer service affects the ability to keep earning revenue. Those involved in customer service are concerned with: setting policies and making decisions relating to customer needs, providing retailers with product availability and pricing data, providing customers with warranty information, and dealing with complaints. Key information needed to perform those duties include some knowledge of customer attitudes as well as costs of servicing customers.

Judging from the above survey, it is clear that the revenue cycle requires the involvement of a variety of departments. In order for managers in such departments to make better decisions, both accounting and non-accounting information is needed. To provide such information, a great deal of data pertaining to revenue function must be made available. This is the function of accounting information systems.

Given the number of steps involved, it is better to treat the revenue cycle as made up of three interrelated sub-cycles: credit sales, cash receipts, and accounts receivable maintenance. Such is the case in this book. We also differentiate between the manual and computerized systems pertaining to the three above-mentioned subsets of the revenue cycle.

## 16.3 DOCUMENTATION OF THE REVENUE CYCLE

In this section, the documentation needed for a manual revenue cycle is described. The revenue cycle source documents typically found in firms that employ manual processing and make product sales include:

*Customer order.*   Often the customer order is the customer's purchase order, and thus not a document prepared by the selling firm. However, it may be a form prepared by the salesperson that lists information that identifies the customer as well as the items ordered, together with their unit prices and discounts provided, if any. The form also identifies the person taking the order.

*Sales order.*   Although similar to the customer order, a sales order has significant differences. First, it is often a more formal, multicopy form that is prepared from the customer order. Secondly, it is prenumbered for more effective control. Finally, it often contains price and extension columns, so that it can be completed to serve as the invoice. In some firms the sales order is called the shipping order, since it provides authorization for the shipping action.

*Order acknowledgment.*   Usually, the order acknowledgment is a copy of the sales order, although it may be a separate form. Sometimes the customer also requires that the selling firm return a signed acknowledgment that has been prepared by the customer.

*Picking list.*   In some cases, a copy of the sales order is sent to the warehouse for use in picking the ordered goods from the bins. Alternatively, a separate picking list document may be prepared. This document specifies the goods to be picked and authorizes their removal. The ordered product data are often arranged in a picking list in accordance with the bin locations in the warehouse. Thus, the picking can be done more efficiently.

*Packing list or slip.*   The packing list is enclosed with the goods when they are packaged. It is generally a copy of the sales order, or it may be the picking list.

*Bill of lading.*   A document such as a bill of lading is relatively uniform from firm to firm. It is intended for the agents of the common carrier that is to transport the products, informing them that goods are legally on board the carrier, that the freight has been paid or billed, and that the consignee is authorized to receive the goods at the destination. In addition to the carrier, both the shipping department and the customer receive copies. Another copy may serve as the freight bill (invoice) and be forwarded to the traffic department (if any) of the customer or seller, depending on who is paying.

*Shipping notice.*   Often a copy of the sales order, when duly noted by the shipping manager, serves as proof that the goods were shipped. When a common carrier is involved, a copy of the bill of lading can serve this purpose. In other cases a separate prenumbered shipping notice, also called a shipping report, may be prepared by the shipping department. This notice is forwarded to the billing department and is to be used for completing the invoice sent to the customers.

*Remittance advice.*   The remittance advice is a counterpart to the sales invoice, since it contains the amount of the cash receipt from the customer and the payor's name. It can be described as a turnaround document, since it represents a portion of the sales invoice that is returned by the customer with the cash. If the customer does not include a cash remittance, the firm must prepare one for use as the posting medium.

*Deposit slip.*   Deposits of cash in the bank must be accompanied by deposit slips. A deposit slip usually contains imprints of both the firm name and the bank name. Coding at the bottom of the slip refers to the account number and the bank code.

*Backorder.*    A form called a backorder is prepared when insufficient quantities are in inventory to satisfy sales orders. It should be prenumbered and contain data concerning the customer for whom the backorder is being placed, the original sales order number, the quantity needed, and the date requested. If the original order is partially filled and the remaining quantities are backordered, the backorder number and relevant data should be entered on the sales invoice. These backorder data has also likely been posted to the inventory records. When the backordered items are received from the supplier, they will be immediately shipped to the customer and the notation will be removed from the inventory records. A new sales invoice will also be prepared for the backordered items and mailed to the customer.

*Credit memo.*    Before ordered goods can be returned or before allowances can be granted, a prenumbered credit memo must be prepared and approved. When filled in, it will state that a customer's account balance will be credited by an indicated amount (e.g., $100). On receipt by the customer, it will serve that firm as the posting medium to reduce the balance in the accounts payable account. A credit memo relating to a return should be approved only on the basis of clear-cut evidence, such as a sales return notice, that lists the physical count of returned goods by the receiving department.

*Other documents.*    A new-customer credit application is useful when customers apply for credit. It should include all of the data pertaining to the applicant's current financial condition and earning level. A salesperson call report may be used to describe each call on a prospective customer and to indicate the result of the call. A delinquent notice may be sent to customers who are past due on their credit account balances. A write-off notice or form is a document prepared by the credit manager when an account is deemed to be uncollectible. A journal voucher is prepared as the basis for each posting to accounts in the general ledger. For instance, a journal voucher may be prepared to reflect one or more account write-offs. Finally, retail firms that make cash sales use receipted sales tickets or cash register receipts to reflect cash received.

## 16.4 DOCUMENTATION FOR COMPUTER-BASED SYSTEMS

All of the above-mentioned hard-copy documents may also be used in computer-based systems. If so, they should be designed to speed the entry of data into the computer system with few errors. Preformatted screens may be used to expedite data entry pertaining to sales orders, sales returns, and cash receipts transactions. Such screens may be designed to handle individual transactions or batches of transactions. They do not differ greatly in format from hard-copy forms. With slight changes, for instance, hard-copy forms could be employed as preformatted screens. Also, on-line computer systems may generate hard copies of documents after data have been entered via preformatted screens.

The on-line entry of data is becoming increasingly common. Note that the underlying computer program aids data entry by performing several functions automatically. Thus the date and batch number are entered when the preformatted screen is accessed. Similarly, data for a number of fields may be retrieved from the open sales invoice and customer files, which the program searches by reference to the entered customer and invoice numbers. Also, the total of all payment amounts entered is computed and placed in the field designated on the screen. In addition, the entered total and previously computed control total are compared by the program, and the difference is displayed.

## 16.5 JOURNAL ENTRIES USED FOR THE REVENUE CYCLE

Sales transactions give rise to their share of general entries. Each sale results in an invoice, and every invoice leads to entries into a sales journal, as well as to individual customer accounts in the accounts receivable subsidiary ledger. Periodically, summary totals in the sales journal are posted to the general ledger accounts as follows:

1. Dr. Accounts Receivable XXX

   Cr. Sales XXX

   To record the total of credit sales.

If the inventory is maintained on a perpetual basis, the following entry would be recorded on a journal voucher to accompany each daily sales entry:

2. Dr. Cost of Goods Sold XXX

   Cr. Merchandise Inventory XXX

   To record the costs of those goods sold during this date.

If necessary, journal vouchers are prepared to reflect sales returns and allowances or write-offs via the following entries:

3. Dr. Sales Returns and Allowances XXX

   Cr. Accounts Receivable XXX

   To adjust customer account balances to reflect returns and allowances on previous sales.

4. Dr. Allowance for Doubtful Accounts XXX

   Cr. Accounts Receivable XXX

   To write off those customers' account balances that are deemed to be uncollectible (assuming that the allowance method for establishing bad debt expense is used).

The flow of cash receipts is also a source of transactions flowing through the accounting cycle. Checks received from customers initiate most cash receipts transactions. The amounts of the checks are entered from accompanying remittance advice into a cash receipts journal and are posted to the accounts receivable subsidiary ledger. Periodically, totals from the journal are posted to the general ledger by the following entry:

1. Dr. Cash XXX

   Cr. Accounts Receivable XXX

   To record daily cash receipts.

Cash may be received from sources other than customer accounts. If so, it may be recorded as a part of the above entry, or it may be reflected in a separate entry like the following:

2. Dr. Cash XXX

Cr. Notes Payable XXX

Cr. Interest or Dividend Income XXX

To record cash received from sources other than customers.

## 16.6 TRANSACTION CODING

Codes are essential for identifying key aspects of sales and cash receipts transactions. Codes may be assigned to customers, sales territories, salespersons, and product types when recording sales transactions. Codes such as these reduce the quantity of data entered and provide unique identifiers. They also facilitate the preparation of sales analyses. Based on the above-mentioned example, sales analyses could show amounts of sales made to customers within each sales territory and amounts of sales made of the various products by each salesperson.

Codes that describe transactions should be group codes, since they encompass several characteristics. In turn, the code for each characteristic could incorporate several features. For instance, a particular customer code might be R9174248, based on the format ABBCCCCD, where:

- A represents the class or type of customer
- BB represents the year the customer became active
- CCCC represents the specific customer identifier
- D represents a self-checking digit

On-line computer system codes that identify transactions are often used. For instance, transaction codes may be entered together with the remaining data for transactions, so that the system knows which application program to employ. Transaction codes may be developed for sales, sales returns or allowances, and other prospective types of transactions. These other transaction codes may not be needed if preformatted screens are accessed via menus, since menu-driven systems designate the suitable application programs through the menu selections.

## 16.7 DATABASE FILES

The database for the revenue cycle contains master files, transaction files, open document files, and other types of files. If appropriate database software is available, it may also contain data structures. Two likely master files employed within the revenue cycle pertain to customers and accounts receivable. If a firm has product sales, a third file is the merchandise inventory master file. It will not be needed if the firm offers services only. The content of these files will vary from firm to firm, depending on factors such as types of customers and markets, variety of desired managerial reports, and degree of computerization. Our survey of these files is necessarily brief.

*Customer master file.* A customer master file contains records pertaining to individual credit customers. Generally, its primary and sorting key will be the customer number. Each record of this file contains customer data such as the shipping and billing addresses, telephone number, past payment performance, credit rating, trade discount allowed, and sales activity. These data items (i.e., elements) are useful in preparing sales invoices and monthly statements, as well as in determining the credit limit.

*Accounts receivable master file.* The records in an accounts receivable master file also relate to credit customers. They represent the computerized counterpart to the accounts receivable subsidiary ledger sheets in a manual system. Two data items are essential, the customer identification (usually the customer account number) and the current account balance. Other data items are optional, and they may include: credit limit, balance at the start of the year, year-to-date sales, and year-to-date payments. For firms using the open invoice method, data concerning specific unpaid invoices are also needed.

The data from both of the above-mentioned files are needed for preparing outputs such as customer monthly statements. If desired, however, the added items could be moved to the customer file. Alternatively, all data items pertaining to a customer, including all transactions for this year, could be consolidated into the accounts receivable file. The most important reason for separating the customer and accounts receivable files is to reduce the size of records during updates and, hence, the processing run times.

*Merchandise inventory master file.* A merchandise file (or finished goods file) is relevant to the revenue cycle for a firm that sells products. Data items that might appear in a record layout include the product (inventory item) number, description, warehouse location code, unit of measure code, reorder point, reorder quantity, unit cost, quantity on order, date of last purchase, and quantity on hand. If inventory is maintained on the perpetual basis, the current balance will also be included. The primary and sorting key is usually the product or inventory item number.

*Transaction and open Document files.* Transactions in the revenue cycle involve sales orders, sales invoices, shipping reports, credit memos, backorders, and cash receipts. Each of the records in related transaction files would contain roughly the data shown in the documents discussed earlier. Three of the major transaction files require closer attention.

*Sales order file.* The content of the sales order files is similar to the data elements contained in accounts receivable master file. When stored in a record on a computerized medium, however, elements such as the customer name and product descriptions may be omitted. They may be drawn from the customer and merchandise inventory files when preparing sales invoices. The record may allow for repeating line items such as product numbers and quantities ordered. Alternatively, these line items may be placed in a separate line item file and cross-referenced. If this file consists of orders that have not yet been shipped and billed, it is known as an open sales order file. The sorting and primary key of the file is usually a preassigned order number. When an order is shipped and billed, it is usually transferred to a closed orders file.

*Sales invoice transaction file.* In a manual system, a sales invoice file consists of a copy of each current sales invoice. Records in this file provide the details of sales transactions posted to the accounts receivable records; by maintaining them in a separate file, the size of the accounts receivable records can be reduced. In a computer-based system a printed copy may or may not be filed. The record stored on magnetic media essentially contains the sales date, customer number, sales order number, terms, product codes, unit prices, quantities, and transaction amount. It likely omits customer names and product data, since the data are available in other files. The primary (and sorting key) is likely the sales invoice number. Each record remains in an open sales invoice file until payment is received from the customer (or until the end of the period if the balance forward method is used).

**Cash receipts transaction file.**   In a manual system a cash receipts transaction file likely consists of a copy of each current remittance advice. In a computer-based system the record layout on magnetic media may contain the customer's account number, the sales invoice number against which the payment is being applied, date of payment, and amount of payment. It also includes a code to identify the record as a cash receipt transaction, and it may be assigned a transaction number.

**Other Files.**   In addition to files described above, some systems may have a shipping and price data reference file. It will be a reference file that may contain such shipping data as freight rates, common carrier routes and schedules, current prices of all products, trade discounts, and so on. Another reference file, especially in manual systems, is a credit file, used to check and approve the credit of customers. Additionally, systems may also have files to track sales history and accounts receivable details.

**Sales history file.**   A sales history file contains summary data from sales order invoices. In a computer-based system, the records pertaining to sales orders and invoices are transferred to this file when they are removed from the open files. These records are retained in the history file for a reasonable period. For instance, a firm may decide to maintain a history file for the five years past. Records older than five years would be purged from the file. However, the firm that employs computer-based processing may retain printed copies of sales orders and invoices for a longer period. Data from this file are used to prepare sales forecasts and analyses. If desired, a cash receipts history file could also be maintained.

**Accounts receivable detail file.**   In addition to creating separate sales invoice and cash receipts transaction files, the application program may accumulate records of the transactions in a detail file. Since it is restricted to the current period's transactions, it is not a replacement for the history files. Instead, its main purpose is to facilitate the preparation of monthly statements for customers. Also, it can serve as an audit trail.

## 16.8 THE WORKING OF THE REVENUE CYCLE

For the sake of convenience, the data flows and processing steps of a revenue cycle are divided into three major subsets: processing of sales transactions, processing of cash receipts transactions, and maintenance of the accounts receivable ledger. Exhibit 16–1 depicts these subdivisions and it also lists related documentation. We will take up the three subsets of a manually processed revenue cycle and then discuss their counterparts in a computer-based system. Understanding the work flows in a manual system will help make clear the processing steps performed by computer.

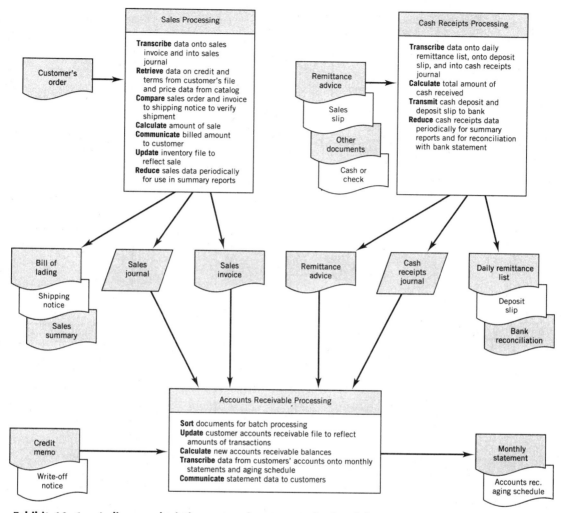

**Exhibit 16–1.   A diagram depicting processing steps and related documents for the revenue cycle.**
Adapted from J. Wilkinson, *Accounting Information System: Essential Concepts & Applications,*
2nd. ed. (NY: John Wiley & Sons).

## 16.9 CREDIT SALES PROCEDURES IN MANUAL PROCESSING SYSTEMS

To help readers get oriented with the procedures involved in the manual processing
system for the revenue cycle, Exhibit 16–2 presents a document flowchart of procedures
involving the credit sales of products. Since the processing is performed manually, the
emphasis is on the flows of source documents and outputs. Although the flowchart is
detailed and appears to be quite complex, it provides an overall view of a sound sales
procedure.

1. The receipt of an order by the order department leads to the preparation of a sales
   order, which should be prenumbered. Customer orders can be received by tele-

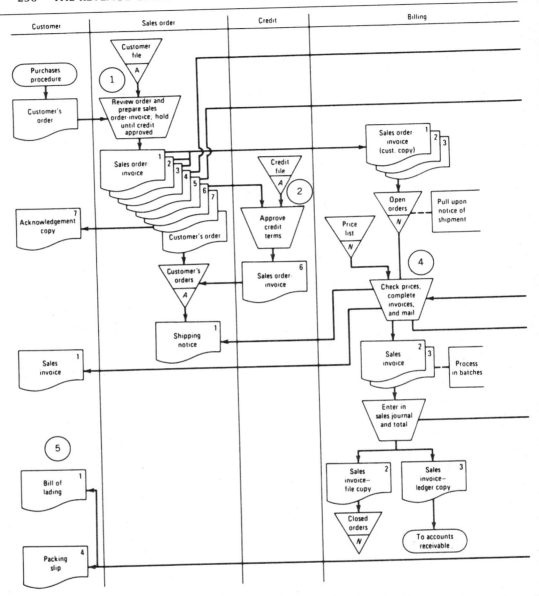

**Exhibit 16–2.** A document system flowchart of a manual credit sales transaction processing procedure. Adapted from J. Wilkinson, *Accounting Information System: Essential Concepts & Applications*, 2nd. ed. (NY: John Wiley & Sons).

phone, mail, sales agent, and electronically. This sales order form gets distributed across many departments. To facilitate such distribution, a sales order can have several copies. Here the form is assumed to consist of seven copies. While preparing the document, the sales order clerk will refer to the customer master file to obtain pertinent information such as credit history.

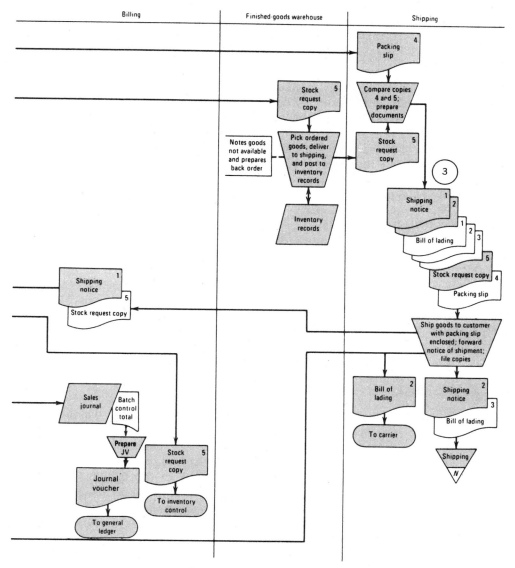

**Exhibit 16–2.** (*Continued*).

**2.** Credit check is the next step. A copy (#6) of the sales order is sent to the credit department. For long-standing customers this is done by utilizing a credit limit check, while for new customers, or for sales above the limit, formal credit approval is needed. If the credit is approved, the sale is authorized. The copy of the sales order is sent back to the sales order department. Copies of the sales order are then distributed as follows:

**A.** Copies #1, 2, and 3 are sent to the billing department to await the notice of shipment before additional steps can be taken with respect to their subsequent use.

**B.** Copy #4 is sent to shipping to allow them to anticipate the arrival of goods from the warehouse.

**C.** Copy #5 of the sales order is sent to warehouse, where it will lead to the preparation of a picking list.

**D.** Copy #6 is made a part of the customer order file.

3. Shipping of ordered goods. This is initiated at the finished goods warehouse, which has custody of the merchandise. After getting the copy #5 of the sales order, the staff in the warehouse will assemble the order. They may use the copy of the order itself or another document, the picking list, to put together the goods needed for the order. If some of the ordered goods are not available, that absence is noted on the order. Subsequently, the merchandise and the copy #5 are delivered to the shipping department. In the shipping department, they will compare the goods delivered to them, copy #5, and copy #4 of the sales order. After such checking, the order is shipped to the customer together with copy #4 of the sales order. In some cases the shipping will also adjust the inventory record, although it is not usually feasible in a shipping department, given the nature of their environment. The shipping department also prepares the shipment-related documentation, such as the bill of lading. The bill of lading can have several copies. Here, it is a document with three copies. The copy #1 is for customer, #2 is for the carrier, and #3 is for the shipping department's records.

4. Billing for goods sold and shipped. After the goods are shipped, the shipping department notifies the billing department. There a clerk will compare the quantities shipped with the quantities ordered, and enter onto the invoice the unit prices from a current pricing list. At times this may be checked by another individual for accuracy. A third individual may then complete the amounts to be billed, and then accumulate enough invoices to form a batch and compute the batch totals. The batch of sales invoices is then sent to the accounts receivable department for posting. There a journal voucher is prepared for posting to the ledger. This procedure is further explained in the section "Accounts Receivable Maintenance Procedure."

5. Receipt of the order by the customer. The customer receives the goods shipped together with a packing slip, actually a copy of the sales order, as well as a copy of the bill of lading. Prior to the receipt of the order itself, customers usually receive a copy of the sales order, which confirms the order placed. Some time after receiving the shipment, the customer will receive an invoice proper listing the amount owed to the seller. It is usual for the customer to compare the invoice, the packing slip, and the receiving report prepared by the customer's receiving department after the shipment arrived.

Before leaving this subset, it may be helpful to understand the disposal of the sales order copies. This will also serve as a useful review of an involved process:

1. Copies #1, 2, and 3 get sent to billing. After additional processing there, #1 becomes the sales invoice and goes to the customer, #2 becomes the sales invoice file copy and remains in the closed orders file, while #3 becomes the sales invoice ledger copy and is sent to accounts receivable.

2. Copy #4 is sent to shipping, often becomes a packing slip and goes to the customer with the order.

3. Copy #5 goes first to the warehouse (stock request copy), then goes back to the billing department, and finally to the inventory control department.

4. Copy #6 goes to the credit department and then becomes part of the customer order file.

5. Copy #7 goes to the customer as an acknowledgement of the order received.

There can be variations. The invoice received by customer, copy #1, could itself have a copy that the customer can keep for his/her records.

## 16.10 CASH RECEIPTS PROCEDURE

As noted earlier, cash receipts are seen as an independent subset of the revenue cycle. Exhibit 16–3 shows a flowchart of a procedure involving the receipts of cash related to

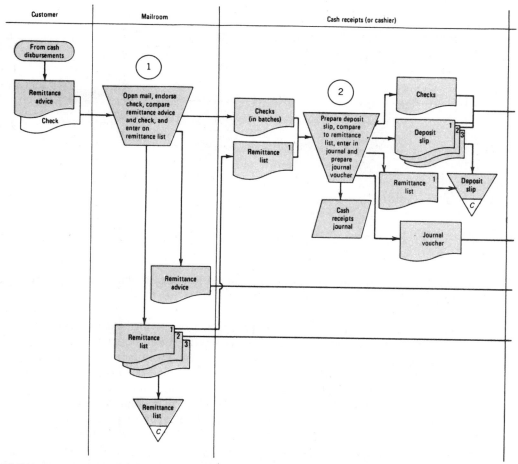

**Exhibit 16–3.** A document system flowchart of a manual cash receipts transaction processing procedure. Adapted from J. Wilkinson, *Accounting Information System: Essential Concepts & Applications*, 2nd. ed. (NY: John Wiley & Sons).

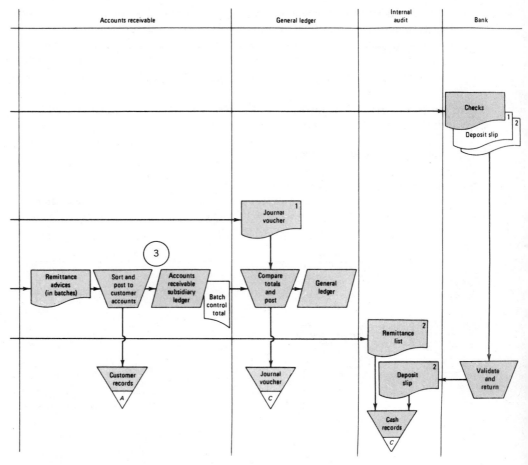

**Exhibit 16–3.**   *(Continued).*

sales. The principal control points involve the receipt of cash, the processing of the cash deposit, and the posting of cash amounts to the ledgers.

1. **Receipt of cash.** The cash receipts procedure begins with the daily receipt of mailed cash and remittance advices from customers. An authorized person, such as a mailroom clerk, compares the checks with the remittance advices and prepares advices when none are received. Then the clerk endorses/stamps the checks "For Deposit Only," and enters their amounts on a prenumbered remittance list, and computes a total of the batch received. One copy of the remittance list (also called a "prelist") is sent to the cashier with the checks; a second copy is sent to the internal audit department (if any) for later reviews; the third copy is filed.

2. **Processing of the cash deposit.** A person who is authorized to handle the cash, such as the cashier, prepares a deposit slip in triplicate. All checks from customers, plus cash received from other sources that day, are listed on the deposit slip. After the cashier compares the computed deposit total with that shown on the remittance list,

he or she delivers the deposit to the bank intact. A cash receipts clerk then enters the total amount of receipts in the cash receipts journal. The clerk prepares a journal voucher, which is sent to the general ledger department for posting. The internal audit department receives an authenticated copy of the deposit slip, which has been stamped and initialed by a bank teller and delivered directly by the bank. This deposit slip is compared to the remittance list, as well as to the deposit slip in the cashier's office and to the general ledger posting.

3. Posting of cash amounts. After preparing the remittance list, the mailroom clerk forwards the remittance advices to the accounts receivable department for posting. The cashier sends a journal voucher to the general ledger department.

## 16.11 ACCOUNTS RECEIVABLE MAINTENANCE PROCEDURE

The third subset of the revenue cycle pertains to updating accounts receivable. The accounts receivable subsidiary ledger is the link between the sales and cash receipts procedures. Exhibit 16–4 illustrates the steps related to posting sales transactions to the accounts receivable ledger. It parallels the figure describing the cash receipts subset. Both figures include the posting step to the general ledger. In addition, Exhibit 16–4 shows the procedures relating to sales returns and allowances and write-offs.

- Posting to accounts receivable accounts. On receiving copies of the sales invoices from the billing department, a clerk in the accounts receivable department posts the credit sales amounts to the customers' accounts. Another clerk verifies the posting and obtains a total of the amounts posted. This clerk then forwards customers' accounts. Sales postings need verification of their accuracy, as do the total of the amounts posted. The batch total is then forwarded to the general ledger department.

- Posting to general ledger accounts. In the general ledger department a clerk compares (for each type of transaction) the total posted to the precomputed batch total amount, as shown on a summary journal voucher. If they agree, the clerk posts the totals to the general ledger accounts. If they disagree, the clerk locates discrepancies, corrects the errors, notifies the accounts receivable clerk of posting errors, and then completes the general ledger postings.

- Processing of sales return and allowances. When goods are returned from customers or if they are to be given allowances, a manager must authorize the transaction. If a return of merchandise previously sold is involved, it is received in the receiving department, where a clerk counts and lists them on a sales return notice. If there was no prior authorization for the return, the receiving department will forward a copy to the credit manager for approval. The approved notice (or a return notification, if issued) is transmitted to the billing department, where the prices are checked against the original sales invoice. Then a prenumbered credit memo is prepared, with copies being sent to the accounts receivable department for posting and to the customer. A clerk in the billing department also prepares a journal voucher for the general ledger department, which posts the sales return transaction. Allowances on sales amounts are granted for damaged goods, shortages, or similar deficiencies. In such cases the sales order department or salesperson settles the amount, which is approved by the credit manager. Sales allowances are then processed in the same manner as sales returns.

- Processing of account write-offs. Another type of accounts adjustment is the write-off of customer account balances. After reviewing account balances and other

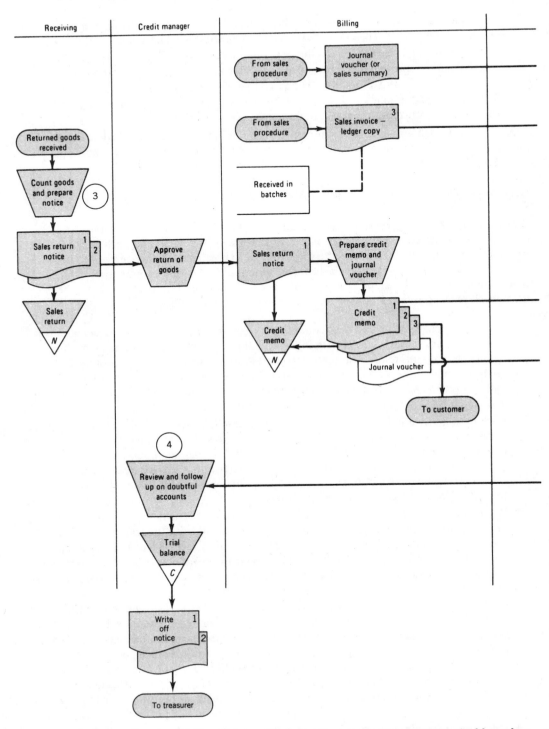

**Exhibit 16–4.   Steps for posting sale transactions to the Accounts Receivable Ledger.** Adapted from J. Wilkinson, *Accounting Information System: Essential Concepts & Applications*, 2nd. ed. (NY: John Wiley & Sons).

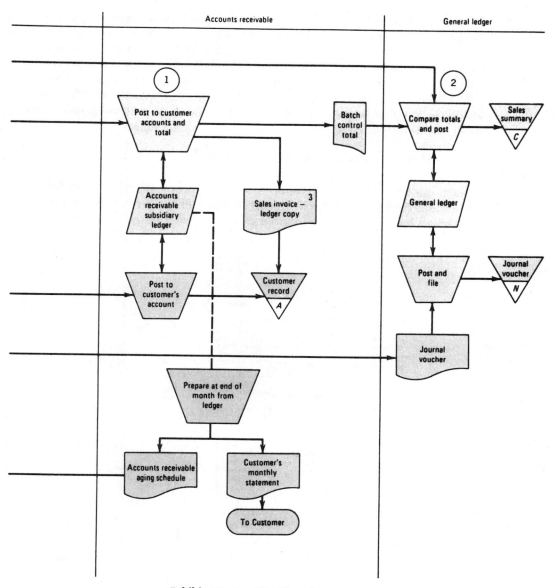

**Exhibit 16–4.** (*Continued*).

available evidence, the credit manager makes a decision concerning the collectibility of accounts. As a result of the review, a write-off notice is prepared. After the treasurer or other designated manager approves the notice, it is processed in the same manner as sales returns and allowances.

## 16.12 COMPUTER-BASED PROCESSING SYSTEMS

When computers are used to process sales and cash receipt sub-cycles, transactions may be processed by either a batch approach, an on-line approach, or a combination of the two. In the batch approach, the transaction data are keyed onto magnetic tape or disk, are sorted according to customer account numbers, and are posted sequentially to the accounts receivable master file. In the on-line approach, the transaction data are entered via a terminal and are posted individually to the master file. In the combined approach, the transaction data are entered via a terminal; however, they are gathered into a batch before being processed, either sequentially or directly, to the master file. The combined approach, described next, is popular, since it aids data entry and editing while retaining batch total controls.

## 16.13 CREDIT SALES PROCEDURE IN COMPUTERIZED SYSTEMS

Exhibit 16–5 presents a system flowchart of the on-line batch sales processing procedure. The flowchart logically divides into three segments: order entry, shipping, and billing.

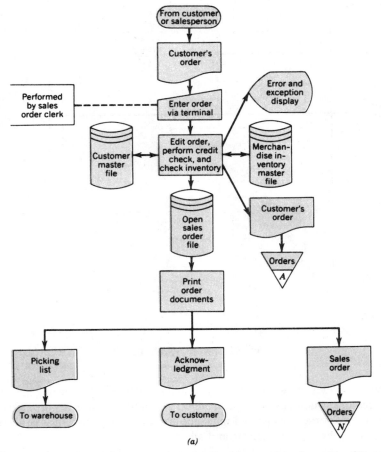

*(a)*

**Exhibit 16–5.** A system flowchart of an on-line/batch computer-based credit sales transaction processing procedure. (a) Order entry. Adapted from J. Wilkinson, *Accounting Information System: Essential Concepts & Applications,* 2nd. ed. (NY: John Wiley & Sons).

From warehouse

Picking list (amended)

Enter data concerning goods delivered to shipping

Performed by shipping clerk

Error and exception display

Edit data concerning picked goods

Picking list (amended)

Shipping file

Ship- ping refer- ence file

Print shipping documents

Packing slip

Bill of lading 1 2

Shipping notice 1 2

Performed by shipping clerk

Pack goods, with packing slip en- closed, and ship

Shipping file

N

To carrier and customer

To billing

*(b)*

**Exhibit 16–5.** (*Continued*). (*b*) **Shipping.**

**1.** Order Entry. Each customer's order is entered when received by means of a terminal in the sales order department. The edit program validates the accuracy of the data, performs the credit check, and verifies that adequate merchandise is available to fill the order. If insufficient quantities of goods are available, the sales order clerk may specify to the computer system that a backorder is to be prepared. If the credit limit is not exceeded, the order is accepted and is placed in the open order file. When the order is ready to be filled (which may be immediately), the

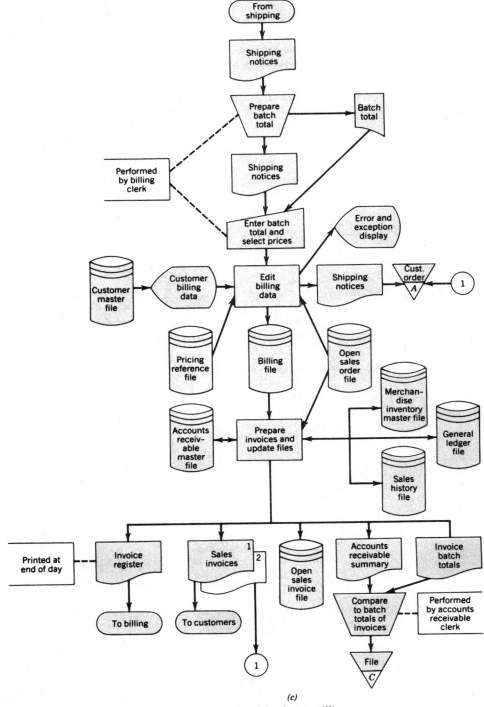

*(c)*

**Exhibit 16–5.** *(Continued).* (c) Billing.

order entry program prints the acknowledgment to the customer, a picking list, and a backup file copy.

2. **Shipping.** After the ordered goods have been picked in the warehouse, the picking list is initialed by the picker and is amended to show any changes (e.g., items out of stock, substitutions). The goods are moved to the shipping department, where a clerk counts the goods and enters the quantities ready for shipment from the picking slip. A shipping program prepares the necessary documents for the shipment. When the goods are packed, they are delivered to the carrier for shipment. A shipping notice (which is, in effect, a copy of the bill of lading) is generated on the billing department's printer concerning the shipment. It shows not only quantities shipped but also the shipping routes, freight charges, and other needed shipping data.

3. **Billing.** Upon receiving the shipping notices for the day, a billing clerk prepares and enters the batch total of quantities shipped. The same clerk also converts each order into an invoice by viewing the order on the terminal screen, selecting product prices, and so on. All entered data are validated by an edit program. Then the data for all the readied invoices are stored temporarily in a billing file until processed. At that time (a) the invoices are printed, (b) each customer's account is debited, (c) the inventory records are reduced by the quantities shipped, (d) the sales order is closed to the sales history file, (e) a new record is created in the sales invoice file, (f) a sales invoice register and summary of accounts receivable are printed, and (g) the total amounts affecting the sales and accounts receivable accounts are posted to the general ledger accounts. Finally, the accounts receivable clerk verifies that the postings to the accounts receivable ledger agree with the batch total.

Other features not shown in the flowchart may be incorporated. The billing file could be sorted by customer number before posting to the accounts receivable master file. Sales returns may be processed together with sales, so that credit memos are produced. An accounts receivable detail file may also be maintained, to serve as an audit trail. If desired, the general ledger processing could be maintained as a separate application. In that case, the billing application would generate a summary entry for later posting to the general ledger.

## 16.14 CASH RECEIPTS PROCEDURE IN COMPUTERIZED SYSTEMS

Exhibit 16–6 presents a system flowchart of the on-line cash receipts procedure. It is divided into two segments: (a) entry and deposit, and (b) end-of-day processing.

**(a) Entry and deposit.**  As checks and remittances are received, one mailroom clerk endorses the checks and prepares a batch total. In place of the batch total the clerk could prepare a remittance list and add up the amounts of remittances. Another clerk enters the batch total and the data (amount, customer number, and sales invoice number) for each payment. If the indicated sales invoice pertaining to an amount is unpaid, and if the customer number is correct, the amount is accepted by the system. If the total of all the individual amounts entered, when computed by an edit program, is shown equal to the precomputed batch total, the batch is accepted. Following this, the application program credits the account receivable records for the customers remitting payments, documents

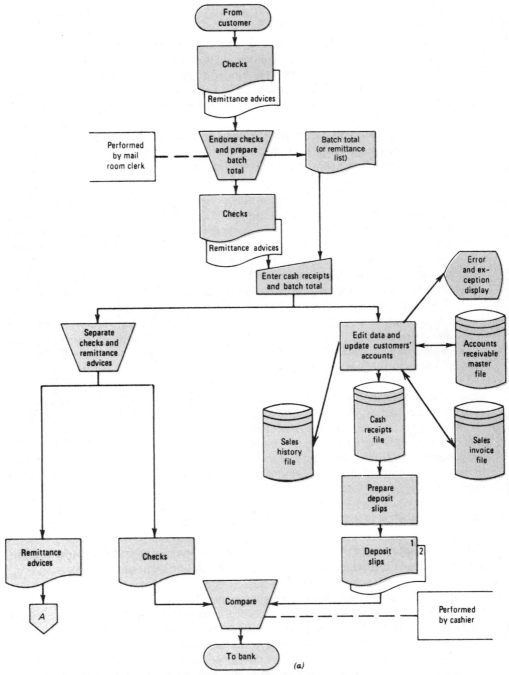

**Exhibit 16–6.** A system flowchart of an on-line computer-based cash receipts transaction processing procedure. (a) Entry and deposit. Adapted from J. Wilkinson, *Accounting Information System: Essential Concepts & Applications*, 2nd. ed. (NY: John Wiley & Sons).

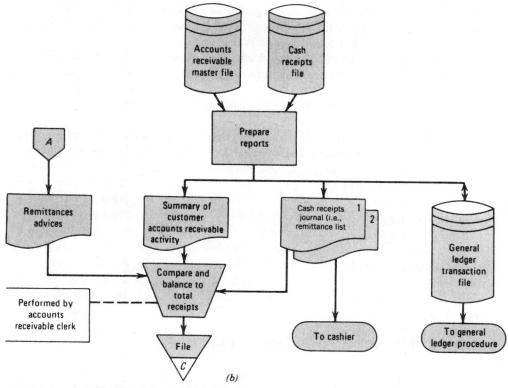

*(b)*

**Exhibit 16–6.** (*Continued*). (*b*) **End-of-day processing.**

that the affected sales invoices are paid, and closes the sales invoices to the sales history file. As in the case of sales transaction processing, the batch of cash receipts could be sorted by the application program prior to the posting step. After the processing is completed, the mail clerk sends the remittance advices to the accounts receivable department and the checks to the cashier. The cashier compares the checks to deposit slips prepared by a print program. In order to verify that all received checks are listed on the deposit slips, the cashier may also (1) receive a computer-prepared listing of cash remittances, or (2) access via a terminal the record of cash receipts for the day. When satisfied that the deposit slips are accurate and complete, the cashier has the checks and deposit slips delivered to the bank. The remainder of the deposit procedure corresponds to the steps in a manual system.

**(b) End-of day processing.**   At the end of the day, two summaries are prepared: a summary of accounts receivable activity and a remittance list (i.e., a cash receipts journal for the day). These summaries, which contain the batch total of receipts, are compared with the remittance advices. This comparison, performed by an accounts receivable clerk, verifies the accuracy of posting to the accounts. As another end-of-day step, a summary journal entry pertaining to the day's total cash receipts, is transferred to the general ledger transaction file. If the open invoice method is employed for posting payments from customers, the foregoing procedure might be modified as follows: instead of entering payments into the computer system, the mailroom clerks would prepare a remittance listing manually. They would then forward a copy to the accounts receivable department. An accounts receivable clerk would then enter the batch total and payment data via an on-line terminal. During the entry of each payment, the clerk would match the cash payments directly against specific sales amounts in the appropriate customer's account.

## 16.15 ADVANTAGES OF COMPUTER-BASED PROCESSING SYSTEMS

Computer-based processing of revenue cycle transactions has several advantages, especially when on-line systems are employed. Processing efficiency is greater, since the relevant files are integrated and can be updated concurrently. Data in the files can be kept more up to-date and can be easily retrieved when needed. Processed data are likely to be more accurate, since the transaction data are validated as entered. Also, batch totals can be employed, regardless of whether processing is performed directly or performed sequentially. Furthermore, a greater variety of control reports and summaries can be prepared automatically by the application programs as a byproduct of processing steps. Finally, needed and useful outputs can be produced more easily. For instance, monthly statements for customers can be generated from data in on-line files. Analyses and reports for managers, such as the accounts receivable aging schedule, can likewise be printed as often as desired.

## 16.16 CONTROLS NEEDED FOR THE REVENUE CYCLE

Transactions within the revenue cycle are exposed to a variety of risks. Exhibit 16–7 lists representative risks and consequent exposures due to these risks. For example, one risk is that payments from credit customers may be lapped when the accounts receivable records are posted. Lapping is a type of embezzlement that involves the theft of cash and its concealment by a succession of delayed postings to customers' accounts. A clerk who undertakes lapping first cashes a check from a customer and keeps the cash. Since the check has not been recorded, the customer's account is in error and is overstated. To cover his or her tracks, the clerk credits the customer's account on receiving a check for an equal or larger amount from another customer. Then the clerk credits the second customer's account with the proceeds from the check of still another customer. This falsifying process continues indefinitely, unless the clerk decides to return the embezzled funds. The major risk exposure to the firm from lapping is clearly a loss of funds received from customers. Another risk exposure, however, is that certain accounts will be overstated, making the amount shown in the balance sheet for accounts receivable overstated.

To minimize the dangers accruing from the kind of risks listed in Exhibit 16–7, various forms of controls may be put in place. In order to have effective measures that counteract its risk exposures, a firm must first clarify the most important control objectives. With respect to the revenue cycle, several key control objectives are to ensure that:

- All customers who are accepted for credit sales are credit-worthy
- All ordered goods are shipped or services are performed by dates that are agreeable to both parties
- All shipped goods are authorized and accurately billed within the proper accounting period

| Risk | Exposure(s) |
|---|---|
| 1. Credit sales made to customers who represent poor credit risks | 1. Losses from bad debts |
| 2. Unrecorded or unbilled shipments | 2. Losses of revenue; overstatement of inventory and understatement of accounts receivable in the balance sheet |
| 3. Errors in preparing sales invoices (e.g., showing greater quantities than were shipped or showing unit prices that are too low) | 3. Alienation of customers and possible loss of future sales (when quantities are too high); losses of revenue (when unit prices are too low) |
| 4. Misplacement of orders from customers or unfilled backorders | 4. Losses of revenue and alienation of customers |
| 5. Incorrect postings of sales to accounts receivable records or postings to wrong accounting periods | 5. Incorrect balances in accounts receivable and general ledger account records (e.g., overstatement of Mary Smith's balance), overstatement of revenue in 1992 and understatement in 1993 |
| 6. Excessive sales returns and allowances, with certain of the credit memos being for fictitious returns | 6. Losses in net revenue, with the proceeds from subsequent payments by affected customers being fraudulently pocketed |
| 7. Theft or misplacement of finished goods in the warehouse or on the shipping dock | 7. Losses in revenue; overstatement of inventory on the balance sheet |
| 8. Fraudulent write-offs of customers' accounts by unauthorized persons | 8. Understatement of accounts receivable; losses of cash receipts when subsequent collections on written-off accounts are misappropriated by perpetrators of the fraud |
| 9. Theft of cash receipts, especially currency, by persons involved in the processing; often accompanied by omitted postings to affected customers' accounts | 9. Losses of cash receipts; overstatement of accounts receivable in the subsidiary ledger and the balance sheet |
| 10. Lapping of payments from customers when amounts are posted to accounts receivable records | 10. Losses of cash receipts; incorrect account balances for those customers whose records are involved in the lapping |
| 11. Accessing of accounts receivable, merchandise inventory, and other records by unauthorized persons | 11. Loss of security over such records, with possibly detrimental use made of the data accessed |
| 12. Involvement of cash, merchandise inventory, and accounts receivable records in natural or man-made disasters | 12. Losses of or damages to assets, including possible loss of data needed to monitor collection of amounts due from previous sales |

**Exhibit 16–7.   A table of risk exposures within the revenue cycle. Adapted from J. Wilkinson,** *Accounting Information System: Essential Concepts & Applications*, **2nd. ed. (NY: John Wiley & Sons).**

- All sales returns and allowances are authorized and accurately recorded and based on actual returns of goods
- All cash receipts are recorded completely and accurately
- All credit sales and cash receipts transactions are posted to proper customers' accounts in the accounts receivable ledger
- All accounting records, merchandise inventory, and cash are safeguarded

To fulfill these control objectives, the firm must then specify and incorporate adequate general controls and transaction controls.

## 16.17 GENERAL CONTROLS

General controls that particularly concern the revenue cycle include those involving organizational independence, documentation, practices and policies, and security measures.

*Organizational independence.*   With respect to sales transactions, the units having custodial functions (i.e., the warehouse and shipping department) should be separate from each other and from those units that keep the records (i.e., billing department, accounts receivable department, inventory control department, general ledger department, data processing department). Also, the sales order and credit departments, whose managers authorize credit sales transactions as well as account adjustments and write-offs, should be separate from all the above-mentioned units. For cash receipts transactions, the mailroom and cashier (who handle cash) should be separate from each other and from the accounts receivable department, general ledger department, and data processing department.

*Documentation controls.*   Complete and up-to-date documentation should be available concerning the revenue cycle, including copies of the source documents, flowcharts, and record layouts. In addition, details pertaining to sales and cash receipts, and edit and processing programs should be organized into separate packages that are meant for use by programmers, computer operators, and systems users. Furthermore, management policies concerning credit approvals, account write-offs, and so forth should be in written form so that all employees have knowledge of them.

*Operating Practices.*   Practices relating to operational matters such as processing schedules, control summaries and reports, and personnel should be clearly established and soundly based. For instance, all employees who handle cash are required to be bonded and are subject to close supervision.

*Security Measures.*   Security should be maintained (for on-line systems) by techniques such as (1) requiring that clerks enter assigned passwords before accessing accounts receivable and other customer-related files, (2) employing terminals with restricted functions for the entry of sales and cash receipts transactions, (3) generating audit reports (access logs) that monitor accesses of system files, and (4) dumping the accounts receivable and merchandise inventory master files onto magnetic tape backups. Security measures for manual and computer-based systems include the use of physically restricted warehouses (for protecting goods) and safes (for holding cash receipts). Also, a lockbox collection system may be considered, where feasible. A lockbox is a postal address that is used solely to collect remittances, which are removed and processed by the firm's bank.

## 16.18 TRANSACTION CONTROLS

The following controls and control procedures are applicable to revenue cycle transactions and customer accounts. They are arranged by input, processing, and output categories.

Input controls include:

- Prenumbered and well-designed documents relating to sales, shipping, and cash receipts.
- Validation of data on sales orders and remittance advices as the data are prepared and entered for processing. In the case of computer-based systems, validation is performed by means of programmed checks built into the system.
- Correction of errors that are detected during data entry. Detected errors should be corrected by an error-correction procedure that is appropriate to the processing approach being employed.
- Batch control totals that are precomputed from sales invoices (or shipping notices) and remittance advices. These precomputed batch control totals are compared with totals computed during postings to the accounts receivable ledger and during each processing run. In the case of cash receipts, the total of remittance advices is also compared with the total on deposit slips.

Processing controls include:

- Issuance of multi-copy sales orders and/or invoices on the basis of valid authorization, such as credit approved customer orders
- Movement of ordered goods from the finished goods warehouse and shipment on the basis of written authorizations, such as picking lists or stock request copies.
- Invoicing of customers only on notification by the shipping department of the quantities that have been shipped.
- Issuance of credit memos and write-off notices only on the basis of prior approvals by designated managers. In the case of sales returns, approval is not granted until the goods being returned have been received.
- Verification of all data elements and computations on sales invoices by a billing clerk other than the preparer or by a computer program before mailing. Another verification consists of comparing the sales invoices against shipping notices and open sales orders, to ensure that the quantities ordered reconcile with the orders shipped and backordered.
- Verification that total postings to the accounts receivable master file accounts agree with the total postings to the general ledger accounts (in the case of a computer-based batch processing system).
- Deposit of all cash receipts with a minimum of delay. This step eliminates the possibility of cash receipts being used to pay employees or to reimburse petty cash funds. It also preserves the audit trail.
- Reconciliation of accounts in the accounts receivable subsidiary ledger with the accounts receivable control account in the general ledger.
- Correction of errors made during processing steps. All such corrections consist of reversing erroneous postings to accounts and reentering correct data. The audit trail concerning accounts being corrected should show the original error, the reversal, and the correction.

Output controls include the following:

- Preparation of monthly statements, which are mailed to all credit customers. Since a customer will likely complain if overcharged, this practice provides control over accidental and fraudulent acts.

- Filing of copies of all documents pertaining to sales and cash receipts transactions by reference to their preassigned numbers, with the sequence of numbers in each file being periodically checked to see if gaps exist. If transactions are not supported by preprinted documents, as often is the case in on-line computer-based systems, numbers are assigned to the documents and they are stored in a transaction file.

- For computer-based systems, preparation of printed transaction listings, such as the cash receipts journal and account summaries, on a periodic basis to provide an adequate audit trail. Also, various outputs that aid control, such as exception and summary reports, should be prepared.

- Reconciliation of all bank accounts should be done monthly by someone not involved in the revenue cycle processing activities. New bank accounts are to be authorized by designated managers.

## 16.19 MANAGERIAL REPORTS

The revenue cycle database can provide a wealth of information to aid managers in making decisions. One example already listed is the accounts receivable aging schedule, which is useful for decision making as well as operational control. Several other useful reports and analyses are worth noting.

*Performance reports.*   These reports reflect results in terms of key measures such as average dollar value per order, percentage of orders shipped on time, and the average number of days between the order date and shipping date.

*Sales analyses.*   Analyses of sales reflect the relative effectiveness of individual salespersons, sales regions, product lines, customers, and markets. One possible report shows a sales analysis that compares the actual sales for three of the above-mentioned factors against established quotas.

*Cash flow statements.*   These reports show the sources and uses of cash for an accounting period, classified by operating profit-directed activities, financing activities, and investment activities. They provide the basis for developing cash forecasts and budgets. Hence, they aid the process of managing the all important cash resource.

Other useful reports are open orders reports, backorders reports, sales analyses by product, and unbilled shipments reports.

## 16.20 CONCLUSION

The revenue cycle permits a firm to trade its products or services for cash. The cycle contains the following steps: receiving orders from customers, checking customers' creditworthiness, processing the orders, obtaining the merchandise from the warehouse, packing and shipping orders to customers, billing customers, receiving and depositing the payments, maintaining the customers' accounts, posting transactions to the general ledger, and finally, informing decision makers by preparing pertinent reports, which could be about products sold, sales force's performance, and the status of accounts receivable and other elements of the revenue cycle.

The revenue cycle is not a localized affair. It directly or indirectly involves every department in the organization. To ensure that the information needs of each department are met and to maintain adequate records, documentation unique to revenue cycle are necessary. This may include sales orders, packing lists, shipping notices, invoices, and credit memos. In the Information Age, preformatted screens are replacing the forms and other types of documentation used in manual transaction processing. The forms or the preformatted screens both feed data into the revenue cycle database, which includes files related to customer accounts, order status, sales history, and other aspects of the cycle.

The revenue cycle is also confronted with an assortment of risks. A firm's exposure to these risks can be minimized through different forms of control. The cycle is represented by various accounting entries, and, given the volume of such transactions, involving both customers and employees is essential to ensure proper procedures and accurate coding.

# CHAPTER 17

# THE EXPENDITURE CYCLE

## 17.1 INTRODUCTION

This chapter describes the expenditure cycle's objectives, functions, and relationships to other organizational units. This is done by identifying data sources, forms of input, and accounting entries. The database used for the cycle is outlined, as well as the steps and approaches used in processing its transaction data flows. The cycle must be safeguarded from a variety of risks, and steps that can be taken to minimize such risks are described.

An organization has expenditures for goods and services. Goods may consist of merchandise, raw materials, parts, subassemblies, supplies, and plant assets. Services may include those provided by outside parties, such as telephone and legal services, as well as the labor provided by the organization's employees. Because the expenditure cycle involves the outflow of funds, it is the counterpoint to the revenue cycle, which provides inflows of funds.

Most acquired goods and services represent resources to the organization. For instance, merchandise, raw materials, and supplies are materials resources, plant assets are facilities resources, and hours worked by employees are human service resources. All such resources also fall within the resources management cycle, which is closely related to the expenditure cycle. This chapter focuses on the goods involved in the materials resource, plus outside services, while the next two chapters examine resources such as employee services and facilities.

The major purpose of the expenditure cycle is to facilitate the exchange of cash with suppliers for needed goods and services. When goods (i.e., merchandise, supplies, or raw materials) are purchased, the functions of the expenditure cycle consist of recognizing the need for the goods, placing the order, receiving and storing the goods, ascertaining the validity of the payment obligation, preparing the cash disbursement, maintaining the accounts payable, posting transactions to the general ledger, and preparing needed financial reports and other outputs. In the case of services, the functions of receiving and storing the goods are replaced by the function of accepting the ordered services. In the case of direct payments by cash (as is done through a petty cash fund), the function of maintaining the payable records is unnecessary. These groups of functions are pictured in Exhibit 17–1.

Exhibit 17–2 presents a data flow diagram that includes the several functions performed as part of the cycle. Each function is viewed in the diagram as a logical process that is linked to other processes and to entities and data stores. The diagram emphasizes the critical role played by each activity within the cycle, while Exhibit 17–3 provides additional insight by depicting relationships within expenditure cycle.

In addition to the functions listed above, some related functions performed by the cycle must also be mentioned. These other related functions include payroll disbursements,

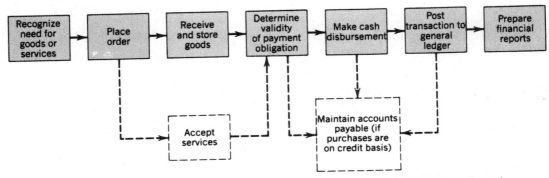

**Exhibit 17–1.** A diagram depicting typical functions of an expenditure cycle. Adapted from J. Wilkinson, *Accounting Information System: Essential Concepts & Applications,* 2nd. ed. (NY: John Wiley & Sons).

expense distribution, purchase returns and allowances, miscellaneous cash disbursements, and petty cash disbursements. Payroll disbursements (together with the distribution of employee labor expenses) are discussed in the next chapter.

## 17.2 ORGANIZATIONAL CONTEXT OF THE EXPENDITURE CYCLE

The above-mentioned expenditure cycle functions are typically achieved under the direction of the inventory management and finance/accounting organizational units of the firm. The expenditure cycle therefore involves both the inventory management information system and the accounting information system (AIS). Moreover, the results attained and information generated by the expenditure cycle further the objectives of these organizational units.

The inventory management function in a merchandising firm is concerned with managing the merchandise that the firm acquires for resale. In addition to planning responsibilities, the function is responsible for purchasing, receiving, and storing the merchandise. In some firms, for example, manufacturing firms, inventory management can be viewed as the major sub-function within a broader logistics function. Also included in the logistics function might be the production and distribution functions. Alternatively, distribution may be assigned to the marketing/distribution function.

Purchasing focuses primarily on selecting the most suitable suppliers or vendors from whom to order goods and services. It makes the selections on the basis of factors such as the unit prices charged for the goods or services, the quality of the goods or services offered, the terms and promised delivery dates, and the supplier's reliability. Together with inventory control, purchasing also ascertains the quantity of goods to acquire. The optimal order quantity is determined by a formula that includes factors such as the expected demand for the goods, the carrying cost, and the ordering cost. However, this formula is normally applied only to high-cost or high-volume goods. Order quantities for low-cost or low-volume goods are more likely to be determined on a rough basis that seeks to avoid stockouts. In some cases a good buying opportunity or a price break determines the quantities to order.

Receiving has the responsibilities of accepting only those goods that were ordered, verifying their quantities and condition, and moving the goods to the storeroom. Storing, or stores, has the responsibility of safeguarding the goods from theft, loss, obsolescence,

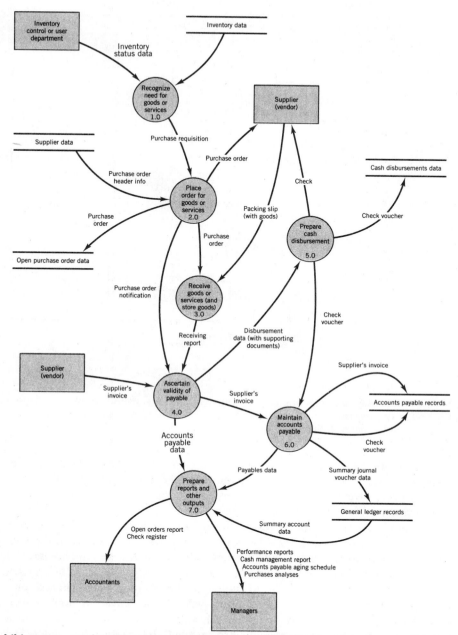

**Exhibit 17–2.   A data flow diagram pertaining to accounts payable activity. Adapted from J. Wilkinson, *Accounting Information System: Essential Concepts & Applications*, 2nd. ed. (NY: John Wiley & Sons).**

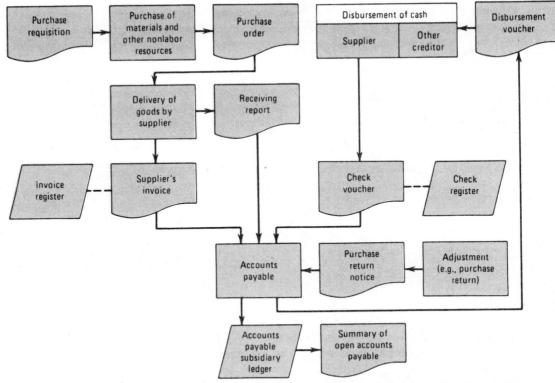

**Exhibit 17–3.   A diagram depicting relationships within the expenditure cycle. Adapted from J. Wilkinson, *Accounting Information System: Essential Concepts & Applications*, 2nd. ed. (NY: John Wiley & Sons).**

and deterioration. The storeroom is also charged with getting the goods together and delivering them when properly authorized requisitions are presented to the staff in charge of the storeroom. Inventory control maintains the records pertaining to inventory balances and initiates the reordering of goods.

The second major player in the expenditure cycle is the team of the accounting department and financial managers. The essence of the cycle relates to a firm spending money to acquire resources, goods, and services. Both accounting and finance have a lot to say about the availability and disbursement of money. The objectives of financial and accounting management relate broadly to funds, data, information, planning, and control over resources. With respect to the expenditure cycle, the objectives are limited to cash planning and control, to data pertaining to purchases and suppliers' accounts, to inventory control, and to information pertaining to cash and purchases and suppliers.

The senior financial managers are often the vice-president of finance, the treasurer, and the controller. One important manager reporting to the treasurer is the cashier. The treasurer's responsibilities include cash disbursements, whereas the controller's responsibilities include accounts payable, inventory control, and the general ledger. Between them, they take care of the actual disbursements pertaining to expenditure cycle. Cash disbursements, a part of the cashier's department, prepares checks for disbursement and maintains the related records. Accounts payable maintains the records of individual suppliers and approves their invoices for payment. General ledger maintains control over all asset, equity, expense, and income accounts.

## 17.3 MANAGERIAL DECISION MAKING

Three department managers having key responsibilities with respect to the expenditure cycle for goods are the inventory manager, the treasurer, and the controller. They must make decisions that influence the way purchases are made and cash is disbursed. To a large extent, the decisions pertaining to expenditures rely on the information that is a byproduct of processing within the expenditure cycle. The inventory manager (or logistics manager in some firms) depends on the inventory management information system to provide needed information for decision making. No matter what the source of information, specific questions requiring answers of the inventory manager (or subordinate managers) of a merchandising firm include the following:

- What levels of inventory should be stocked?
- When should particular inventory items be reordered?
- What quantities of particular inventory items should be reordered?
- When should long-term purchase contracts be obtained for particular inventory items?
- From which suppliers should particular inventory items be ordered?
- What procedures should be followed in receiving and storing merchandise inventory?
- What organizational units are to be included in the inventory management and logistics functions?
- What logistics plans and budgets are to be established for the coming year?

Like the inventory managers, those working for the controller and treasurer make decisions pertaining to expenditures. Among the questions requiring answers of the controller and treasurer (or subordinate managers) are:

- What policies concerning purchase terms and discounts should be established?
- What level of services should departments be allowed to acquire?
- What accounts payable records are to be maintained concerning amounts owed to suppliers?
- What financial plans and budgets are to be established for the coming year?
- What sources of funds are to be employed?

In order to make sound decisions in the context of the expenditure cycle, the senior managers need the assistance of accountants, purchasing analysts, auditors, and systems developers. Managerial accountants aid in developing appropriate decision models and in determining what information is needed. Purchasing analysts are specialists who aid in making purchases and inventory decisions and in analyzing their effects. Auditors examine and evaluate the controls incorporated in the expenditure cycle; they also review the financial statements and related financial outputs for reliability. Systems developers participate in the analysis and design of the systems used to process purchases and cash disbursements transactions; thus, they aid in specifying outputs that provide information to managers.

## 17.4 DOCUMENTATION NEEDED FOR THE CYCLE

Documentation used in the expenditure cycle is mainly based on inputs from the inventory records and from suppliers. The inventory records are the primary source of most purchase transactions, whereas suppliers invoices are the source of payable/disbursement transactions. Other sources are the records of department heads, buyers, supplier history files, receiving and stores departments, and (for manufacturing firms) the production departments. If payroll disbursements are included within the expenditure, such sources as the personnel and payroll records are to be added.

## 17.5 DOCUMENTATION FOR MANUAL SYSTEMS

The expenditure cycle source documents typically found in firms that employ manual processing include the following:

*Purchase requisition.*    The initiating form in the expenditure cycle is the purchase requisition. It both requests and authorizes the purchasing department to place an order for goods or services. Key items of data that it conveys are the quantities and identifications of the goods to be purchased, plus the date needed, the name of the requestor and department, and the approver's name or initials. Optional data include the suggested supplier (vendor), suggested unit prices, and shipping instructions. The request after approval by the purchasing department on the confirmation copy is returned to the requester.

*Purchase order.*    This is the formal request, signed by an authorized buyer or purchasing manager, to an outside supplier. If the supplier agrees to all stated terms and conditions on the order, it is binding on the issuing firm. In addition to the signature area at the bottom of the form, a typical purchase order has two sections: a heading and a body. The heading contains the supplier's name and address, as well as shipping instructions. The body contains one or more line items, with each line item pertaining to a single item of merchandise or material being ordered. Although unit prices for the various line items are included, cost extensions are not normally provided because the unit prices are tentative.

*Receiving report.*    A prenumbered document prepared by a clerk in the receiving department is called a receiving report (or record, memorandum, or ticket). This document states that the listed quantities of goods have been received and indicates their overall condition. It can be used to reflect the receipt of any goods, including goods on consignment or goods returned by a dissatisfied customer. However, most receipts of goods—and receiving reports—relate to issued purchase orders. A copy of the receiving report usually accompanies the goods to the storeroom and then is forwarded to accounts payable.

*Supplier's invoice.*    To the buying firm, a supplier's (vendor's) invoice is a response to a previously issued purchase order. To the supplier, it is a sales invoice. The invoice is the source document used as the posting medium into the accounting cycle. Neither the purchase order nor receiving report can serve this purpose, since they do not reflect the amount of the obligation incurred.

*Disbursement voucher.*   This document, also called a payment voucher, authenticates a liability for an expenditure and authorizes payment. It is prepared when the widely popular voucher system is used. A voucher is generated when one or more invoices are received from a supplier. It is entered in a voucher register, which serves as a journal. Then the voucher may be used as the medium for posting to the accounts payable ledger, if one is maintained. Summary totals from the voucher register are posted to the general ledger. A disbursement voucher offers several advantages. It allows several invoices to be accumulated, thereby reducing the number of checks to be written. Because it is prenumbered, the voucher provides numerical control over the payment. Finally, it provides a convenient means for grouping the vouching documents together in a package and reflecting the approval for its payment.

*Check voucher.*   A disbursement check is the final document in the expenditure cycle. Usually, the check has an attachment, which in effect is a copy or abbreviated version of the voucher. In some systems the check voucher is prepared in lieu of the disbursement voucher. The check is signed by an authorized signer. In some instances, as when the amounts are large, it may be countersigned. Then a copy of the check, containing the attached voucher section (and hence called the check voucher), is entered in a cash disbursements journal or check register.

*Debit memorandum.*   When a purchasing firm decides to reject received goods or desires to obtain adjusted terms, it first requests that the selling firm authorize a return or allowance. When granted, the purchasing firm prepares a debit memorandum. In effect, this document notifies the selling firm and the accounts payable department.

*Additional documents.*   A new supplier (vendor) form is useful in the selection of new suppliers. It contains data such as prices or rates, types of goods or services provided, experience, credit standing, and references. A request for proposal (RFP), or for quotation, is a form used during a competitive bidding procedure. It lists the various goods or services needed and provides columns in which the bidding suppliers enter their proposed prices, terms, and so on. A bill of lading generally accompanies received goods. Finally, a journal voucher is prepared as the basis for each posting to the general ledger. For instance, a journal voucher is prepared by reference to debit memoranda that have been issued during a period.

## 17.6 DOCUMENTATION NEEDED BY COMPUTER-BASED SYSTEMS

All of the aforementioned hard-copy documents may also be used in computer-based systems. In some cases, however, they may be generated automatically upon the entry of data via computer. Moreover, the source documents can be designed to aid the entry of data into the computer system with fewer errors. Preformatted screens may be used to enter data concerning purchases, payments, purchase returns, and cash disbursement transactions. These screens, like simplified entry forms, may be designed to handle individual transactions or batches of transactions.

## 17.7 ACCOUNTING ENTRIES

In a manual system, data from a supplier's invoice are entered into a purchases' journal and posted to individual supplier accounts in the accounts payable subsidiary ledger.

Otherwise, invoices are used to prepare a disbursement voucher, which then becomes the source of input into the voucher register. Periodically, the summary total from the journal or register is posted to the general ledger. Under a periodic inventory system, this particular entry will be a debit to purchases and a credit to accounts payable. Under a perpetual inventory system, the entry will be a debit to inventory and a credit to accounts payable.

Entries must also be made to reflect purchases returns and allowances. Usually, journal vouchers are prepared to reflect such returns and allowances and the entry made involves a decrease in the accounts payable through a debit entry with a credit to purchase returns account. Exhibit 17–4 shows the flow of purchases transactions through the accounting cycle, beginning with the supplier's invoice.

Exhibit 17–5 portrays the flow of cash disbursements through the accounting cycle. Checks are prepared for suppliers for their invoices or disbursement vouchers. Copies of such checks provide the amounts and other details to be entered into the cash disbursement journal and then posted to individual supplier records in the accounts payable subsidiary ledger. Periodically, totals from the cash disbursements journal are posted to the general ledger as a debit to accounts payable and credit to cash. If there are any cash discounts involved for prompt payments, they will form a part of the entry as credit.

The entries discussed above could be different if the payment was recorded under the net value method instead of the gross value method. In such a case, the entry will be different if the discount criteria were not met. The discounts lost will be added to the purchase price.

Cash can be expended for reasons other than purchase of merchandise. In such cases, the entries will be a debit to the given account, with credit going to cash. Certain purchased goods and services are consumed over the course of more than one accounting period. At the end of each period the consumed portion should be reflected in the accounts. In the case of supplies, which were set up in accordance with the perpetual method when acquired, an end-of-period adjustment would be a debit to supplies expense account and credit to supplies inventory. The notation for such an adjusting entry will explain that the entry records the amount of supplies used during the period.

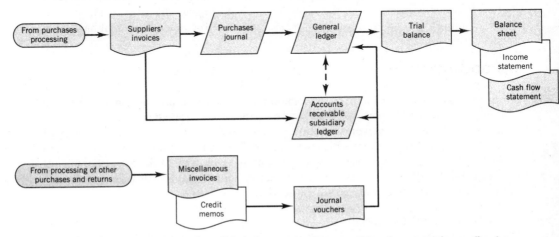

**Exhibit 17–4.   A diagram describing the transactions resulting from purchases flowing through the accounting cycle. Adapted from J. Wilkinson, *Accounting Information System: Essential Concepts & Applications*, 2nd. ed. (NY: John Wiley & Sons).**

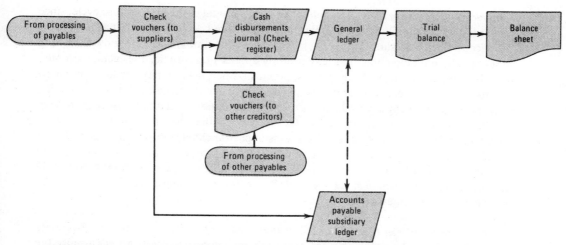

**Exhibit 17–5. A diagram depicting the flow of cash disbursements through the accounting cycle. Adapted from J. Wilkinson,** *Accounting Information System: Essential Concepts & Applications*, **2nd. ed. (NY: John Wiley & Sons).**

## 17.8 TRANSACTION CODING

Since codes were discussed in detail in Chapter 15, which dealt with the general ledger cycle as well as the accounting system, the discussion here will be very brief. Codes can aid the entry of purchases and cash disbursements transactions. In addition to codes for general ledger accounts, codes can be entered for suppliers, purchase orders, products, buyers, purchase contracts, and so on. When such codes are used, various purchase analyses can be performed. Group codes are useful when recording transactions such as purchases, since they encompass several dimensions. They may also be used to represent entities such as products, suppliers, and buyers. A product code for a battery sold by a retail chain, for instance, might include these fields: the coded item is numbered 28-B-485-34-D, with 28 representing the product line, such as auto accessories, B representing the product subgroup, in this case a battery, 485 referring to the vehicle for which the battery is to be used, 34 denoting the size of the battery, and D representing the battery type, such as dry cell.

## 17.9 DATABASE FOR THE EXPENDITURE CYCLE

The database for the expenditure cycle contains master files, transaction files, open document files, and other types of files. If appropriate database software is available, it may also contain data structures.

**(a) Master Files.** Two likely master files employed within the expenditure cycle pertain to suppliers and accounts payable. If a firm purchases merchandise for resale, a third file is the merchandise inventory master file. The content of these files will vary from firm to firm, depending on factors such as types of suppliers and sources of services, variety of desired managerial report, degree of computerization, and structure of the database.

*Supplier (vendor) master file.*   A supplier master file is vital to the expenditure cycle, since it specifies where the checks for suppliers are to be mailed. In many firms, it also serves as the accounts payable subsidiary ledger by showing the amount currently owed to each supplier. The data elements that may usefully appear in a supplier file include account number, supplier name, mailing address, phone number, fax number, credit terms, year-to-date purchases, year-to-date payments, and the current account balance. Generally, the primary and sorting key of the file is the supplier number. An important concern with this file, as with the inventory master file, is keeping the records up-to-date. When a supplier obtains a new mailing address, for instance, this change must be quickly reflected in the record. When a new supplier is added to the approved supplier list, a new record should appear in the file; when a supplier is dropped, the corresponding record should be deleted also. These changes can be made during the processing of invoices, or they may be made at other times as they arise.

*Accounts payable master file.*   The records in the accounts payable master file also relate to suppliers. Two data elements are essential: the supplier identification (usually the supplier account number) and the current account balance. If this file is used, the account balance would be removed from the supplier master file. A separate accounts payable master file is useful, since records containing fewer fields can be processed faster during file update runs.

*Merchandise inventory master file.*   The merchandise inventory master file is essentially the same file discussed in Chapter 16 for revenue cycle. If the firm is a manufacturer, the file will be called the raw materials inventory master file. Some firms also maintain a separate supplies inventory master file.

**(b) Transaction and Open Document Files.**   Transactions in the expenditure cycle involve purchase requisitions, purchase orders, receiving reports, suppliers' invoices, debit memos, and cash disbursements. If a voucher system is in effect, vouchers are also included. Each of the records in related transaction files contains roughly the data shown in the documents discussed earlier. Three major transaction files deserve closer attention.

*Purchase order file.*   The content of the purchase order file is similar to the content of a purchase order, discussed earlier. When stored on a computerized medium, the data are often separated into two records: a header record and one or more detailed line item records. The header record consists of the purchase order number, supplier number, purchase order date, and expected delivery date. Each line item record contains the purchase order number, line item number, inventory item number, quantity ordered, and expected unit price. If the supplier's product number differs from the inventory item number, the former should also be listed. Other data elements, such as the supplier's name and address, can be obtained from the supplier's master file when preparing vouchers and checks. The purchase order records pertaining to purchases not yet approved for payment are stored in an open purchase order file. The sorting and primary key of the file is the preassigned order number. When an order is approved for payment, its record is usually transferred to a closed orders file.

*Supplier's invoice transaction file.*   In a manual system a supplier's invoice file consists of a copy of each current supplier's invoice. Records in this file provide the details of purchases transactions posted to the accounts payable records. In practice the file may be separated into two files: a pending invoice file of those invoices not yet vouched, and an open invoices file of those invoices not yet paid. Each invoice may also be assigned a new number, or the supplier's number may be used for reference. If a voucher system is in use, the open invoices file may be replaced by an open vouchers file. In a computer-based system only a few key data elements would be stored on magnetic media. For a voucher, the data elements would include the voucher number (the sorting key), each invoice number and date, each invoice amount, and the payment discount date. All of the associated invoices would be linked to the voucher record and would cross-reference the related purchase order, requisition, and receiving report. Records in the open documents file are closed when payments are made

**Check disbursements transaction file.**    In a manual system this file consists of a copy of each current check voucher, arranged in check number order. In a computer-based system the record layout on magnetic media may contain the supplier's account number, related purchase order number(s), date of payment, and amount of payment. It also may include a code to identify the record as a cash disbursement transaction.

**Other files.**    The supplier reference and history file is one of the other files used. Considerable data are needed concerning suppliers, so that buyers can make informed decisions when placing purchases. Such information can be maintained in a reference file. The accounts payable detail file is another useful file. In addition to creating separate suppliers' invoice and cash disbursement transaction files, the application program may accumulate records of the transactions in a detail file. Since it is restricted to the current period's transactions, this detail file is not a replacement for the history files. Instead, its main purpose is to facilitate the preparation of purchases reports and analyses. Also, it can serve as an audit trail.

## 17.10 DATA FLOWS AND PROCESSING FOR MANUAL SYSTEMS

Within the expenditure cycle, the data flows and processing steps can be divided into three major subsets: processing of purchases transactions, establishment of accounts payable, and processing of cash disbursements. Each of these subsets can be examined through system flowcharts for manual and computer-based processing procedures.

**(a) Manually Processed Purchases System.**    Exhibit 17–6 presents a document flowchart of a procedure involving the purchases of goods on credit. Since the processing is performed manually, the emphasis is on the flows of source documents and outputs. Although the flowchart is detailed and appears to be quite complex, it provides an overall view of a sound purchases procedure. The flowchart, together with the following narrative, will present a clear picture of the data flows and processing. Also, an understanding of the document flows will clarify the processing steps performed by computer-based systems.

The narrative description will be assisted through the use of reference numbers, which have been placed on the flowchart at key spots. These numbers designate several control points within the purchases procedure.

1. Determination of a need for goods. The procedure for purchasing goods usually begins in the inventory control department. There an accounting clerk refers to the

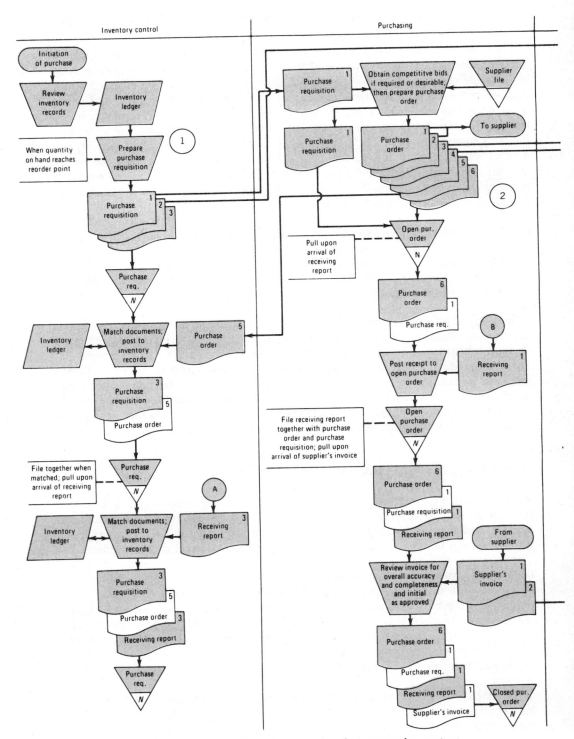

**Exhibit 17–6. A flowchart of a manual purchases transaction processing system.** Adapted from J. Wilkinson, *Accounting Information System: Essential Concepts & Applications*, 2nd. ed. (NY: John Wiley & Sons).

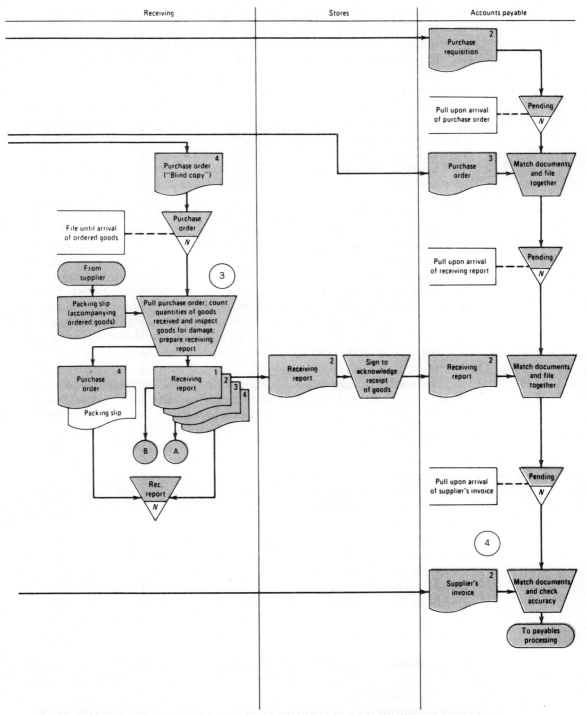

**Exhibit 17–6.** (Continued). Adapted from J. Wilkinson, *Accounting Information System: Essential Concepts & Applications*, 2nd. ed. (NY: John Wiley & Sons).

inventory ledger records to locate those items whose on-hand quantities are below a preestablished reorder point. Those items that need to be reordered are listed on a prenumbered and well-designed purchase requisition form. For each item the clerk specifies a precomputed economic order quantity. On approval of the requisition, perhaps by the inventory manager, copies are sent to the purchasing department and receiving department.

2. **Preparation of the purchase order.** When the purchase requisition is received in the purchasing department, a buyer is assigned by the purchasing manager to handle the purchase transaction. No purchasing action can take place without authorization by a requisition. If the goods or circumstances are non-routine, competitive bids are obtained. If the needed goods are routine or after bids have been evaluated, the buyer selects the most suitable supplier from an approved supplier file and prepares a prenumbered purchase order. When the purchase order has been checked for prices and terms and signed by an authorized person, such as the purchasing manager, the copies are distributed. Two copies are mailed to the supplier. Other copies are forwarded to the inventory control, receiving, and accounts payable departments. The copy sent to inventory control (which may actually be an amended copy of the requisition) is used to post ordered quantities to the inventory records. The copy for the receiving department (which has the quantities blanked out, i.e., is "blind") is used later to verify the authenticity of the received goods. Thus it serves an important control purpose. The copy sent to accounts payable provides prior notification that an invoice is soon to be received. The last copy is filed in the open purchase order file to await the arrival of the invoice.

3. **Receipt of ordered goods.** When the ordered goods arrive at the receiving dock, the "blind" copy of the purchase order is matched to the packing slip, in order to verify that the goods were ordered. Next, the receiving clerk inspects the goods for damage and counts the quantities received. Then he or she prepares a prenumbered receiving report on which the findings are recorded. The original copy of this report accompanies the goods to stores, where the storekeeper or warehouseman signs the copy (to acknowledge receipt) and forwards it to accounts payable. Other copies of the receiving report are sent to the purchasing department (to update the open purchase order) and to the inventory control department (to update the inventory ledger records).

4. **Receipt of the supplier's invoice.** When the supplier's invoice arrives shortly after the ordered goods, it is routed to the purchasing department for comparison with the documents relating to the purchase. If found to be proper and complete, the invoice is forwarded to the accounts payable department for more extensive processing. Invoices pertaining to services are first routed to the using departments, where they are approved for payment by the managers responsible for incurring the expenditures. Then they are forwarded to accounts payable.

**(b) Manually Processed Accounts Payable Procedure.**  Since accounts payable is an accounting department not directly involved in purchasing and receiving goods, it is the most suitable department to examine the supplier's invoice and to trace its content to the supporting documents (i.e., to vouch the invoice). Exhibit 17–7 illustrates the vouching step and the processing steps that follow. On receiving a supplier's invoice in the accounts payable

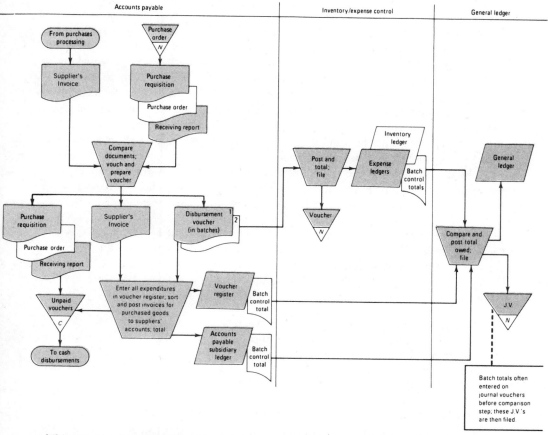

**Exhibit 17–7.** A system flowchart of a manual payables processing procedure, with an emphasis on document flows. Adapted from J. Wilkinson, *Accounting Information System: Essential Concepts & Applications*, 2nd. ed. (NY: John Wiley & Sons).

department, a clerk pulls the supporting documents from a file. Then the clerk performs the various comparisons and checks that constitute vouching. These verifications are intended to determine that (1) the purchase has been authorized, (2) the goods or services listed in the invoice have been duly ordered, (3) the goods or services have been received in full, (4) the unit prices applied are in conformity with the purchase order, or are satisfactory to the purchasing department, (5) the terms and other specifications are in agreement with the purchase order, and (6) all computations are correct. After finishing, the clerk initials an audit box (stamped either on the invoice or on another document, such as a voucher) to acknowledge that the verifications have been performed and that the supplier's invoice is approved for payment. Any differences must be settled, however, before a supplier's invoice can be approved for payment. For instance, if only part of the order is received, the purchase order should be so marked and returned to the file.

Assuming that the voucher system is used, a disbursement voucher is prepared on the basis of one or more approved suppliers' invoices. Then the voucher is entered in a voucher register. Batch control totals are computed from the columns in the voucher register, including the total amount of payment, the total merchandise cost, the total selling expense, and so on. A journal voucher is prepared from the totals.

A clerk posts the vouchers to the suppliers' accounts in the accounts payable subsidiary ledger. Batch totals are computed of the posted credits. Also, copies of the vouchers are forwarded to accounting departments that maintain the ledgers relating to the various expenditures (e.g., inventory control). Clerks in these departments post debits to inventory, supplies, plant assets, selling expense, and administrative expense ledgers. Batch totals are computed of the posted debits. Then the batch totals of the posted debits and credits are compared with the journal voucher previously prepared. If all amounts agree, the entry is posted to the accounts in the general ledger.

Finally, the originals of the vouchers, together with the supporting documents, are filed in a file arranged by payment due dates, that is, a "tickler" file. There, the unpaid vouchers remain until ready for use in cash disbursements processing.

(c) **Manually Processed Cash Disbursements Procedure.**    Exhibit 17–8 shows a flowchart of a procedure involving the disbursements of cash related to purchases on credit. The principal control points involve assembling the unpaid vouchers, preparing the checks, signing the checks, processing the cash disbursement records, and posting the cash amounts.

1. Assembly of the unpaid vouchers. The cash disbursements procedure begins in the accounts payable department with the unpaid voucher file. Each day a clerk extracts the unpaid vouchers due to be paid. She or he reviews each voucher "package" to see that it contains all of the supporting documents including the suppliers' invoices. After computing the total amount to be paid and posting the payment amounts from the vouchers to the appropriate suppliers' accounts, the clerk forwards the vouchers and supporting documents to the cash disbursements department.

2. Preparation of the checks. A cash disbursements clerk inspects each voucher for completeness and authenticity and then prepares a prenumbered check. When finished, the clerk forwards the original checks to an authorized check signer, together with the supporting documents. Then the check copies are entered into the records.

3. Signing of the checks. In many firms, the authorized check signer is the treasurer, although the cashier may be authorized to sign checks below a designated amount. The signer first reviews the supporting documents. Then he or she signs each check that is properly supported and sends the checks directly to the mailroom. From the mailroom the checks are delivered to the post office.

4. Processing of the cash disbursement records. The amounts and other key data concerning the checks are entered in a check register, and the total of the paid amounts is computed. One copy of the check voucher is filed in the cash disbursements department numerically. The other copy of the check voucher is stapled to the supporting documents, which are stamped as paid. The number of each check and the date are entered in the voucher register, and the package is filed alphabetically by supplier in the accounts payable department. Firms that process large volumes of invoices often find that the bulky voucher packages consume much storage space and are awkward to retrieve. Thus, they may decide to microfilm the documents after processing and then destroy the documents. The microfilm images may be arranged by voucher number and cross-referenced to supplier names.

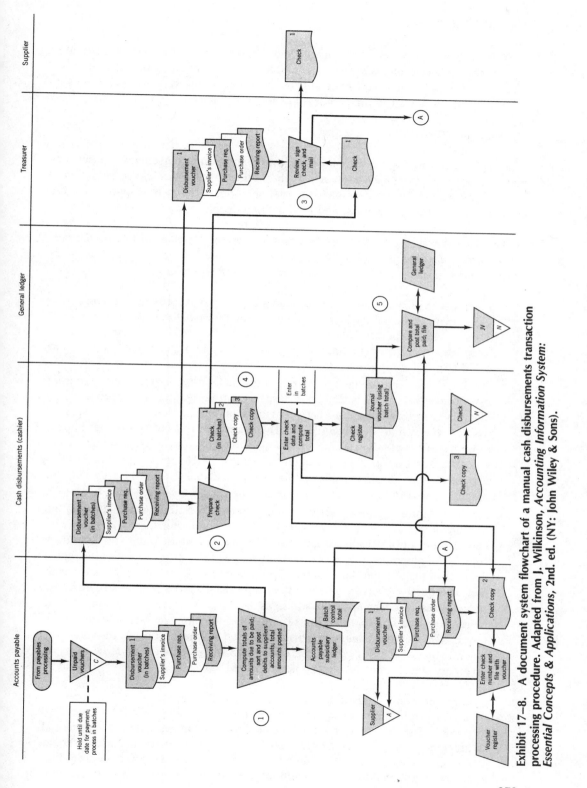

**Exhibit 17–8. A document system flowchart of a manual cash disbursements transaction processing procedure.** Adapted from J. Wilkinson, *Accounting Information System: Essential Concepts & Applications*, 2nd. ed. (NY: John Wiley & Sons).

5. Posting of cash amounts. A journal voucher is prepared on the basis of the total of prepared checks and sent to the general ledger department. If the amount in the journal voucher agrees with the total debits posted to the accounts payable ledger, the entry is posted to the accounts in the general ledger.

## 17.11 COMPUTER-BASED PROCESSING SYSTEMS

When computer-based systems are used, purchases and cash disbursements transactions may be processed by the batch approach, by the on-line approach, or by a combination of the two approaches. In our examples, the on-line approach is used for purchases, the combined approach for payables, and the batch approach for cash disbursements.

(a) **Purchases Procedure.** Exhibit 17–9 presents a system flowchart of the on-line purchases processing procedure. The flowchart logically divides into three steps: inventory requisitioning, purchase order processing, and purchase approving.

1. Inventory requisitioning. In the first step the inventory records are checked to find those items whose on-hand quantities have been drawn below their reorder points. This step may be performed when batches of sales transactions are being processed to update the inventory master file. If backorders are prepared and processed, they could also be an input to this step.

2. Purchase order processing. The output from this first step is an inventory reorder list, which is, in effect, a batch of purchase requisitions. After being approved by an inventory control manager, the list is sent to the purchasing department. Buyers are assigned the various items on the list. By making inquiries of an on-line supplier reference file containing evaluation data, the buyers select suitable suppliers and enter them on the list. Then they enter the data needed to prepare purchase orders into their video display terminals. On being validated by an edit program, the purchase order data are stored temporarily in a purchase transaction file.

3. Purchase approving. In this step a purchase order is printed, using data from the transaction file as well as the supplier and inventory master files. Also, the computer system automatically assigns a number to the purchase order and dates it. A copy of the purchase order is placed in the on-line open purchase order file, and a notation of the order is placed in the appropriate record(s) of the inventory master file. The purchasing manager, or other authorized signer, reviews the purchase order. When he or she approves and signs the order, it is mailed to the supplier. If revisions are necessary, the buyer retrieves the purchase transaction data via the terminal, makes the changes, and prints a revised purchase order. At the end of the day a listing of the day's purchase orders is printed. Other reports, such as an inventory status report, may also be generated.

(b) **Receiving Procedure.** Exhibit 17–10 shows a system flowchart of an on-line receiving procedure. A clerk in the receiving department first counts and inspects the received goods. Then he or she keys the count and inventory item numbers into a departmental terminal, together with the related purchase order number listed on the packing slip. A receiving program checks the entered data against the on-line open purchase order file. Any differences between the ordered quantities and the counted quantities are displayed

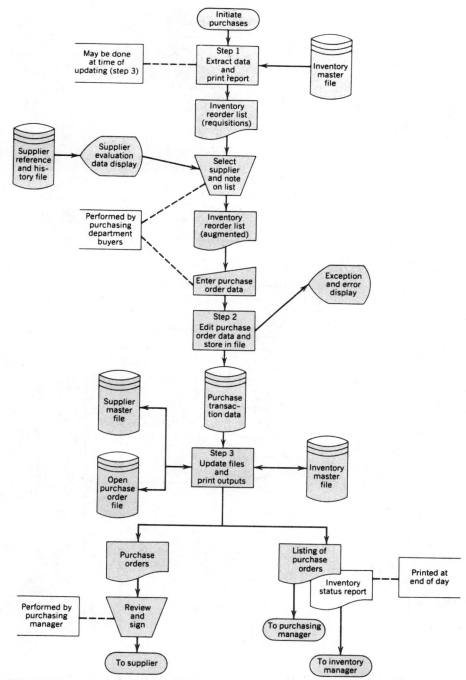

**Exhibit 17–9.** A system flowchart of an on-line computer-based purchases procedure. Adapted from J. Wilkinson, *Accounting Information System: Essential Concepts & Applications*, 2nd. ed. (NY: John Wiley & Sons).

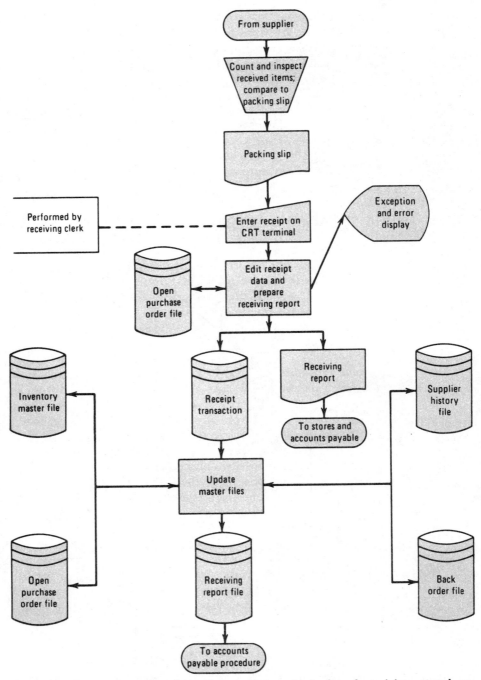

**Exhibit 17–10.   A system flowchart of an on-line computer-based receiving procedure.**
Adapted from J. Wilkinson, *Accounting Information System: Essential Concepts & Applications*, 2nd. ed. (NY: John Wiley & Sons).

on the terminal screen. Also, if no matching purchase order number is found in the on-line file, an alerting message is displayed.

Assuming that the goods are accepted, the program prints a prenumbered receiving report. A copy of this report accompanies the goods to the stores department, is signed, and is forwarded to the accounts payable department. The receiving program also (1) updates the inventory master file, increasing the on-hand quantity and eliminating the quantity on order, (2) notes the quantity received in the open purchase order file, and (3) notes the date of receipt in the supplier history file. If a backorder is involved, the backorder record is flagged.

**(c) Accounts Payable Procedure.** Exhibit 17–11 portrays a system flowchart of a combined on-line/batch payable procedure. As invoices are received from suppliers, an accounts receivable clerk performs a visual check for completeness, pulls the related receiving reports, and then computes a batch control total based on the invoice amounts. He or she enters into a terminal the batch total, plus key data elements from each invoice. An edit program validates the entered data, checks the quantities against those in the open purchase order file, recomputes the batch totals, and displays any differences. Then the vouched transaction data are stored until a designated processing time (e.g., end of the day). When the processing time arrives, an accounts payable posting program updates each record in the supplier master file (i.e., accounts payable subsidiary ledger) that is affected by a supplier's invoice. No sorting is necessary, since the program accesses and retrieves each supplier record directly from its location in the file. Then the program adds the amount of the supplier's invoice to the balance in the account. If all of the quantities pertaining to the related purchase order have been received, it also closes the purchase order and transfers its record to the supplier history file.

A variety of outputs are generated during this posting of the supplier master file. Prenumbered disbursement vouchers are printed, with one copy being filed together with the supporting documents in an unpaid vouchers file. A voucher register is printed to provide a key part of the audit trail. Summaries of the vouchers are added to the on-line open vouchers file. The debits from the summary vouchers are accumulated by account number; the totals are then compiled by the program into a sequentially numbered journal voucher and added to the journal voucher transaction file or posted immediately to the accounts in the general ledger. A total of the invoice amounts is also computed by the program and displayed on a terminal in the accounts payable department. A clerk verifies that this total agrees with the precomputed batch total.

**(d) Cash Disbursements Procedure.** Exhibit 17–12 shows a system flowchart of a batch cash disbursements procedure. Our illustrated procedure begins with the approved disbursement vouchers that have been printed during the payable procedure and filed with the supporting documents. This procedure has been widely used, and it provides a contrast to the on-line procedures discussed above. Also, it facilitates the prior computation of batch control totals and the review of documents by the check signer. However, it is important to recognize that the alternative on-line procedure offers significant benefits. For instance, it can use as transaction data the open vouchers stored in an on-line file, thereby omitting the need to key data from the vouchers. The vouchers in that file are sorted by due date at the end of each day, and the vouchers due to be paid on the next payment date are extracted.

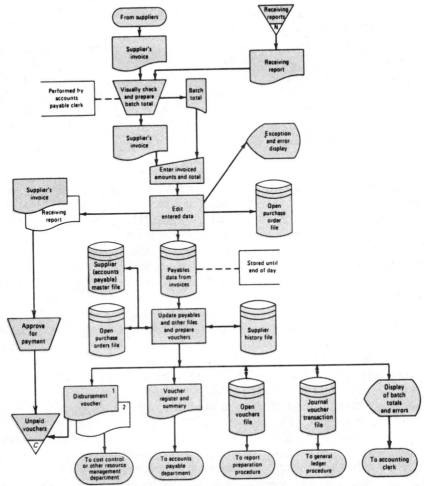

**Exhibit 17–11.** A system flowchart of an on-line computer-based payables procedure. Adapted from J. Wilkinson, *Accounting Information System: Essential Concepts & Applications*, 2nd. ed. (NY: John Wiley & Sons).

The batch procedure shown in the flowchart can be described as follows. On each payment date, the vouchers due to be paid are pulled from the unpaid vouchers file. As in all batch processing procedures, the first step is to compute one or more batch control totals. Then the data needed for preparing checks are keyed from the vouchers onto a magnetic disk. The data are edited, either during the keying or during a separate edit run. Next, the data are sorted, in run 1 by supplier account numbers. Finally, in run 2 the voucher amounts are posted to the accounts payable ledger. The program also prints the checks and a check register, as well as a summary of disbursements. It removes the paid vouchers from the open vouchers file (if one is maintained on-line) and adds the data concerning disbursements to the supplier history file. Finally, it accumulates the amounts disbursed and prepares a sequentially numbered journal voucher, which it adds to the journal voucher transaction file.

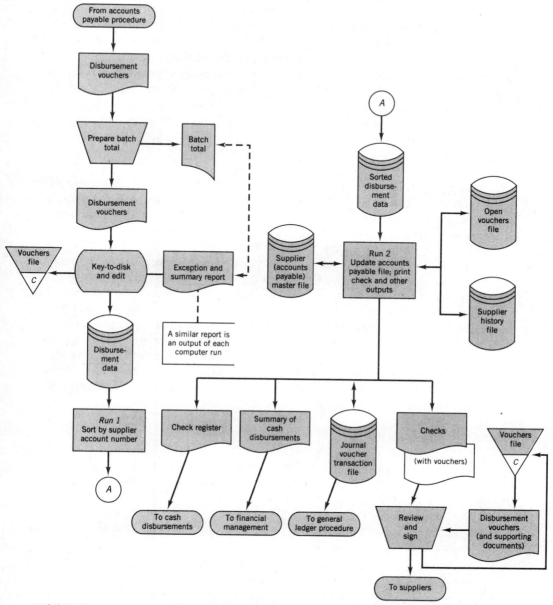

**Exhibit 17–12.** A system flowchart of a batch computer-based cash disbursements procedure. Adapted from J. Wilkinson, *Accounting Information System: Essential Concepts & Applications*, 2nd. ed. (NY: John Wiley & Sons).

The person authorized to sign reviews the checks and supporting documents. After signing, he or she forwards the checks to the mailroom. Alternatively, the checks may be signed automatically by the program in run 2, using a signature plate. If this is done, the checks are later reviewed to see that they pertain to authorized obligations.

Computer-based processing of expenditure cycle transactions offers several advantages, especially when on-line systems are employed. Processing efficiency is greater, since the relevant files are integrated and can be updated concurrently. Data in the files can be kept more up-to-date and can be easily retrieved when needed. Processed data are likely to be more accurate, since the transaction data are validated when entered. Also, batch totals can be employed regardless of whether processing is performed directly or performed sequentially. Furthermore, a greater variety of control reports and summaries can be prepared automatically by the application programs as a byproduct of processing steps. Finally, needed and useful outputs can be produced more easily.

## 17.12 CONTROLS NEEDED FOR THE EXPENDITURE CYCLE

Like transactions within other cycles, those within the expenditure cycle are also exposed to a variety of risks. Exhibit 17–13 lists representative risks and consequent exposures due to these risks. An adequate set of general and transaction (application) controls is necessary to counteract risks due to the above-mentioned exposures and to realize the objectives of the expenditure cycle.

Typical risks, for example, are that goods may be ordered that are not needed, or more goods may be ordered than are needed. Goods that are not ordered may be received or, conversely, goods may be ordered that are not received. These risks expose the acquiring firm to excessive inventory and storage costs or to possible losses due to goods being unavailable for sale. Another risk is that checks may be kited. Kiting is a type of embezzlement that involves transfers of checks among bank accounts. The purpose is generally to cover cash shortages or to inflate the assets. Transfers typically take place near the end of the month, so that float causes the checks not to be recorded until the following month. Such activity, besides causing the loss of cash through embezzlement, can lead to an overstating of cash balances in the financial statement.

## 17.13 CONTROL OBJECTIVES

In order to counteract exposures to the kind of risks listed in Exhibit 17–13, a firm must first clarify the control objectives that the management sees as crucial. With respect to the expenditure cycle, several key control objectives may be as follows:

- All purchases are authorized on a timely basis when needed, and are based on economic order quantity calculations.
- All received goods are verified to determine that the quantities agree with those ordered and that they are in good condition.
- All services are authorized before being performed and are monitored to determine that they are properly performed.
- All suppliers' invoices are verified on a timely basis with respect to accuracy and conformance with goods received or services performed.
- All available purchase discounts are identified, so that they may be taken if economical to do so.
- All purchase returns and allowances are authorized and accurately recorded and based on actual returns of goods.
- All cash disbursements are recorded completely and accurately.

| Risk | Exposure(s) |
|------|-------------|
| 1. Orders placed for unneeded goods or more goods than needed | 1. Excessive inventory and storage costs |
| 2. Receipt of unordered goods | 2. Excessive inventory and storage costs |
| 3. No receipt of ordered goods | 3. Losses due to stock-outs |
| 4. Fraudulent placement of orders by buyers with suppliers to whom they have personal or financial attachments | 4. Possibility of inferior or overpriced goods or services |
| 5. Overcharges (with respect either to unit prices or to quantities) by suppliers for goods delivered | 5. Excessive purchasing costs |
| 6. Damage of goods en route to the acquiring firm | 6. Possibility of inferior goods for use or sale (if undetected) |
| 7. Errors by suppliers in computing amounts on invoices | 7. Possibility of overpayment for goods received |
| 8. Erroneous or omitted postings of purchases or purchase returns to suppliers' accounts payable records | 8. Incorrect balances in accounts payable and general ledger account records |
| 9. Errors in charging transaction amounts to purchases and expense accounts | 9. Incorrect levels (either high or low) for purchases and expense accounts |
| 10. Lost purchase discounts due to late payments | 10. Excessive purchasing costs |
| 11. Duplicate payments of invoices from suppliers | 11. Excessive purchasing costs |
| 12. Incorrect disbursements of cash, either to improper or fictitious parties or for greater amounts than approved | 12. Losses of cash and excessive costs for goods and services |
| 13. Improper disbursements of cash for goods or services not received | 13. Excessive costs for goods or services |
| 14. Fraudulent alteration and cashing of checks by employees | 14. Losses of cash |
| 15. Kiting of checks by employees | 15. Overstatement of bank balances; possible losses of deposited cash |
| 16. Accessing of supplier records by unauthorized persons | 16. Loss of security over such records, with possibly detrimental use made of data accessed |
| 17. Involvement of cash, merchandise inventory, and accounts payable records in natural or man-made disasters | 17. Loss of or damage to assets, including possible loss of data needed to monitor payments of amounts due to suppliers within discount periods |

Exhibit 17–13.  A table of risk exposures within the expenditure cycle. Adapted from J. Wilkinson, *Accounting Information System: Essential Concepts & Applications*, 2nd. ed. (NY: John Wiley & Sons).

- All credit purchases and cash disbursements transactions are posted to proper suppliers' accounts in the accounts payable ledger.
- All accounting records and merchandise inventory are safeguarded.

To fulfill these control objectives, the firm must specify and incorporate adequate general controls and transaction controls.

General controls that particularly concern the expenditure cycle include those involving organizational independence, documentation, practices and policies, and security measures. Each is briefly discussed below.

*Organizational independence.*   With respect to purchases transactions, the units having custodial functions (i.e., the warehouse or receiving department) should be separate from each other and from those units that keep the records (i.e., inventory control department, accounts payable department, general ledger department, data processing department). Also, the purchasing department, whose manager authorizes purchases and returns, and the user departments, whose managers authorize services, should be separate from all the above-mentioned departments. Of course, this separation is not always possible with respect to services, since every department requires some types of services. For cash disbursements, the cash disbursements department and treasurer (who prepare and sign checks) should be separate from the accounts payable department, general ledger department, and data processing department.

*Documentation.*   Complete and up-to-date documentation should be available concerning the expenditure cycle, including copies of the documents, flowcharts, record layouts, and reports available. In addition, details pertaining to purchases and cash disbursements edit and processing programs should be organized into separate books or "packages" that are directed to programmers, computer operators, and system users. Furthermore, management policies concerning purchase discounts, purchase returns, and so forth should be in written form.

*Operating practices.*   Practices relating to such operational matters as processing schedules, control summaries and reports, and personnel should be clearly established and soundly based. For instance, all employees who handle cash should be required to be bonded and subjected to close supervision.

*Security measures.*   Security should be maintained (in the case of on-line systems) by techniques such as (1) requiring that clerks enter assigned passwords before accessing supplier and inventory files, (2) employing terminals with restricted functions for the entry of purchases and cash disbursement transactions, (3) generating audit reports (access logs) that monitor accesses of system files, and (4) dumping the supplier and inventory files onto magnetic tape backups. Security measures for manual and computer-based systems include the use of physically restricted stores areas (for protecting goods) and safes (for holding stocks of blank checks).

**(a) Transaction Controls.**   The following controls and control procedures are applicable to expenditure cycle transactions and supplier accounts. They are arranged by input, processing, and output categories.

Input controls include the following:

- Prenumbered and well-designed documents relating to purchases, receiving, payable, and cash disbursements.
- Validation of data on purchase orders and receiving reports and invoices as the data are prepared and entered for processing. In computer-based systems, validation is performed by means of the kinds of programmed checks listed in Exhibit 17–14.

| Type of Edit Check | Typical Transaction Data Being Checked | | Assurance Provided |
|---|---|---|---|
| | Purchases | Cash Disbursements | |
| **1.** Validity check | Supplier account numbers, inventory item numbers, transaction codes | Supplier account numbers, transaction codes | The entered numbers and codes are checked against lists of valid numbers and codes that are stored within the computer system. |
| **2.** Self-checking digit | Supplier account numbers | Supplier account numbers | Each supplier account number contains a check digit that enables errors in its entry to be detected. |
| **3.** Field check | Supplier account numbers, quantities ordered, unit prices | Supplier account numbers, amounts paid | The fields in the input records that are designated to contain the data items (listed at the left) are checked to see if they contain the proper mode of characters (i.e., numeric characters). If other modes are detected, an error is indicated. |
| **4.** Limit check | Quantities ordered | Amounts paid | The entered quantities and amounts are checked against preestablished limits that represent reasonable maximums to be expected. (Separate limits are set for each product.) |
| **5.** Range check | Unit prices | None | Each entered unit price is checked to see that it is within a preestablished range (either higher or lower than an expected value). |
| **6.** Relationship check | Quantities received | Amounts paid | The quantity of goods received is compared to the quantity ordered, as shown in the open purchase orders file; if the quantities do not agree, the receipt is flagged by the edit program. When an amount of a cash payment is entered as a cash disbursement transaction, together with the number of the voucher or invoice to which the amount applies, the amount in the open vouchers (or invoices) file is retrieved and compared with the entered amount. If a difference appears, the transaction is flagged. |
| **7.** Sign check | None | Supplier account balances | After the amount of a cash disbursement transaction is entered and posted to the supplier's account in the accounts payable ledger (thereby reducing the account balance of the supplier), the remaining balance is checked. If the balance is preceded by a negative sign (indicating a debit balance), the transaction is flagged. |
| **8.** Completeness check[a] | All entered data elements | All entered data elements | The entered transactions are checked to see that all required data elements have been entered. |
| **9.** Echo check[a] | Supplier account numbers and names, inventory item numbers and descriptions | Supplier account numbers and names | After the account numbers for suppliers relating to a purchase or cash disbursements transaction (and also the product numbers in the purchase transaction) have been entered at a terminal, the edit program retrieves and "echoes back" the related supplier names (and product descriptions in the case of purchase transactions). |

[a]Applicable only to on-line processing systems.

**Exhibit 17–14. A table of programmed edit checks useful for validating data entered into the expenditure cycle.** Adapted from J. Wilkinson, *Accounting Information System: Essential Concepts & Applications*, 2nd. ed. (NY: John Wiley & Sons).

When data are keyed onto a computer-readable medium, key verification is also recommended.

- Correction of errors that are detected during data entry. Detected errors should be corrected by an error correction procedure that is appropriate to the processing approach being employed.
- Use of control totals (when feasible) that are precomputed from amounts on suppliers' invoices and vouchers due for payment. These precomputed batch control totals are compared with totals computed during posting to the accounts payable ledger and during each processing run. In the case of cash disbursements, the precomputed total of vouchers is also compared with the total of the check register.

Processing controls include the following:

- Issuance of multiple-copy purchase requisitions, purchase orders, disbursement vouchers, checks, and debit memoranda on the basis of valid authorizations.
- Verification of all data elements and computations on purchase requisitions and on purchase orders by persons other than the preparers. Another verification consists of counting the quantities received and comparing the counted quantities against the ordered quantities.
- Vouching of all data elements and computations on suppliers' invoices, including comparisons with the corresponding purchase orders and receiving reports (in the case of goods) by an accounts payable clerk or computer program.
- Monitoring of all open transactions, such as partial deliveries and rejected goods. Also, all transactions in which one or more supporting documents are missing are investigated.
- Issuance of debit memoranda only on the basis of prior approval of the purchasing manager.
- Reconciliation of accounts in the accounts payable subsidiary ledger and the expense ledgers (if any) with control accounts in the general ledger.
- Verification that total posting to the accounts payable master file accounts agree with the totals posted to the general ledger accounts (in batch processing systems).
- Monitoring of discount terms relating to payments, in order to ensure that all purchase discounts are taken (if economical).
- Review of evidence supporting the validity of expenditures and the correct posting of amounts prior to the signing of checks.
- Use of check protectors to protect the amounts on checks against alteration before the checks are presented to be signed.
- Countersigning of checks over a specified amount by a second manager.
- Verification of all inventories on hand by physical counts at least once yearly, and reconciliation of the counted quantities with the quantities shown in the inventory records. The inventory taking should be performed under close supervision, and adjustments should be made when necessary to reflect the actual quantities on hand.
- Use of an imprest system for disbursing currency from petty cash funds, with the funds being subject to surprise counts by internal auditors or a designated manager.

- Establishment of purchasing policies that require competitive bidding for large and/or non-routine purchases and that prohibit conflicts of interest, such as financial interests by buyers in current or potential suppliers.

Output controls include the following:

- Establishment of clear-cut receiving and payable cutoff policies, so that the balances for inventories and accounts payable are properly measured at the end of each accounting period.
- Establishment of budgetary control over purchases, with periodic reviews of actual purchase costs and key factors such as inventory turnover rates.
- Comparison of monthly statements from suppliers with the balances appearing in the suppliers' accounts in the accounts payable ledger.
- Filing of copies of all documents pertaining to purchases and cash disbursements by number, including voided documents such as checks. The sequence of numbers in each file should be checked periodically to see if gaps exist. If transactions are not supported by preprinted documents, as often is the case in on-line computer-based systems, numbers are assigned to generated documents and the documents are stored in transaction files.
- In the case of computer-based systems, preparation of transaction listings (e.g., check registers) and account summaries on a periodic basis to provide an adequate audit trail.
- Reconciliation of all bank accounts monthly by someone who is not involved in expenditure cycle processing activities. New bank accounts should be authorized by designated managers only.

## 17.14 REPORTS AND OTHER OUTPUTS

Outputs generated by the expenditure cycle include financial and nonfinancial reports, purchases-related and cash-related reports, daily and weekly and monthly reports, and hard-copy and soft-copy reports. These reports are classified as operational reports, managerial reports, and inquiry screens.

**(a) Operational reports.** Various registers and journals help to maintain the audit trail. The invoice or voucher register is a listing of invoices received from suppliers or a listing of the vouchers prepared from the invoices. The check register is a listing of all the check issued. It is sometimes called a cash disbursement journal. Each day's listing is accompanied by a summary of the gross amount of payable reduced, the discounts taken, and the net amount paid.

Another group of reports deals with so-called open documents. The open purchase order report shows all purchases for which the related invoices have not yet been approved. The open invoice report lists invoices approved but not yet paid. Another report is the inventory status report, which lists quantities received, shipped, and on hand for each merchandise item. The receiving register contains a listing of all incoming shipments. The overdue deliveries report pinpoints those purchases that are yet to be received.

**(b) Inquiry screens.**   If part or all of expenditure cycle is on-line, a variety of reports can be produced interactively within seconds. Many programs have software packages that permit inquiries to be made of the system concerning orders, vendors, payments, order receipts, and other elements of the expenditure cycle.

**(c) Managerial reports.**   The expenditure cycle database contains considerable information that can aid managers in making decisions. This information can be used to generate a number of reports, some of which, such as the cash flow statement, overlap with the revenue cycle, while others focus on purchases, suppliers, and disbursements. Purchase analyses are useful reports that show the levels of purchasing activity for each supplier, inventory item, and buyer. For instance, a typical analysis may show the number and dollar amount of purchases placed with each supplier this year, as well as the average dollar expenditure. This analysis shows the degree to which purchases are concentrated with certain selected suppliers. An analysis of the number of purchases placed by each buyer indicates the relative productivity of the buyers. The most important report relating to suppliers is known as a vendor performance report. This report describes the performance of suppliers (vendors) in terms of on-time shipments, quality of goods, unit prices, level of service, and condition of goods delivered. Discounts on purchases are important to most firms. Thus, a purchase discounts lost report can highlight the relative effectiveness of a firm in paying invoices promptly.

## 17.15 CONCLUSION

The expenditure cycle facilitates the exchange of cash with suppliers for needed goods and services. Functions of the cycle (in the case of goods) are to recognize the need, place the order, receive and store the goods, ascertain the validity of the payment obligation, prepare the cash disbursement, maintain the accounts payable, post transactions to the general ledger, and prepare needed financial reports and other outputs. These functions are often achieved under the direction of the inventory management, as well as finance and accounting managers.

Most of the documentation used in the cycle arises from suppliers' or inventory records. Documents typically employed are the purchase requisition, purchase order, receiving report, supplier's vendors invoice, disbursement voucher, and check voucher. Preformatted screens may be used in on-line systems to enter purchases and cash disbursements data. The database includes files such as the supplier (vendor) master file, merchandise inventory master file, open purchase order file, cash disbursements transaction file, and supplier reference and history files. The expenditure cycle is exposed to a number of control risks. Steps can be taken to minimize such risks through organizational independence, segregation of duties, and written procedure documentation, among other controls.

# CHAPTER 18

# THE RESOURCES MANAGEMENT CYCLE: EMPLOYEE SERVICES

## 18.1 INTRODUCTION

This chapter and the next two introduce readers to the treatment of organizational resources within the accounting information system. For reasons of convenience, resources management activity will be viewed as being a single cycle, consisting of four sub-cycles or systems. These are employee services management, facilities management, inventory management, and funds management. We will deal with the first, employee services management in this chapter, while inventory management will be treated in Chapter 19. Facilities management and funds management will be dealt with together in Chapter 20.

## 18.2 RESOURCES MANAGEMENT: AN OVERVIEW

An organization is a composite of its resources, such as employees, facilities, and inventories. On the surface, an organization is no more than the sum of its resources, but the synergy resulting from effective organizational management can make it into something much more than the sum of its parts. The survival, not to mention the success and ongoing prosperity, of every organization depends on the manner in which it manages the resources available to it in the context of an ever-changing environment. Every organization uses resources that may be categorized as inventory, human and technical services, facilities, funds, and data. Each type of resource must be managed.

## 18.3 OBJECTIVES OF EMPLOYEE SERVICES MANAGEMENT SYSTEM

Employee services are the major component of the human services resource employed by organizations. In certain organizations, such as service-oriented firms and governmental agencies, this resource is responsible for the largest share of the operating expenditures. In other organizations, such as manufacturing firms, this resource can be a major cost in the conversion of raw materials into finished goods. Given the increasing automation in the industrial and service sectors, the complexity of products being manufactured, and changing demographics, labor-related costs are starting to represent something less than a prime share of the total product costs, even while the so-called manufacturing overhead increases its share.

Functions related to employee services include hiring, training, transferring, terminating, classifying, adjusting pay levels, establishing safety measures, maintaining employee benefits programs, reporting to governmental agencies, and payroll. The major purpose of the employee services management system is to facilitate the exchange of cash with employees for needed services. Because it centers on payments to employees, it is also known as the payroll system. The label is not fully descriptive of what management of human resources entails in the 1990s, and, for this reason, we will not use it here. Nevertheless, the key discussions here pertain largely to payroll functions.

Employee services functions discussed here deal with establishing pay status, measuring and recording the services rendered, preparing paychecks, issuing and distributing paychecks, distributing labor costs, and preparing required reports and statements. Exhibit 18–1 presents a data flow diagram that includes the several functions just described. Each function is viewed in the diagram as a logical process that links to other processes and to entities and data stores. The diagram emphasizes the parallel flows involving the distribution of paychecks and distribution or allocation of payroll-related costs. It also shows the several recipients of information based on employee services activity and pay-related reports.

## 18.4 ORGANIZATIONAL CONTEXT OF EMPLOYEE SERVICES

The functions described in Exhibit 18–1 are typically performed under the direction of the personnel and finance/accounting organizational units of the firm. The employee services management system therefore involves the interaction of the personnel information system and the accounting information system (AIS). In addition, every department or other organizational unit is involved, since the employees are located throughout the organization. Exhibit 18–2 shows the relations between key departments and the system functions.

(a) **Personnel Management's Role.**   Personnel management has the primary objective of planning, controlling, and coordinating the issues pertaining to employees—the internally employed human resource—within an organization. The personnel function itself may be under the direction of a vice-president of personnel, while managers who might report to this senior officer are those in charge of employment and personnel planning, safety and benefits, industrial relations, employee development, and human resource administration. Employment and personnel planning is concerned with recruiting and testing prospective employees, hiring selected employees, assuring sound promotion and termination procedures, and determining future personnel needs (in terms of both skills and management levels). Safety and benefits is responsible for employees' safety and health and for providing pleasant working conditions. The industrial relations unit is responsible for dealing with unions and other labor-related organizations. The employee development unit is concerned with training employees and with the improvement of executive skills in managers. Human resource administration is responsible for salary compensation plans, group insurance, and related programs; it also administers and maintains the records of all employees and related personnel actions. Thus, the human resource administration unit, at times called personnel administration, is the most closely related to the payroll activity and of most concern in this chapter.

(b) **Finance/Accounting.**   The objectives of financial and accounting management relate broadly to funds, data, information, planning, and control over resources. Organizational units that are within this function and are involved in the management of employee services include timekeeping, payroll, accounts payable, cash disbursements, cost distribution, and

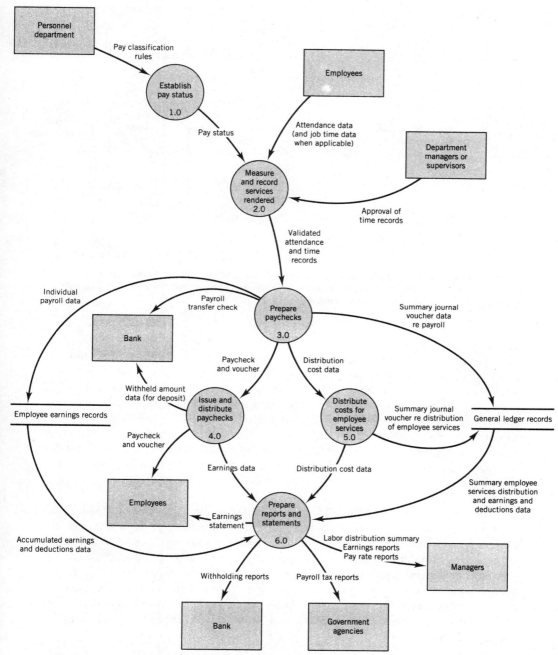

**Exhibit 18–1. A data flow diagram pertaining to employee services activity. Adapted from J. Wilkinson, *Accounting Information System: Essential Concepts & Application*, 2nd. ed. (NY: John Wiley & Sons).**

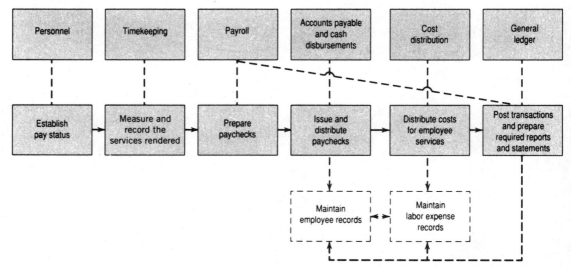

**Exhibit 18–2.** A flowchart of typical functions and related organizational units of an employee services system. Adapted from J. Wilkinson, *Accounting Information System: Essential Concepts & Applications* 2nd. ed. (NY: John Wiley & Sons).

general ledger. Timekeeping maintains control over the time and attendance records of hourly employees. Payroll prepares paychecks, maintains the payroll records, and prepares required reports and statements. Accounts payable, in the context of employee services, approves the disbursement voucher pertaining to employee services. Cash disbursements, together with the cashier, signs and distributes the paychecks. Cost distribution maintains the records reflecting detailed costs of employee services, as allocated to responsibility centers and products. General ledger maintains control over all asset, equity, expense, and income accounts. Note that the timekeeping and cost distribution units are more typically found in manufacturing firms than in other types of organizations.

**(c) Managerial Decision Making.** Three managers who have key responsibilities with respect to the employee services management system are the personnel manager, the treasurer, and the controller. They must make constantly make and remake decisions that influence acquiring, managing, and paying employees and keeping them motivated. The personnel manager depends most heavily on the personnel information system to provide needed information for decision making. Many of the decisions are not based, even in part, on routine accounting transactions. However, some of the payroll-related decisions do draw on information from the AIS. Some of the typical decisions most closely bound to the payroll activity are:

- What organizational units are to be included within the personnel function?
- Which applicants should be hired as new employees?
- Which employees should be promoted, transferred, given pay raises, or terminated?
- What benefits of financial value should be made available to employees?

The controller and treasurer depend primarily on the AIS for needed decision making information. Much of the information is a byproduct of processing within the employee services management system. Among the decisions requiring answers of these managers (or their subordinates) are:

- What employees' earnings records are to be maintained concerning amounts paid to employees?
- Which payroll deduction plans, for example, United Way, are to be made available to employees?
- What type of payroll bank accounts are to be established?
- Who is to sign paychecks, and how are pay amounts to be distributed to employees?
- What pay periods (e.g., weekly, biweekly) are to be established?

Among the information needed to make these decisions are qualifications of applicants, evaluations of employees, wage and salary scales (both for the firm and for the industry), expected costs of benefit plans, and educational and work experience histories of employees.

## 18.5 DOCUMENTATION NEEDED FOR EMPLOYEE SERVICES

Source documents typically used in the management of employee services include the following:

*Personnel action form.* A personnel action form serves to notify interested parties of actions concerning employees. These actions may pertain to hiring, changing of status, evaluating job performance, and so on. A personnel action form notifies the payroll department of a situation or change affecting the status of an employee's pay. Another category of personnel actions concerns deductions. Some of these forms are issued by the firm and some by government agencies. An example of the latter is the W4 Form, Employee Withholding Allowance Certificate, provided by the Internal Revenue Service and informing the employer about the number of exemptions being claimed by the employee. Other documents provide employees with the total amount of taxes withheld from their earnings during the fiscal year.

*Time and/or attendance form.* The timecard, also known as a clock card, documents the actual hours spent by hourly employees at their work locations. It contains an employee's name and number, plus the dates of the applicable pay period. Usually, the bottom of the card has a space for the supervisor's signature attesting that the hours recorded are accurate. Attendance forms and other forms of timecards are not limited to hourly employees, time sheets are also used by salaried employees, such as those working for public accounting firms.

*Job-time ticket.* In contrast to the timecard, which focuses on attendance at the work site, the job-time ticket focuses on specific jobs or work orders. Each time an hourly employee, such as a production worker, begins and ends work on the job, he or she records the time on the card. As in the case of the timecard, the means of entering the times may be a timeclock or terminal. If appropriate to the employees' tasks, spaces are provided for entering the productivity in terms of pieces completed during the elapsed periods.

*Paycheck.*  A paycheck, with voucher stub, is the final supporting document in the employee services management cycle. The stub shows all necessary details, including overtime pay and deductions.

## 18.6  ACCOUNTING ENTRIES

Exhibit 18–3 shows the flow of payroll transactions through the accounting cycle. The transactions begin with records reflecting times worked by employees. These source documents, when expressed in dollar terms after applying rates of pay, are used in the two steps comprising payroll processing. As the first step, the data from these source documents are entered into a summary that distributes the costs for employee services to activities such as selling goods, performing administrative tasks, and manufacturing goods. The totals from this summary are posted to the general ledger accounts as follows, to record the gross payroll costs of employee services for the period:

Dr. Selling Expense Control XXX
Dr. Administrative Expense Control XXX
Dr. Direct Labor or Work-in-Process XXX
Dr. Manufacturing Overhead Control XXX
Cr. Wages and Salaries Payable XXX

Details can also be posted to appropriate expense subsidiary ledgers for manufacturing overhead control, selling expense control, and administrative expense control, if such ledgers are maintained.

The second step consists of using the time data to prepare payroll checks. Copies of these paychecks, together with attached earnings statements, are the source documents

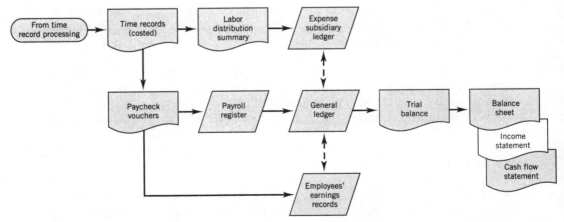

**Exhibit 18–3.  A diagram showing the flow of payroll transactions through the employee services system. Adapted from J. Wilkinson, *Accounting Information System: Essential Concepts & Applications*, 2nd. ed. (NY: John Wiley & Sons).**

from which pay data are entered into the payroll register. They are also used as the basis for posting the same data to the earnings records of individual employees. Totals from the payroll register are then posted to the general ledger accounts as follows, to record the payroll for the period:

Dr. Wages and Salaries Payable XXX
Cr. Federal Income Taxes Withholding Payable XXX
Cr. FICA Tax Payable XXX
Cr. Other Accrued Deductions Payable XXX
Cr. Cash XXX

A payroll clearing account may be used in place of the wages and salaries payable account, since the effect is to clear the costs of employee services. Either account can also serve as a control over the total of the gross amounts of pay for all employees.

## 18.7 TRANSACTION CODING

Codes are useful for identifying data needed in payrolls and labor cost distributions. In addition to codes for general ledger accounts, codes can be used to identify employees, departments, production jobs, and skills. Many firms use an employee's social security number, since it is unique and familiar to the employees. Other firms may use group codes that include factors such as pay category (i.e., hourly or salaried), department number, skill, date of hire, and self-checking digit.

## 18.8 DATABASE NEEDED FOR EMPLOYEE SERVICES

A considerable amount of data are needed in order to process the payroll, as well as to take care of other functions associated with employee services. To facilitate the use of such data, they are arranged in assorted files. Among the files needed in managing employee services are the employee payroll master, personnel reference and history, time record transaction, paycheck transaction, compensation reference, and personnel planning files. These are described below.

*Employee payroll master file.*   An employee payroll master file contains the earnings records of the employees. It is updated to show the amounts received from paychecks at the end of each pay period. Typical items listed in this file are: employee's name, identification number, marital status, number of exemptions claimed, department where employed, pay classification, pay rate, overtime rate, deduction rate, year-to-date withholding, and year-to-date gross pay. Generally, the primary and sorting key is the employee number. The accumulated amounts shown in the record serve various purposes. For instance, the year-to-date gross pay has the purpose of determining the maximum levels for deductions pertaining to unemployment benefits and social security.

A major concern with respect to this file is keeping the records and their permanent data up-to-date. If an employee marries and obtains a new last name, for instance, this change should appear quickly in her record. When a new employee is hired, a record must be established before the end of the pay period. On the other hand, when an employee is terminated, the record should not be discarded until after the end of the year. Certain year-end reports require data concerning all employees who were active during any part of the year.

*Personnel reference and history file.*    As the main source of personnel data in the firm, the personnel reference and history file complements the payroll master file. It contains a variety of nonfinancial data as well as financial data concerning each employee. For instance, it might contain the employee's address, skills, job title, work experience, educational history, performance evaluations, and even family status. This file may be consolidated and maintained in the personnel department. Alternatively, it may be split into several reference files, which may be located in the payroll and/or data processing departments as well as the personnel department. A related file is the skills file, which provides an inventory of job capabilities required by the firm and the employees who currently possess each skill. This type of file enables a firm to locate qualified candidates when an opening or new need arises.

*Time record transaction file.*    A time record transaction file consists of copies of all the time cards or sheets for a particular pay period. In computer-based systems they are likely stored on magnetic media for use in processing the payroll.

*Paycheck transaction file.*    In a manual system the payroll transaction file consists of a copy of each current paycheck, arranged in check number order. In a computer-based system the record layout on magnetic media may appear similar to a record in the check disbursements transaction file.

*Compensation reference file.*    A table of pay rates and salary levels for the various job descriptions serves as a compensation reference file.

*Personnel planning file.*    In order to provide the basis for planning for future personnel needs, a firm may maintain a collection of information relating to current and past trends as well as projections. It might show the number of employees in each department during the past ten years, for instance, as well as the turnover for each department.

## 18.9 DATA FLOWS AND PROCESSING FOR MANUAL SYSTEMS

Exhibit 18–4 presents a document flowchart of a procedure involving the payment of hourly operations-type employees (e.g., production employees) who also work directly on specific jobs. The numbers on the flowchart designate the various control points within the procedure and parallel the functions described earlier.

1. Establishment of pay status. This beginning function takes place in the personnel department, where all of the personnel actions and changes are prepared and then transmitted to the payroll department.
2. Measurement of the services rendered. The time records are prepared in the operational (e.g., production) departments and timekeeping areas. Employees clock in and clock out under the eye of a timekeeper. The job-time tickets are available right at the work site. Employees either punch the tickets on a clock or mark them manually under the eye of their supervisor. At the end of each day, the

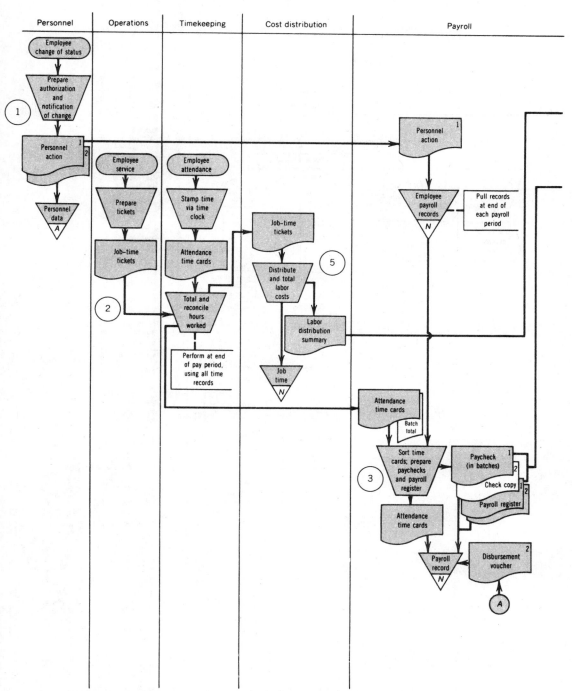

**Exhibit 18–4.** A document system flowchart of a manual employee payroll transaction processing procedure. Adapted from J. Wilkinson, *Accounting Information System: Essential Concepts & Applications*, 2nd. ed. (NY: John Wiley & Sons).

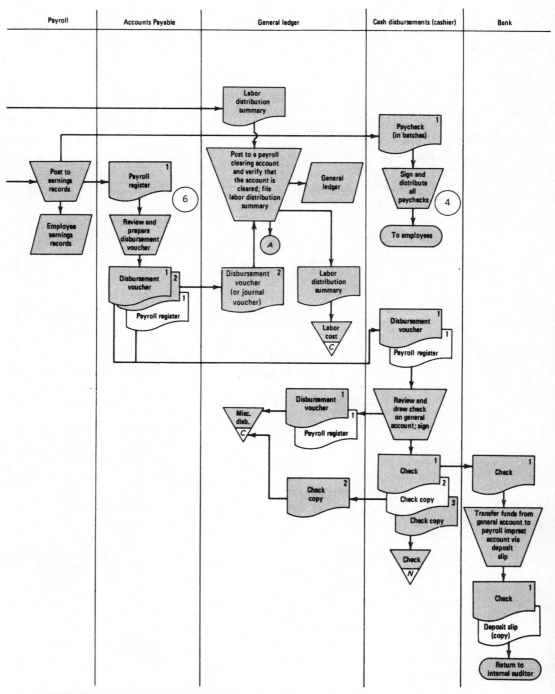

**Exhibit 18–4.    (Continued).**

job-time tickets are collected and approved by the employees' supervisor. Then the supervisor forwards the tickets to the timekeeper. At the end of the pay period the timekeeper compares the total hours shown on each employee's job-time ticket with the total hours shown on his or her attendance timecard. If the two sets of total hours are approximately equal (allowing for breaks, lunch, etc.), the time records are said to be reconciled. Then the timekeeper sends the attendance timecards to the payroll department (together with the total of hours worked) and the job-time tickets to the cost distribution department.

3. Preparation of the paychecks. In the payroll department, a clerk prepares a paycheck and voucher stub for each employee, based on data from the time card and from the employee's payroll reference file. Next, the clerk enters the relevant information from the paycheck and voucher stub (i.e., gross pay, deductions, net pay, overtime premium) on the payroll register. Another clerk then posts the information to the employee's earning record (i.e., the payroll master). Still another clerk verifies that the hours used in preparing the payroll register equal the total hours on the timecards and that the total payroll amount entered into the register equals the total amount posted to all employees' earnings records. The paychecks and attached voucher stubs are sent, in a batch, to the cash disbursements department (or cashier).

4. Issuance and distribution of the paychecks. On receiving a copy of the payroll register, an accounts payable clerk verifies its correctness and then prepares a disbursement voucher. A clerk in the cash disbursements department draws a check on the firm's regular bank account and gives it to the cashier for signing. The signed check is delivered to the bank and deposited in the special imprest payroll account. Then the cashier signs all the paychecks. A paymaster (a designated person not otherwise involved in personnel or payroll procedures) distributes the paychecks.

5. Distribution of the labor costs. Meanwhile, a clerk in the cost distribution department spreads or distributes the costs of services incurred by the operational personnel (e.g., production employees) to the various jobs in progress. The clerk next reports the costs, via a labor distribution summary or a journal voucher, to the general ledger department. Then the general ledger clerk debits the amounts to the various labor-related accounts (e.g., direct factory labor) and credits a payroll control account (e.g., wages and salaries payable). Subsequently, the general ledger clerk clears the payroll clearing account by reference to the disbursement voucher (or related journal voucher) prepared by accounts payable. That is, he or she debits the payroll clearing account and credits the cash account and the accounts with liabilities for payroll deductions. Since the total of gross pay from both sources (labor cost distribution and disbursement voucher) should be equal, the payroll clearing account will be cleared to zero if processing has been correct. Note that this clearing procedure is a partial substitute for the computation of formal batch totals in the timekeeping department. When attendance time records are not accompanied by time records related to jobs, it is highly desirable to compute batch totals at the point where the time records are assembled.

6. Preparation of required outputs. Numerous payroll-related reports and statements and other outputs are prepared. The only outputs shown in the flowchart are the labor distribution summary (mentioned in the previous step) and the payroll register. Additional reports can also be extracted from the data available.

## 18.10 COMPUTER-BASED PROCESSING SYSTEM

Although the on-line approach may be used in computer-based processing, the batch approach is better suited to the payroll procedure. Since all of the records in the employee payroll master file are affected, sequentially accessing and updating the records is the more efficient alternative. Exhibit 18–5 therefore shows a system flowchart in which the batch approach is applied. The flowchart has three segments: time date entry, paycheck preparation, and pay status changes.

*Time data entry.*   Attendance timecards are first gathered in a batch by the timekeeper and transmitted to the payroll department. In the system being described, the employees do not prepare job-time tickets. Therefore, to enhance control the timekeeper (or a payroll clerk) computes batch totals based on the time records. One total is based on the hours worked, a second total is based on the employee numbers, and a third total is based on a count of the timecards.

*Paycheck preparation.*   The batch of timecards, prefaced by a batch transmittal sheet, is forwarded to the data processing department. There, the time data are keyed onto a magnetic disk and edited. In the first computer processing run the data are sorted by employee numbers. In run 2 the time data are processed to produce paychecks (and voucher stubs). The program also updates the employee payroll master file and prints the payroll register. The paychecks and a copy of the register are sent to the cashier, where the paychecks are signed and distributed. (The transfer of funds from the regular account may be included if desired.) The program in run 2 also adds a journal voucher concerning the payroll transaction to the general ledger transaction file.

*Pay status changes.*   The left side of the flowchart portrays changes made in the pay status of employees. Clerks in the personnel department enter all personnel actions via departmental terminals. Since the employee payroll master file is stored on a magnetic disk, the actions (e.g., a change in pay rate) can be entered into the affected employee records promptly by direct access. Thus, all actions can be effected during the pay period and before the payroll processing begins.

## 18.11 CONTROLS NEEDED FOR EMPLOYEE SERVICES

Transactions within the employee services system are exposed to a variety of risks. Exhibit 18–6 lists representative risks and consequent exposures due to these risks. With respect to the employee services system, several key objectives are to ensure that:

- All services performed by employees, including hours worked on specified tasks such as production jobs, are recorded accurately and in a timely manner.
- All employees are paid in accordance with wage contracts or other established policies.
- All paychecks are calculated accurately, with due allowance for authorized payroll deductions and approved benefit programs.
- All costs for employee services are distributed to accounts in accordance with clearly established accounting policies.

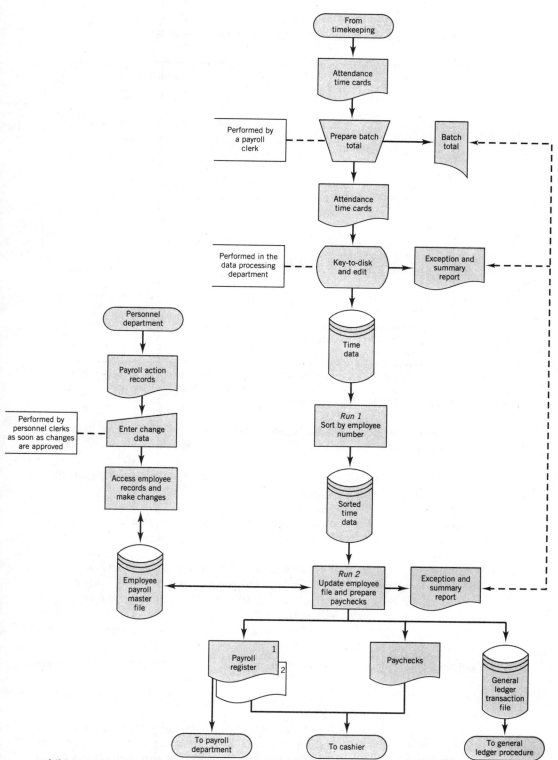

**Exhibit 18–5.** A system flowchart of a batch computer-based employee payroll transaction processing procedure. Adapted from J. Wilkinson, *Accounting Information System: Essential Concepts & Applications*, 2nd. ed. (NY: John Wiley & Sons).

| Risk | Exposure(s) |
|------|-------------|
| 1. Employment of unqualified persons | 1. Lessened productivity and higher training costs |
| 2. Employment of larcenous persons | 2. Possibility of loss of assets and circumvented policies and controls |
| 3. Errors or omissions in time records | 3. Incorrect payroll records and labor distribution summaries |
| 4. Errors in payments to employees | 4. Possibility of overpayments and/or adverse effects on employee morale; erroneous quarterly statements sent to federal and state agencies |
| 5. Incorrect disbursements of paychecks to fictitious or terminated employees, or diversions of valid paychecks to unentitled employees | 5. Excessive wage and salary costs |
| 6. Errors in charging labor expenses or in stating payroll liabilities | 6. Incorrect levels for expense and liability accounts |
| 7. Violation of government regulations and laws, with regard to payments or reporting requirements | 7. Possibility of penalties and fines being assessed |

**Exhibit 18–6. A table of risk exposures within the employee services management system. Adapted from J. Wilkinson** *Accounting Information System: Essential Concepts & Applications,* **2nd. ed. (NY: John Wiley & Sons).**

- All required reports are accurately and completely prepared in accordance with prescribed laws and regulations and submitted by their scheduled dates.

To fulfill these control objectives, the firm must then specify and incorporate adequate general controls and transaction controls.

**(a) General Controls.**   General controls include those involving organizational independence, documentation, security measures, and practices and policies.

- Within the organization, the persons and units that have custodial functions (i.e., the cash disbursements department, cashier, and paymaster) should be separate from those units that keep time records (i.e., timekeeping) and that prepare the payroll documents (i.e., the payroll department). Also, the personnel department, which authorizes personnel actions, and the departmental supervisors, who approve the time records, should be separate from all the above-mentioned units.
- Complete and up-to-date documentation should be available concerning the management of employee services.
- Security should be maintained (in on-line systems) by such techniques as (a) requiring the clerks enter assigned passwords before accessing employee payroll files, (b) employing terminals with restricted functions for the entry of personnel actions, (c) generating audit reports (access logs) that monitor accesses of system files, and (d) dumping the employee files onto magnetic tape backups. Physical security measures include the use of safes for holding stocks of blank paychecks and signature plates.

- Operating policies and practices relating to processing schedules, reports, changes in programs, and other matters should be clearly established.

**(b) Transaction (Application) Controls.** The following controls and control procedures are applicable to employee services transactions and employee records:

- Documents relating to payables and cash disbursements are prenumbered and well designed. Also, timecards, and job-time tickets where applicable, are preprinted with the employees' names and numbers.
- Data on time records are validated (and key-verified if suitable) as the data are prepared and entered for processing. In the case of computer-based systems, validation is performed by means of programmed checks such as (a) validity checks on employee numbers, (b) limit checks on hours worked, (c) field checks on key identification and amount data, and (d) relationship checks on employee numbers and related departments to which employees are assigned.
- During payroll processing, the results, such as net pay, are validated. In the case of computer-based systems, validation is performed by means of programmed checks such as the sign check and cross-foot balance check. Thus, the net-pay field of each paycheck is verified by the former to see that the sign is positive and by the latter to see that the amount equals the gross pay, less all deductions.
- Errors detected during data entry or processing are corrected as soon as possible by means of an established error-correction procedure. A part of this procedure may involve the printing of suitable exception and summary reports during edit runs. In the case of the payroll application, processing of paychecks may be delayed until all transaction data errors and discrepancies have been corrected.
- Personnel actions (such as new hires and pay rate changes) and paychecks are issued promptly on the basis of valid authorizations.
- Timecards, and job-time tickets where applicable, are approved by the employees' supervisors.
- Where job-time tickets are used, the hours that they reflect for each employee are reconciled with the hours shown on the attendance timecards.
- Batch control totals are precomputed on hours worked, as reflected by timecards, and on net pay amounts, as shown in the payroll register; these batch totals are compared with totals computed during paycheck preparation and during postings to the employee payroll master file, respectively.
- Paychecks are drawn on a separate payroll-imprest bank account.
- Voided paychecks are retained, in order that all paycheck numbers can be accounted for.
- Unclaimed paychecks are traced back to the time records and employee payroll master file, to verify that they belong to actual, current employees.
- In the case of computer-based systems, a preliminary payroll register is reviewed before the paychecks are printed, to determine that all errors have been corrected. Payroll account summaries are also printed periodically to enhance the audit trail.

In addition, all controls pertaining to cash disbursements also apply to the issuance of paychecks.

## 18.12   REPORTS AND OTHER OUTPUTS

The reports made available to managers and other users by the employee services can be either to meet operational needs or to help managerial decision making.

(a) **Operational reports.**   One of the most used outputs is the payroll register. In essence, it is the journal in the payroll procedure, and it lists the key payment data concerning each employee for a single pay period, ranging from gross pay to net pay. A related output is the deduction register, which provides a detailed breakdown of the deductions for each employee. The cumulative earnings register shows amounts earned year-to-date, and possibly quarter-to-date, for each employee. Various control reports are also needed. One example is a report that shows the number of checks printed and the total amounts of the checks.

Required governmental reports include those pertaining to withholdings of social security and federal income taxes, plus a variety of others. Some are due during the month following the end of each quarter; others are due during the month following the end of each year.

(b) **Managerial reports.**   Various analyses are of interest to managers, such as those pertaining to absenteeism, overtime pay, turnover, sales commissions, and indirect labor costs. One useful analysis is a projection of salaries for the upcoming months of the year. Other reports that are often helpful include surveys of average pay rates per occupational category, compared with similar firms, and personnel strength reports, showing levels of staffing and changes during the past month.

The labor distribution summary can serve two purposes: as the basis for accounting entries and as an analysis for management. In essence, the summary shows the amounts of employee services costs to be distributed to various accounts. However, it can also include details concerning the costs incurred by individual employees and responsibility centers for various tasks. For instance, it might show that Ray Valdez incurred $350 this week with respect to Production Job 301 and $420 with respect to Production Job 318. In addition, it separates the costs between direct labor and indirect labor.

## 18.13   CONCLUSION

Resources allow an organization to function, and they must be managed. The employee services function consists of hiring, training, assigning, compensating, evaluating, and when needed, terminating employees. The focus of the chapter is on compensating employees, or payroll function, even though employee services or human resources comprise of much more. Functions surveyed include establishing pay status, measuring the services rendered, preparing paychecks, issuing and distributing paychecks, distributing employee service costs, and preparing required reports and statements. This system involves the personnel, timekeeping, payroll, accounts payable, cash disbursements, cost distribution, and general ledger units. Key inputs are the personnel action form, time record, job-time ticket, and paycheck. The database includes the employee payroll master and personnel reference and history files. Outputs include the payroll and other registers, plus analyses of labor cost components.

# CHAPTER 19

# THE RESOURCES MANAGEMENT CYCLE: INVENTORY MANAGEMENT

## 19.1 INTRODUCTION

This chapter continues the coverage of systems within the resources management cycle by surveying inventory management. Inventory refers to goods held for the purpose of sale. The goods held in the inventory by a merchandiser like Sears are sales-ready, finished products. But, in the case of a manufacturer of automobiles, like Ford, it will include raw materials, work in process, finished goods, component parts for factory machines, and supplies. Inventory represents a firm's reason for being. Both manufacturers and merchandisers are in business to sell, and without inventories there will be nothing to sell. Furthermore, inventories can represent a major portion of a firm's working capital. Not only do inventories tie up capital, but their storage and safe-keeping also consumes considerable resources.

   The discussion of inventory management is split into two sections: Inventory management for merchandisers is discussed first, followed by a discussion of inventory management for manufacturers. The discussion of the merchandise inventory system is relatively brief, since various facets concerning the inventory management system for a merchandising firm were included in the earlier discussions of interfacing cycles and systems. For example, Chapter 16, which focused on the revenue cycle, discussed the system considerations governing the sale of inventory, while Chapter 17, on the expenditure cycle, dealt with payments for the merchandise acquired. In Chapters 16 and 17, the emphasis was on those firms that acquire merchandise for resale. Given this, the discussion of merchandise inventory in this chapter will fill in the gaps. After discussing the inventory resource for merchandisers, we will go on to discuss inventory management for firms that manufacture products.

## 19.2 MERCHANDISE INVENTORY MANAGEMENT SYSTEM

The major purpose of the merchandise inventory management system is to facilitate the inflows, storage, and outflows of the merchandise needed by a firm that obtains revenues from the merchandise's resale. Key objectives within this purpose are to (1) assure that an adequate quantity of merchandise is available to meet the demands of customers, (2) reflect accurate values for the merchandise on hand, (3) safeguard the merchandise from theft and other various losses, and (4) handle returns and other adjustments in a prompt and judicious manner.

   Major functions related to merchandise inventory consist of acquisition, receipt and storage, shipment, and adjustment. In each of these functions, the merchandise inventory management system tracks the flows of inventory items and generates information to aid in making decisions. We have already referred to most of the concerns and actions that pertain to the management of merchandise inventory. However, to help readers better visualize the central place of the inventory resource in such firms, we provide an overview of the merchandise inventory management system in this section. Exhibit 19–1 presents a data flow diagram that includes the major functions related to merchandise inventory: acquiring inventory, receiving and storing it, shipping it to purchasers, and adjusting inventory valuation. These functions are discussed below.

(a) **Acquiring Merchandise.**   In order to restock merchandise, a firm must first determine when and how much to reorder. This information is often provided by reference to a reorder point and reorder quantity, whose values are computed in accordance with inventory control principles described in Chapter 17. The information concerning reorders may be provided by a purchase requisition. Alternatively, it may appear on a merchandise reorder report, if one is available.

Data concerning the various items of merchandise inventory are typically maintained in merchandise inventory ledger records. In a manual system, these records are usually kept on ledger cards, one card per merchandise inventory item. Included on a ledger card will likely be a header data, such as the item number, the description of the item, its unit cost, and its location in the warehouse. The body of the ledger card will show the data for each transaction, plus the balances after each transaction.

Before leaving this function, we should observe its effects on a computer-based inventory processing system. As Exhibit 19–2 shows, the ordering procedure might consist of entering data from the merchandise reorder report, via a terminal, into the system. Purchase orders are printed, with a copy being stored in the on-line open purchase orders file. Also, the merchandise inventory master file is updated to reflect the ordered quantity. This merchandise inventory item record, which corresponds to the merchandise inventory ledger card in a manual system, contains a quantity on-order field.

(b) **Receiving and Storing Merchandise.**   When ordered merchandise items arrive, the quantities received are posted to the merchandise inventory ledger records, leading to an increase in the quantities on hand. Exhibit 19–2 shows this posting for a computer-based system. It also indicates that the related purchase order is removed from the open purchase orders file upon its receipt and a copy of the receiving report is added to the receipts transaction file. If a portion of the received quantities are to be allocated to particular customers, perhaps because of backorders, the affected quantities should be noted.

For tracking inventory, most firms that have computer-based inventory systems employ the perpetual method. Under this method, the merchandise inventory account in the general ledger is debited when the payable relating to a purchase has been established. If the inventory account in the general ledger is to be used as a control account, costs pertaining to the various items comprising the purchase must also be posted to the merchandise inventory item records. In a subsequent section, we discuss the basics of inventory costing. The amount posted to each inventory item record is roughly equal to the unit cost of the item times the quantity received. However, allowance may need to be made for costs pertaining to freight and handling.

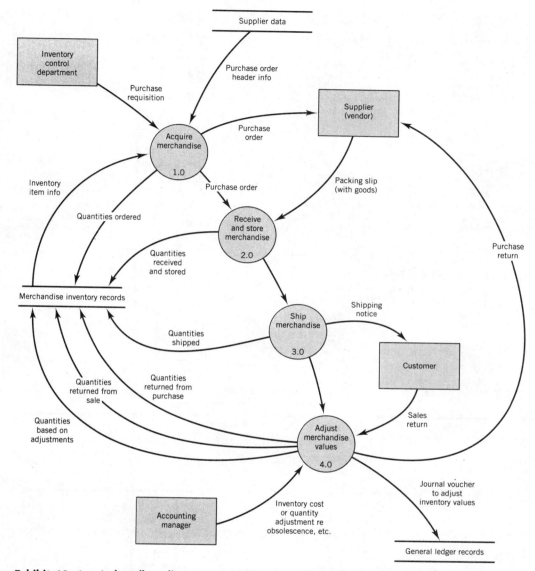

**Exhibit 19–1.  A data flow diagram pertaining to merchandise inventory activity.
Adapted from J. Wilkinson, *Accounting Information System: Essential Concepts &
Applications*, 2nd. ed. (NY: John Wiley & Sons).**

**(c) Shipping Sold Merchandise.**    When merchandise is sold, it is normally shipped a short
time later. Then the quantities sold are removed from the affected inventory records. If
the perpetual method is used, the value of the inventory sold on a given day, for instance,
April 1, would be posted to the inventory account in the general ledger (after billing) by
the following entry, to record the value of inventory sold on April 1, 19XX:

Dr. Cost of Goods Sold        XXX
Cr. Merchandise Inventory    XXX

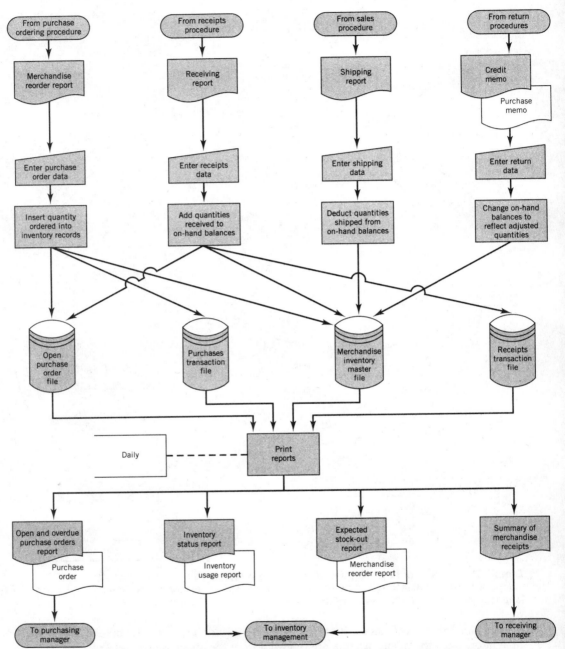

**Exhibit 19–2. System flowchart of an on-line computer-based inventory management system. Adapted from J. Wilkinson, *Accounting Information System: Essential Concepts & Applications*, 2nd. ed. (NY: John Wiley & Sons).**

The measured cost valuation of each inventory item involved in a sale must also be reduced. In determining the amounts to be reduced, the firm multiplies the quantity sold of each item times its unit cost. The unit cost, in turn, is based on the particular inventory valuation method employed by the firm.

**(d) Adjusting Value of Merchandise.**   A firm may accept returns of merchandise it has sold, and it may also return merchandise to its suppliers. A firm may also write down the recorded value of stored inventory, due to obsolescence, pilferage, or other reasons. Such adjustments, which are quite frequent, are posted to the merchandise inventory master file, as Exhibit 19–2 shows.

## 19.3 INVENTORY VALUATION METHODS

Before discussing inventory valuation methods, it is best to clarify the various costs associated with inventory. The costs directly related to inventory are:

- Purchase costs refer to the value of the inventory itself. This is the amount paid to the vendor in exchange for the goods purchased.
- Ordering costs define the managerial and clerical expenses incurred to prepare the purchase order, or in the case of manufacturing firms, the production order.
- Carrying costs are those incurred for inventory's storage, handling, insurance, pilferage, spoilage, obsolescence, taxes, and the opportunity cost of capital tied in the inventory. The last one, the opportunity cost, does not appear in the accounting statements, but managers who ignore it are putting their firms' future in danger.
- Stockout costs are incurred when the needed item is not in stock. It may take the form of lost sales revenue, the cost of delayed production, or penalties incurred for not meeting customers' deadlines. Again, this is included as a part of the accounting reports.
- Set-up costs occur because different products require equipment to be reset to meet the specific needs of each batch of the product. In addition to equipment setups, making a batch of product different from the one previously made involves obtaining the raw materials from the warehouse, moving out the previous stock at the workstations, and filing paperwork. While setups are being changed, workers involved in operating the equipment may be idle, though they are still being paid. The set-up costs are non-value-adding to the firm; they can be minimized by redesigning the production process and simplifying the product design.

The value of inventory on hand must be determined periodically, along with the value of other assets, to enable a company to determine its cost of goods sold and to state the firm's financial position. The basis used for inventory valuation has a significant effect on the costs and earnings reflected in the operating statements and on the asset value shown on the balance sheet.

A variety of methods can be used to value inventory. Income tax computation, accounting principles, ease of application, management reporting, and control considerations all influence the approach used. From a management point of view, the evaluation method that is selected should take advantage of benefits that may affect the operating statement, but should also be easy to use.

The principal methods of cost determination for financial statement preparation begin with standard costs and/or average costs. The development of standard or average unit costs allows the extension of physical units by these costs to determine the ending inventory value, which, when subtracted from the beginning inventory plus materials purchases and conversion expenses for the period, equals the cost of goods sold for the period. Ending inventory values and, correspondingly, costs of goods sold may be further adjusted by the use of FIFO (first in, first out), LIFO (last in, first out), or cost-of-specific-lot methods. For retail and distribution companies, which do not incur conversion costs, the gross margin retail method is the commonly accepted valuation method.

**(a) Standard costs.** A simply administered technique for determining inventory cost is provided by standard costs. To facilitate its use, standard costs representing the expected costs of purchase and conversion must be established for materials, labor, and other expenses directly related to each inventoried unit. Those manufacturing expenses (overhead or burden) not directly identified with a product or inventory item are usually assigned to each unit of production on a realistic basis. Thus, supervision might be assigned on the basis of the labor hours required to make a product, whereas depreciation, insurance, and taxes might be on the basis of machine hours required to make a product.

When actual costs are not equal to the standards, variances result and must be recorded. If these variances are substantial, they are accumulated and reallocated to all the production of the period and the inventory on hand. If the variances are minor, they are generally taken as an expense as they occur. To remain accurate and effective, standard costs must be revised at least annually by correcting for changes in costs and methods; otherwise, the values derived lose validity either for valuing inventory or for controlling purchasing or operations.

**(b) Average cost.** Many companies use an average unit cost to determine inventory value. Two methods are used for this purpose: an average calculated by dividing the total cost of beginning inventory and purchases by the total number of units represented; and a moving average, calculated after each new purchase is made. Where the time lapse between valuations is long or price changes are rapid, the two methods produce somewhat different results. It is important to note that the moving average requires the maintenance of perpetual records, while the weighted average may be computed periodically. In either case, the use of averaging tends to spread the effect of short-range price changes and level their effect on profit determination.

**(c) FIFO.** The FIFO method assumes that the goods are sold in the order in which they were received or manufactured. With the use of FIFO it is not necessary to identify lots or physically segregate items in the order of purchase. Under this method, the final inventory is priced by determining the costs of the most recent purchases or using the most recent standard costs. Generally, this is done by working back from the most recent invoices of units manufactured until the quantity on hand has been covered.

**(d) LIFO.** The LIFO method assumes that the last goods received or manufactured were sold first. In a period of rising prices, this method reduces profits and postpones tax payments because the highest-cost items are used to compute the cost of sales. This method assumes that all items in inventory, or in a segment of inventory, are homogeneous. This assumption allows the use of price indices to convert the final inventory priced at current prices to the same inventory priced in terms of the base year (the year in which LIFO was adopted).

The LIFO method allows a firm to postpone the payment of taxes and thus receive an "interest-free" loan from the government. Generally, any company with the following characteristics should adopt the LIFO method: inventories that are significant relative to its total assets (a manufacturer or distributor); inventory costs that are increasing; an effective tax rate that is not expected to increase significantly; and/or inventory levels that are expected to remain stable or increase. Conversely, if inventory levels or unit costs are expected to decline or the tax rates increase, a company should not use the LIFO method because LIFO could increase rather than decrease the company's future tax liability.

**(e) Cost of specific lots.** For capital-goods industries and some other industries, it is usually desirable to maintain the identity and actual costs of specific lots or items. This is done by recording the actual raw materials purchase price and conversion costs for each lot or item in inventory. The application of this method is usually limited to inventories containing high-value, low-quantity items or to those situations where such records are practical or mandatory.

**(f) Gross-margin method.** The gross-margin method of pricing inventory has been developed in the distribution and retailing industries to facilitate inventory valuation by having items in inventory priced at an adjusted selling price rather than cost. The basis for computing this ratio is an established average margin, or markup, which is added to the cost to arrive at the selling price. Assuming that the goods in inventory are representative, in terms of their markup, of goods purchased during the recent operating period, their cost is easily determined by using the ratio to recompute the inventory value. Using this method, the ending-inventory value is determined by extending the physical units by their selling prices and then reducing the extended amount by the average markup percentage. As with other methods, cost of goods sold is calculated by subtracting the ending-inventory value from beginning inventory plus purchases.

**(g) Selecting a method.** The size, age, and type of inventory must be considered in selecting the method used for valuation. The techniques chosen should provide for the requirements of financial and tax reporting as well as relate to the methods used by management for measuring its performance in inventory management and control. For consistency in financial reporting, it is mandatory that the method of inventory valuation remain constant. If at any time a change in method is made, the effect on the financial results must be clearly stated.

## 19.4 ORGANIZATIONAL CONTEXT OF INVENTORY MANAGEMENT

While specific managers within a firm are directly responsible for the inventory management function, inventory-related functions have impact throughout the organization. Indeed, the functions of the inventory management system overlap with the revenue and expenditure cycles, and span organizational units such as inventory management (or logistics), marketing, finance, and accounting. Departments typically involved are inventory control, purchasing, receiving, sales order, credit, shipping, billing, and general ledger. Collectively those involved with inventory seek to answer questions such as: What should be ordered? Who should it be ordered from? Where should it be delivered? When should it be ordered? How much should be ordered? In addition, the inventory managers also need to keep track of what has been ordered and received.

**(a) What Should Be Ordered.**    The inventory to be ordered or manufactured depends on sales forecasted. In the case of merchandisers, the quantities of various items to be ordered are unrelated to each other: the number of appliances to keep in stock is not related to the number of shirts on hand. This is described as independent demand. On the other hand, for a manufacturer, the number of tires to be ordered is related to the number of automobiles being manufactured. When goods ordered are so related, their demand is called dependent.

While a manufacturer is more likely to order goods that have dependent demands, it can also order items that are independent of the decision to manufacture a certain quantity: the amount of supplies and the number of machine parts ordered may not be directly related to the number of cars being made.

**(b) Who Should It Be Ordered From.**    In the past, the job of purchasing managers was to shop around among a large sample of vendors for the best value. But old ways are changing. A manifestation of the change involves the trend toward bypassing the middle link in the chain—the wholesalers—in favor of suppliers. This thinking means that the number of suppliers that buyers deal with can be much smaller. It has also meant long-term, close relationships with a few vendors. Vendors are expected to be more quality conscious and make deliveries on time. In return, they are assured business.

One consequence of the new thinking has been a reduction in ordering costs. Advances in information technologies, combined with fewer, more reliable vendors, has allowed changes in accounts payable procedures: payments are made faster with less paperwork and the clerical staff that processes purchase orders and receiving reports has been reduced. Bar codes allow arriving shipments to be identified sooner; such identification of shipments received can then allow automatic bank transfers to vendors' accounts.

**(c) Where Should It Be Delivered.**    For manufacturers, the need for reducing inventory on hand has led to having vendors deliver goods directly to work centers on the factory floor as needed. However, not everyone is able to do this, since the production facilities may not have been built to accommodate such deliveries. In any case, manufacturers may be able to have material delivered to the factories, but retailers have more of a problem, since their warehouses and stores may be far apart. Instead of storing inventory in their facilities, the mega-retailers can have the vendors deliver merchandise to their stores directly. This trend is increasing in popularity.

**(d) When Should It Be Ordered.**    The issue of when to order inventory is handled differently, depending upon whether the goods are demand independent or dependent. For dependent demand, the issue revolves around a certain point in time when desired goods must be ordered, or in many cases, reordered. That point in time is described as the reorder point. For most firms, the reorder point is reached when the quantity on hand of a certain item falls below a certain level. At this level there is still some quantity on hand, termed safety stock, to last until the reordered stock reaches the warehouse. The quantity of safety stock is dependent on the lead time needed for stock ordered to be delivered. The reorder points, when drawn graphically, resemble the teeth of a saw, hence they are called the sawtooth model.

One measure of change in retailing has come about through the point of sales computers that link the sales registers to the vendors, alerting them to items that need to be replaced. Such electronically determined reorder points can be of benefit to both the vendors and the buyers.

**(e) How Much Should Be Ordered.**   In the past, the quantities being ordered were related not only to the sales projected but also to a decision to have safety stock on hand. Managers provided safeguards for variation in the deliveries of goods ordered, and took advantage of the economies of purchasing in bulk. More recently, the rules of the game dictate that the burden of carrying inventory be shifted to the supplier from the seller or manufacturer. This has brought on the popularity of just-in-time production, which does away with work-in-process inventories. There is also a greater emphasis on matching production to market requirements, thus greatly reducing finished goods inventory and the costs associated with taking care of it.

## 19.5 COMPONENTS OF MERCHANDISE INVENTORY MANAGEMENT SYSTEM

In this section the various components of the merchandise inventory management system are reviewed, such as documentation, inputs, files, control procedures, the nature of the database, and reports. Since readers were exposed to most of these elements in Chapters 16 and 17, the survey here will be brief.

*Documentation.*   The key data entry inputs include the purchase requisition (or merchandise reorder report), purchase order, receiving report, shipping report, credit memo, and purchase memo.

*Database.*   The critical file is the merchandise inventory master file. Exhibit 19–2 indicates that the open purchase order and receipts files are additionally involved. An inventory history file and standard cost reference file can also be useful.

*Controls.*   Most of the suitable general and transaction controls are listed under the revenue and expenditure cycles. Among those controls especially relevant to the inventory management system are up-to-date and accurate merchandise inventory item records, sound reorder algorithms, and periodic physical inventories followed by reconciliations with inventory records.

*Reports and other outputs.*   Exhibit 19–2 portrays several outputs. A fundamental report is the inventory status report. The merchandise reorder report, which is used as an input to the purchasing procedure, is also an output of the inventory management system. The summary of merchandise receipts is, in effect, a transaction listing, as is an open purchase orders report. A report that shows overdue purchase orders can be used for control purposes. A report showing inventory items expected to be out of stock, either in coming days or weeks, can be used for planning purposes.

## 19.6 RAW MATERIALS MANAGEMENT (PRODUCT CONVERSION) SYSTEM

As noted earlier, in certain industries, such as manufacturing and construction, the inventory resource is varied. In place of just one inventory consisting of merchandise for resale, these firms maintain three distinct inventories: raw materials, work-in-process, and finished goods. These inventories help trace the conversion of raw materials into finished goods.

Raw materials are converted, via a production process, into finished goods. Thus, a system whose purpose is to manage the raw materials resource can also be described as a system whose purpose is to facilitate the conversion of raw materials (and parts) into finished goods (products). Key objectives within this product conversion or raw materials management system are to ensure that adequate raw materials and other resources are available; that production orders are processed and completed in a timely and cost-efficient manner; that the products are of specified quality; that the finished goods are warehoused or shipped on schedule; and that the costs for each order are accumulated and processed.

The functions related to the foregoing objectives are (1) planning and initiating the production process, (2) moving work-in-process through production operations, (3) costing the work-in-process, (4) transferring completed products into the finished goods inventory, and (5) preparing financial reports that report on the production process and the subsequent sale of finished goods.

Basically, production of goods is performed according to one of these five patterns and their corresponding cost tracking systems:

- Process production, which involves the continuous production of standardized products (e.g., petrochemicals, cement, steel). Costs are usually accumulated in process cost systems in terms of processes or work centers.

- Mass production, which involves the production of discrete and relatively similar products on assembly lines (e.g., automobiles).

- Operations-oriented production, which involves the production of discrete and relatively similar products through a mix of standard operations performed automatically and manually (e.g., suits, shoes).

- Batch production, which involves the production of batches or job lots of distinctly differing products; for example, a metal products firm may manufacture bearings in one batch and gears in another. Costs are usually accumulated by job order cost systems.

- Custom production, which involves the production of uniquely individual products, often of a complex nature (e.g., special machine tools). Costs are accumulated in custom cost systems in terms of individual orders.

Rather than discussing all of the production patterns, the discussions will focus on batch or job lot production, which also subsumes the production of customized products.

## 19.7 PRODUCTION LOGISTICS

A key element that underscores production function is logistics. All of the units and flows shown within the dashed box in Exhibit 19–3 can be viewed as closely related to the production logistics function. The sales order entry and shipping units, however, may alternatively be grouped under the marketing/distribution function. Since the roles of several of the production logistics units were described in Chapter 17, only the remaining units will be considered in this section.

Engineering design determines the specifications by which the products are to be manufactured. Production planning determines the quantities of products to be manufac-

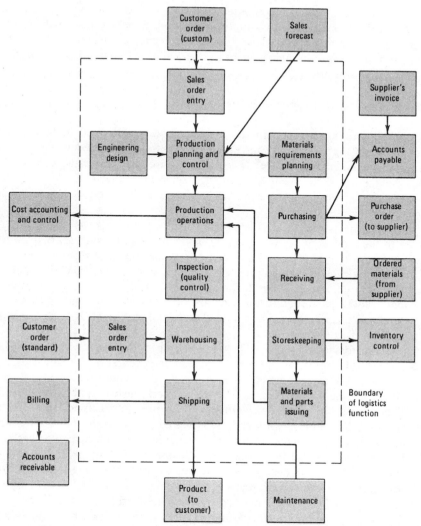

**Exhibit 19–3.** A flowchart of organizational units that direct the functions involved in the product conversion system and the filling of customers' orders. Adapted from J. Wilkinson, *Accounting Information System: Essential Concepts & Applications*, 2nd. ed. (NY: John Wiley & Sons).

tured, the production schedule, and the resources to use; production control dispatches an order into the production process, monitors the order, and takes corrective actions when necessary. Materials requirements planning (MRP) assures that the proper quantities of materials, parts, and subassemblies will be available at the proper time to manufacture the scheduled orders.

## 19.8 EMERGING ORGANIZATIONAL CONTEXT OF PRODUCTION

Manufacturing firms have greatly improved the quality of their production planning processes in recent years. The improvements have been made possible through the availability of more powerful computer hardware and software. At the heart of these improvements, broadly called computer-integrated manufacturing (CIM), are several sophisticated planning approaches. Two of the most useful approaches with special significance for inventory management are known as manufacturing resource planning (MRP) and just-in-time (JIT).

### 19.9 MATERIAL REQUIREMENTS PLANNING

MRP is a planning system concerned with the acquisition of dependent demand goods. Its focus is on knowing the quantity and timing of inventory items needed in accordance with the master production schedule. It seeks to ensure that materials, components, and subassemblies needed are available in the right quantity. When used effectively, MRP can reduce various inventory costs, as well as improve scheduling effectiveness and response to changing market conditions.

The details pertaining to MRP are highly involved, but at its very basic level, MRP is similar to the kind of planning needed to have all the ingredients for a given recipe on hand, before cooking. A cake is a discrete item, and the acquisition of the ingredients (materials requirement planning) to coincide with actual baking is easily done. But it is immensely more complex when dealing with the continuous production of goods that have many models, that require hundreds and often thousands of parts, components, and subassemblies, and that have different lead times and vendors. MRP helps to address the logistics of such complexity.

The key factors in an MRP system are: a master production schedule (MPS), a bill of materials (BOM), an inventory profile, and lead times of all inventory items. Using these factors, computer algorithms can help schedule the acquisition of inventory and its subsequent release for production. The same algorithms can quickly respond to unexpected changes in the customers' demands for its products. Closed-loop MRP includes feedback from both internal production sources and vendor reports. Regenerative MRP systems use batch processing to replan the requirements on a regular basis, perhaps monthly or even weekly. In contrast, net-change MRP systems are on-line and can respond continuously to transactions pertaining to production schedule, vendor reports, and inventory status of various items.

The success of MRP is closely tied to making effective use of a firm's production capacity. Capacity requirements planning (CRP) is to personnel and equipment what MRP is to materials. The acquisition of materials on schedule is of no use unless capacity is available to subsequently convert them into finished products as planned. Because of this, CRP must be effectively integrated with MRP.

In recent years, it has become clear that requirements planning cannot be limited to material and capacity. A firm does not live by material and capacity alone; other company functions such as purchasing, marketing, management, and accounting must also be involved if the firm is to endure. Material requirements do not effectively describe a planning system that involves all the company functions. To describe the system that involves all the resources of a manufacturing firm in requirements planning, a new term was coined: manufacturing resource planning (MRP II). It is a system that allows all the departments of a firm to work according to the same plan. A number of application software packages are available that allow implementation of MRP (II), among them IBM's MAPICS, Arthur Andersen's MAC-PAC, Hewlett-Packard's FMS, and Martin Marietta's MAS II.

At present, MRP is mostly limited to manufacturing environments. However, potential exists to use MRP in the service sector. Hospitals and academic institutions can use it to schedule equipment, space, and personnel.

Despite the considerable usefulness of MRP, there have been problems in its implementation. Problems can occur because of an insufficient management commitment or a failure to realize that MRP is only a computerized scheduling tool; it is only a part of the system, not a replacement for the system. In some instances, MRP is taken to mean no more than meeting the schedule: such a limited perception defeats the spirit of MRP and its potential rewards. One can argue that MRP's effectiveness is ultimately constrained by the personnel using it and the environments in which it is used.

An alternative to MRP is synchronized manufacturing. It seeks to optimize the use of resources such as production time, inventory, and operating expenses by focusing on the actual and potential bottlenecks in the production process.

In short, the MRP procedure consists of "exploding" the materials requirements by multiplying the quantity of products scheduled by the number of materials and parts required per unit. It then leads to requisitioning of needed materials and parts from the storeroom and requesting that purchasing order any additional materials and parts not in house. On the issuance of materials and parts into production, line production departments actually perform the conversion of raw material into work-in-process and finally to finished goods. In the past, the inspection department sought to provide assurance that the products met established quality standards. The current philosophy is that the production crews do their own inspection, and that they actually do the right job so that there is no need for inspection. In the jargon of activity-based costing, inspection is a non-value-adding chore that adds costs to production.

## 19.10 INVENTORY IN THE JIT ENVIRONMENT

The JIT system is characterized by goals such as minimizing inventory investment, uncovering quality problems, reacting faster to demand changes, and shortening the time needed to process a batch. JIT insists on the production or acquisition of the units when and where they are needed. Under JIT, the production line is run on a demand-pull basis. As a result, the work at each station is authorized by the demand of the station following it. Production is stopped when problems occur, such as absent parts or defective work.

The underlying philosophy of JIT is constant improvement sought by ongoing simplification of the production process and elimination of non-value-adding activities. The elimination of nonessential activity, the underlying basis of what has come to be known as activity management, can extend to the staffing needs of accounting department. Under JIT, bookkeeping is reduced to two entries: one made at the start of the process and one made when finished goods exit production and are received by the warehouse.

The Hewlett-Packard Company of Palo Alto, California, is one of an increasing number of firms that has decided to employ the JIT approach. As a beginning step, it implemented the JIT approach in its plant, which produces computer data storage devices. Hewlett-Packard redesigned its production line for smoother flows. The firm also eliminated its raw materials storeroom and provided space beside the beginning point of its production line. Raw materials are now delivered directly to the production line and placed almost immediately into production. The finished goods warehouse has also been reduced in size.

Several benefits have been achieved at Hewlett-Packard since the introduction of JIT. The actual cost per unit for direct material has been reduced by one third. Warehouse space, as noted, has been reduced by 50 percent. Investments in both raw materials and finished goods have been reduced significantly. The accounting procedures have also been simplified. On the other hand, the firm has not as yet significantly reduced its production time per order. Also, the labor cost per unit has increased somewhat, because of added support personnel and training time. Overall, Hewlett-Packard is pleased with the results and expects the benefits to increase as production lines in other plants are converted.

## 19.11 PRODUCTION-RELATED MANAGERIAL DECISION MAKING

Strategic decisions affecting the inventory management in a manufacturing environment are usually made by a group of higher-level managers, including the production manager and marketing manager. They rely primarily on the logistics and marketing information systems to provide key information, such as available production levels and sales forecasts. Some of the decisions that must be made at the strategic level include:

- Which products and product components should be manufactured?
- How much added production capacity should be acquired in each year for the next several years?
- What method of physical production should be employed?
- What materials should be included in each unit product, and what steps should be followed in the production process?
- From what sources should raw materials and direct labor be obtained?
- Where should the production facilities be located?

Other decisions must follow the resolution of these strategic matters. Some of the decisions primarily concern the accounting department. Among such decisions are:

- What type of costing system should be employed?
- If standard costing is selected, what standard costs should be assigned to units of material, hours of direct labor, and applied manufacturing overhead?
- What budget levels should be established for the production departments?

Similarly, decisions related to production include:

- What quantity of goods should be produced for inventory?
- When should production jobs be scheduled to begin and end?

## 19.12 DOCUMENTATION FOR PRODUCT CONVERSION

Data used in the product conversion system are based on a variety of inputs from documents generated by customers and departments involved in the logistics of production. Other sources are the various finance and accounting files pertaining to raw materials purchases, receivables, and payables.

Source documents typically used in the product conversion system (other than those already discussed) include:

- Bill of material. A bill of material, also called a product specification, lists the quantities of materials and parts to be used in a particular product.

- Operations list. An operations list specifies the sequence of operations to be performed in fashioning and assembling the materials and parts required for a particular product. In effect, it is a "recipe" or guide for production employees. The list may mention the work centers at which the specified operations are to take place, as well as machine requirements and standard time allowances.

- Production (work) order. The document identifying a specific job that is to flow through the production process is known as a production order to work order. It incorporates key data from the initiating customer order (or from a sales forecast, in the case of standard products), from the relevant operations list, and from the production schedule.

- Materials issue slip. A materials issue slip, also known as a materials requisition, directs the storekeeper to issue materials or parts to designated work centers, jobs, or persons. It may list the costs of that material or the costs may be entered later by the inventory control department.

- Move ticket. A move ticket, also called a traveler, authorizes the physical transfer of a production order from one work center to the next one listed on the order. It also records the quantities of items received and shows the date received, thereby enabling progress to be tracked.

- Other source documents. Additional source documents used in the logistics activity include the purchase requisitions, purchase orders, receiving reports, and labor job-time tickets. In the case of custom production, the sales orders from customers are also needed.

## 19.13 ACCOUNTING IN A MANUFACTURING CONTEXT

The objectives and most of the units of the finance/accounting function have been discussed in Chapters 16 and 17. But in production environments, the cost accounting and control unit appears for the first time. It accumulates material, labor, and overhead costs incurred in the production process and prepares cost variance reports. This unit also develops labor and material cost standards and assigns values to work-in-process inventory.

Product costs consist of direct materials costs, direct labor costs, and manufacturing overhead costs. Included as manufacturing overhead are costs of indirect labor, supplies, utilities, small tools, depreciation of machines, and other costs related to the production process. Manufacturing overhead is essentially all production-related costs that are not identified as direct material or labor. These product costs are accumulated to reflect the value of the jobs flowing through production operations. Periodically, the costs are posted to the work-in-process records; the total of the costs for all current jobs is then posted to the general ledger from an entry having the following form, to record costs applicable to current work in process:

| | |
|---|---|
| Dr. Work-in-Process | XXX |
| Cr. Raw Materials Inventory | XXX |
| Cr. Labor or Payroll Payable | XXX |
| Cr. Overhead Applied | XXX |

When a production order or job has flowed through all required production operations, and thus is completed, the costs for the finished goods are posted from the following entry, to record the completion and transfer of work-in-process to finished goods:

Dr. Finished Goods           XXX
Cr. Work-in-Process Inventory  XXX

## 19.14 DATABASE FOR PRODUCT CONVERSION

The database for the product conversion system of a manufacturing firm interfaces closely with the revenue and expenditure cycle databases. Its master files pertain to raw materials, work-in-process, and finished goods. The principal transaction file is the open production order file. Other files might contain data concerning the product specifications, production order schedule, work center schedule, machine loadings, standard costs, overhead rates, production order history, and so forth.

Since the finished goods master file is similar in content to the merchandise inventory master file, we will focus on the raw materials and work-in-process inventory master files. Also, we will look at the production order file.

*Raw materials master file.* Each record in the raw materials master file reflects the receipts, issues, orders, and on-hand balances pertaining to a particular material, part, or subassembly. All the records in this file comprise a subsidiary ledger supporting the raw materials inventory account in the general ledger.

*Work-in-process master file.* Each record in the work-in-process master file (i.e., job cost sheet) summarizes the costs incurred—for raw materials, parts, direct labor, and overhead—with respect to a current (open) production order. All of the job-oriented records in this file comprise a subsidiary ledger supporting the work-in-process inventory account in the general ledger.

*Open production order file.* Like the work-in-process master file, the focus of this file is the individual production order. Thus, the two files are closely related. For instance, both use the production number as the primary and sorting key. However, the open production order file does not contain costs. Instead, it shows the current status of each production order in the production process. The data contained in a typical production order are listed in Exhibit 19–4.

## 19.15 DATA FLOWS AND PROCESSING

Exhibit 19–5 presents a document-oriented flow diagram that emphasizes the flows of source and output documents among organizational units. In effect, it connects the pertinent documents to the units portrayed in Exhibit 19–6. Note that the diagram places the product conversion process in perspective. That is, it includes the documents and flows for sales order entry, purchasing of materials, and shipping of finished goods as well as those involved in production. The description of data flows and processing will be in accordance with the listed functions. The focus will be on on-line computer-based information systems, since they offer the following benefits to the product conversion system:

Production order number (primary key)
Customer number (if a special order)
Customer name (if a special order)
Customer order number (if a special order)
Date of order
Date started (or to be started) in production
Date to be completed
Product number (or line number)
Product description
Quantity
Weight (if applicable)
Operation number[a]
Operation description[a]
Date scheduled[a]
Time started[a]
Time ended[a]
Machine number[a]
Special instructions (such as tools needed)[a]
Work center numbers[a]
Inspection results

**Exhibit 19–4.   A table of data elements within a production order. Adapted from
J. Wilkinson,** *Accounting Information System: Essential Concepts & Applications,* **2nd. ed.
(NY: John Wiley & Sons).**

- Better integration of highly interdependent activities
- Improved production efficiency and use of resources resulting from better scheduling of orders
- Immediate editing of data when entered on-line
- Better control over orders due to up-to-date files and timely reports

**(a) Planning and Initiating the Production Process.**    Exhibit 19–7 shows a computer system flowchart of production planning. One input into the planning process is a set of specifications for special orders, as prepared by engineering design. These specifications are added to the product specifications and operations files. At the end of each day, revised sales forecasts and backorders are also entered into the system.

A production planning program can perform several complex steps. First, it prints prenumbered production orders that are needed to initiate the production of custom-ordered and standard products, based on the listed inputs. It also updates the open production orders file. Second, the program prints move tickets and material issues (requisitions) to accompany the new production orders. Third, it prints bills of material and operations lists for each product shown on the production orders, so that they may be reviewed by customers and used by planning personnel. Fourth, the program prints a revised production schedule, showing starting dates for the new orders, and updates the schedule in the file. Fifth, it "explodes" materials requirements and prepares a report showing the materials and parts needed to fill the production orders. It then updates the materials file to reserve materials in the current raw materials inventory file for specific production orders. Finally, it updates the machine loading file to assign machines as needed for the new orders.

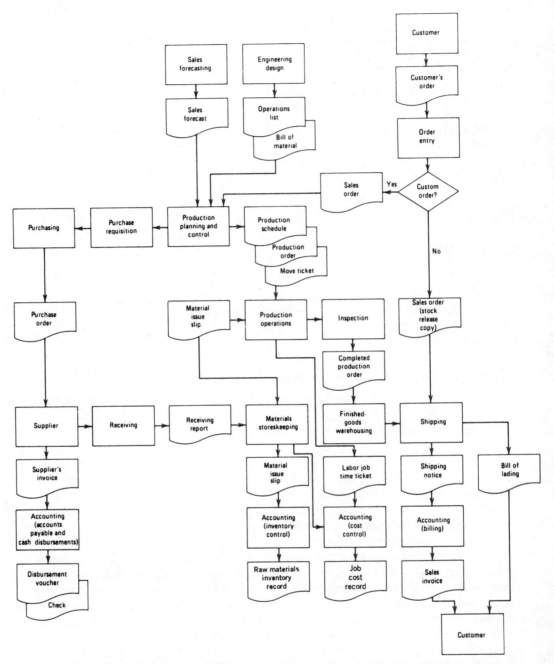

**Exhibit 19–5.**   A diagram depicting the flow of documents among the organizational units involved in product conversion and logistics. Adapted from J. Wilkinson, *Accounting Information System: Essential Concepts & Applications*, 2nd. ed. (NY: John Wiley & Sons).

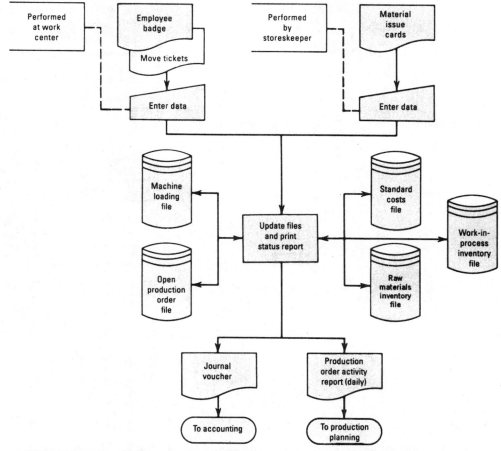

**Exhibit 19–6.** A computer system flowchart of processing steps during production operations. Adapted from J. Wilkinson, *Accounting Information System: Essential Concepts & Applications*, 2nd. ed. (NY: John Wiley & Sons).

**(b) Moving Work-in-Process Through Production Operations.** Exhibit 19–6 shows a computer system flowchart pertaining to production operations. When a new production order is scheduled to start into production, raw materials and parts are issued by the storekeeper. They are delivered to the first work center and accompanied by the first move ticket. Employees and machines are assigned in accordance with data in the production order. As employees begin work on the order, they enter the employee number and order number. When they finish the operation involving the order, they repeat the entry.

Based on these variously entered inputs, the production operations program performs several steps. First, it updates the materials inventory file to reflect the issued materials and parts. Second, it updates the open production order file to show that the operation at the work center is completed. In addition, the program prints a production order activity report to show the current status of all open production orders. Third, it updates the machine loading file to show that the machine is free for the next scheduled operation. Fourth, the program updates the work-in-process file to show the accumulation of production costs. These costs are computed by reference to the standard unit costs and

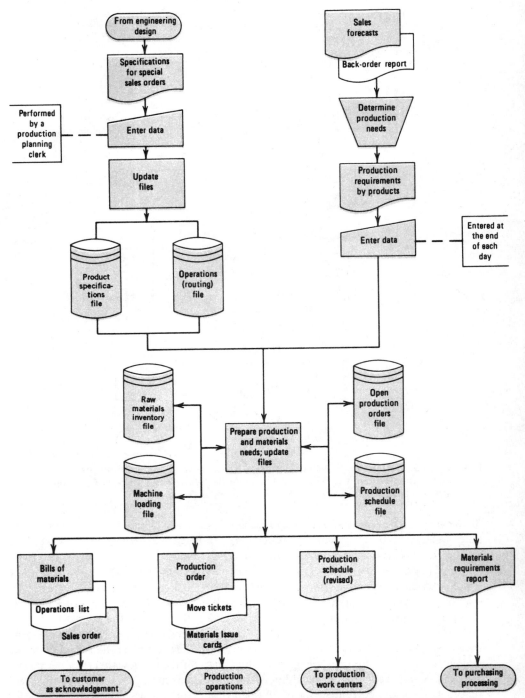

Exhibit 19–7.   A computer system flowchart of production planning. Adapted from J. Wilkinson, *Accounting Information System: Essential Concepts & Applications*, 2nd. ed. (NY: John Wiley & Sons).

overhead rates in a standard costs file. Finally, it prints a journal voucher that reflects the costs to be posted to the work-in-process inventory account in the general ledger.

(c) **Costing the Work-in-Process Inventory.**   Accumulating costs related to production orders is a very important step to accountants. Although this cost accounting step was mentioned in the previous paragraph, it deserves special attention. To see the step more clearly, it is helpful to return briefly to a manual processing system. Exhibit 19–8 shows the postings of production costs in a job order cost accounting system. Raw materials costs pertaining to a particular production order are posted to a work-in-process cost record from materials issue slips (requisitions). Direct labor costs are posted to the cost sheet from job-time tickets. Manufacturing overhead is applied on the basis of direct labor hours. Periodically, a department such as cost accounting and control prepares a journal voucher that summarizes these cost accumulations for the current jobs in production.

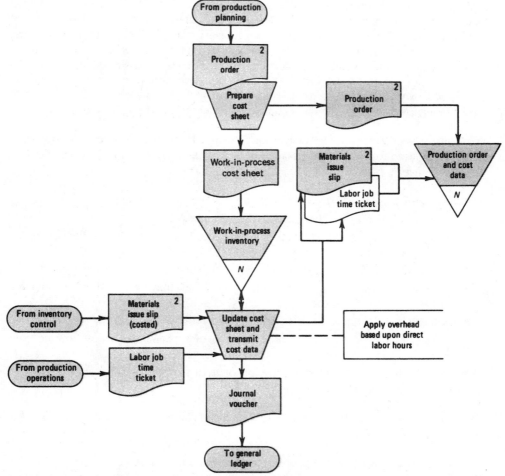

**Exhibit 19–8.   A document system flowchart showing the posting of production costs by cost accounting and control. Adapted from J. Wilkinson,** *Accounting Information System: Essential Concepts & Applications,* **2nd. ed. (NY: John Wiley & Sons).**

The journal voucher is transmitted to the general ledger department for posting. The procedure just described can be identified as an actual job order cost accounting system, using normal costing for manufacturing overhead. In contrast, the standard costs file shown in Exhibit 19–3 depicts a system that employs standard costs. In a standard system, the unit prices and quantities for materials, hourly cost rates and hours allowed for direct labor, and manufacturing overhead rate and base level are carefully preestablished. These standards are applied consistently over a reasonable period (e.g., a year). Standard cost systems enable useful manufacturing cost variances to be computed. Thus, they help production managers to monitor and control costs more effectively. Consequently, the efficiency of production operations can be increased while costs are monitored and hopefully controlled. An alternative type of cost allocation system is centered on activities and was discussed in Chapter 8, which examined activity-based costing.

**(d) Completing the Production Process.** Exhibit 19–9 shows a computer system flowchart relating to the completion of production operations. As work is completed on a production order, the finished goods are inspected; the results are entered into a terminal to update the production order. Those orders that pass inspection are transferred either to the warehouse or to the shipping department. At either destination, data are entered from the final move ticket. This notification causes the production order to be closed, transferred to a history file, and listed in a report. It also causes the work-in-process cost record to be totaled and transferred to the finished goods inventory. Finally, a journal voucher is printed that transfers incurred costs for the day's completed orders to finished goods inventory.

**(e) Controls Needed for Product Conversion System.** Product conversion inventory systems are exposed to a variety of risks. In designing the procedures to be implemented, attention must be paid to risks:

- Raw materials, finished goods, or scrap may be lost or stolen
- Incorrect costs may be charged to work-in-process inventory
- Quantities of finished goods exceeding the quantities specified by production orders may be produced
- Recorded quantities of work-in-process or finished goods may be inaccurate
- Insufficient or excessive quantities of raw materials may be issued into production with respect to production orders
- Production orders may be lost during production operations
- Inventories may be incorrectly valued because of improper use of cost flow methods or charges to wrong accounts
- Inventories may be inflated because obsolete or slow-moving items have not been written down in amount

**(f) Control Objectives.** With respect to the product conversion system, several key control objectives are to assure that:

- All production orders are properly authorized and scheduled
- All needed raw materials and other resources are assigned to production orders promptly and accurately, and the related costs are accumulated fully in accordance with the established system of cost accounting

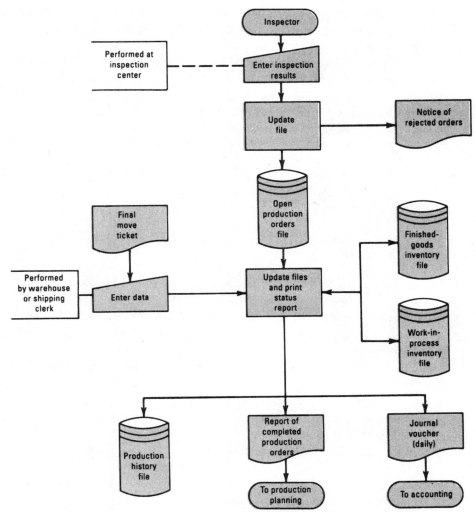

**Exhibit 19–9.** A computer system flowchart of processing steps after completion of productions operations. Adapted from J. Wilkinson, *Accounting Information System: Essential Concepts & Applications*, 2nd. ed. (NY: John Wiley & Sons).

- All movements of production orders through the production process are reflected by acknowledgments at the various work centers
- All finished goods are valued properly
- All inventories are adequately safeguarded

**(g) General Controls.** The following general controls can serve to counteract the risks due to exposures:

- With respect to the organization, the production operations should be separated from those units with custodial functions (i.e., receiving, materials storeskeeping, finished goods warehousing), and both should be separated from those units having recording functions (i.e., production planning, inventory control, cost accounting and control, data processing, general ledger).

- Complete and up-to-date documentation should be available concerning production planning and operations.
- Security should be maintained (in the case of on-line computer systems) by techniques such as (a) requiring clerks to enter passwords before accessing production-related files, (b) employing terminals with restricted functions, (c) generating audit reports (access logs) that monitor accesses of system files, and (d) dumping the production files onto magnetic tape backup. Physical security measures should include locked enclosures for materials and finished goods.
- Operating practices relating to processing schedules, reports, changes in programs, and other matters should be clearly established.

**(h) Transaction (Application) Controls.**   The following controls and control procedures are applicable to the product conversion system:

- Prenumbered and well-designed materials requisitions and production orders, as well as carefully established production schedules
- Required authorizations to issue materials and place orders into production
- Validation of input data by means of edit programs containing a variety of programmed checks
- Verification and transfer of responsibility at every step through the production process
- Up-to-date work-in-process cost records
- Reconcilations of work-in-process cost records and the balance of the control account in the general ledger
- Periodic physical inventories of materials, parts, and finished goods, with the results being reconciled with the balances in the inventory records
- Prompt preparation of reports relating to work-in-process costs, finished goods transfers, batch controls, errors, and so on

## 19.16 REPORTS AND OTHER OUTPUTS

Several listings can be provided on a frequent basis, including the materials requirements report, production activity report, completed production orders report, raw materials status report, and finished goods status report. Inquiries may involve individual production orders, costs of individual materials and parts, and activity at individual work centers.

Among the production-oriented reports of interest to managers are the:

- Work-in process cost report, which compares the actual accumulated costs of production orders against the budgeted costs
- Employee efficiency report, which compares the outputs of production employees against established quotas
- Production overdue report, which shows those production orders that are behind schedule and the number of days that each is behind

- Equipment utilization report, which shows the percentage of time that each piece of production equipment is utilized during a work week
- Work center performance report, which shows the utilization and efficiency of each work center
- Production waste report, which shows the percentages of scrap, rework, and rejects that result from production during a week

## 19.17 CONCLUSION

The inventory management system can be separated into the merchandise inventory management system and the raw materials management system. The merchandise inventory management system integrates the transactions related to merchandise inventory, including ordering, receiving, shipping, and returns. Its main file is the merchandise inventory master file, and the key organizational unit is the inventory control department. Useful outputs include the stock status report, merchandise reorder report, and inventory items expected to be out of stock.

The raw materials management system, also called the product conversion system, includes the functions of planning and initiating the production process, moving orders through production operations, costing the work-in-process inventory, and completing the production process. Key inputs include the production (work) order, materials requisition, job-time ticket, move ticket, bill of materials, and operations list. The database includes the production order file and the work-in-process inventory file. Outputs include the materials requirements report, work-in-process cost report, and raw materials inventory status report.

# CHAPTER 20

# THE RESOURCES MANAGEMENT CYCLE: FACILITIES AND FUNDS

## 20.1 INTRODUCTION

This chapter continues the coverage of resources management, focusing on two kinds of resources, facilities and funds, also known as fixed assets and working capital, respectively. The chapter first describes functions, relationships, and components of the system involved in the management of the facilities resource. It then describes functions and relationships of the system involved in the management of the funds resource.

## 20.2 THE FACILITIES MANAGEMENT CYCLE

The facilities resource concerns the plant assets, also known as fixed assets, or property, plant, and equipment. Within the wide range of plant assets are buildings, machines, furniture, fixtures, vehicles, and other items requiring capital expenditures. The useful lives of these assets extends beyond one year, which subjects them to depreciation—the need to spread the costs incurred in acquiring them over their useful lives. The costs of acquiring these assets is relatively high. While these assets represent a relatively large portion of the total asset value of a firm, they also comprise the ultimate source of organizational revenue. Without its investment in property, plant and equipment, a firm may not be in a position to realize revenue and, by implication, income.

The major purpose of the facilities management system is to facilitate the acquisition of the plant assets, safeguard and maintain them over their economic lives, account for their use over their lives in financial statements, and then see to their disposal. Specifically, the system's objectives are to (1) ensure that all acquisitions are properly approved and recorded and exchanged for cash or equivalents, (2) safeguard the plant assets in assigned locations, (3) reflect depreciation expense properly and consistently in accordance with an acceptable depreciation method, and (4) ensure that all disposals are properly approved and recorded.

## 20.3 THE CAPITAL BUDGETING CONCEPT

To appreciate the working of facilities management cycle, it is best to start by understanding the capital budgeting both as a concept and also as a process in organizations, both public and private. Accordingly, an overview of the nature and working of capital budgeting follows.

Capital budgeting must be seen as a resource investment, since it involves committing resources in the present in the hope of receiving future rewards. The decision to invest is made in order to purchase new equipment, introduce new product lines, modernize plant equipment, or allow for cost reduction or productivity enhancements. In recent years, investments made in capital assets may well be to help meet the regulatory requirements.

Spending for capital needs involves major amounts. Moreover, such spending ties up a major part of organizational resources, especially the monetary assets. Money for one project often is obtained by taking it away from some other use. Rare is the organization that has unlimited resources for undertaking all projects that seem desirable to its managers. These decisions affect the long-term financial health of the organization. The choices are made even more difficult because managers cannot predict the future. Managers in the American automobile industry might not have so whole-heartedly ignored the threat posed by foreign-made, fuel-efficient automobiles in the 1960s if they could have predicted the high price and the fluctuating availability of gasoline in the decades that followed.

Given the high stakes that capital budgeting decisions represent, firms use elaborate processes in order to make them. Such processes are meant to help them make informed decisions. They do so by examining a proposed project to see if it meets preestablished organizational criteria, such as a return of 15 percent on investments involving cost reduction. Or instead of such screening by means of preestablished standards, firms may decide to select a project from among several competing courses of action. Firms may decide to use both screening and selection in order to choose capital projects.

To help make informed decisions, firms seek to use objective means to screen and/or rank projects. The texts used for teaching accounting and finance courses devote considerable space to mathematical models for capital budgeting decisions that use discounted cash flows. The use of discounted cash flows or time value of money better measures the costs and the resulting benefits that will accrue over the life of the project. Since inflation is a constant fact of life, such consideration to the time value of money is highly justified.

The two models that incorporate time value of money in their calculations are the net present value method and the time-adjusted rate of return, otherwise known as the internal rate of return method. Under the present value method, the discounted value of all cash inflows is compared with the present value of all cash outflows that the project will require. The difference between the present values of proposed projects will help determine their relative acceptability. The internal rate of return method seeks to measure the rate of return promised by a project over its useful life. It is the rate of return that will render the net present value of a project to be zero, that is to say the discounted outflows and inflows will break even. It is based on forecasting the future cash flows, while the net present value method is based on the selection of a discount rate.

Both techniques are only that, techniques. They must not become a surrogate for strategic planning. All the more so because the predicted cash flows and the chosen discount rate are subject to manipulation. Managers will be better served if they focus on improving their strategic assumptions and operating considerations instead of relying entirely on discounted cash flows for their capital budgeting decisions. The firms will be better served if their screening and/or selection criteria is broad-based and relies upon a number of factors.

Strict and sole reliance on rate of return has been shown to be counterproductive as a measure of managerial performance. Projects need time and infusion of money before they reach their profit potential. Such infusion of money, when invested in projects, initially leads to a lowered rate of return for the firm or the division involved. Once the project is operational, rates tend to increase. But this may not be enough or be timely for managers whose performance are measured by the current returns. For them a project that leads to a lowered return in the short term, even while promising future increases, is not rewarding enough and they refuse to undertake it, even when strategic considerations call for its acceptance.

## 20.4 THE CAPITAL BUDGETING PROCESS

The capital expenditure process begins when a manager perceives that his or her department, division, or other organizational unit needs an additional plant asset or needs to replace an asset. For instance, a shipping manager may learn from his drivers that certain delivery trucks need replacement. This need should be substantiated through formal capital investment analyses. Such analysis requires that expected benefits and costs be gathered for the economic lives of the new plant assets, as well as for factors such as the expected disposal or salvage values of the current assets. Furthermore, these benefits and costs must be discounted to the present time by a factor (i.e., desired rate of return or opportunity cost of capital) that management specifies.

The manager places a formal request for the needed plant asset. Senior-level management must then approve such a request. The larger the amount involved, the higher the request must ascend for approval. On receiving approval, the request follows a procedure similar to the acquisition of merchandise. That is, a copy of the request is sent to the purchasing department or, in the case of highly technical equipment, the engineering department. Bids are requested, a supplier is selected, and a purchase order is prepared. When the plant asset arrives, a receiving report is completed, and the asset is delivered to the requesting organizational unit. On the receipt of the supplier's invoice, a disbursement voucher is prepared (if the voucher system is in effect). On the due date a check is written and mailed.

## 20.5 MAINTENANCE AND DISPOSAL OF CAPITAL ASSETS

While capital budgeting itself is generally limited to the acquisition of major assets, the facilities management cycle goes beyond the acquisitions to include their subsequent maintenance and disposal. These capital assets usually represent valuable property. In order to safeguard and maintain each acquired plant asset, all relevant details are generally recorded. Included are all acquisition costs, the estimated salvage value, the estimated economic life, and the location. If a plant asset is transferred to a new location, this move is recorded. If costs are incurred during the life of the asset that increase its service potential or extend its economic life, these are added to the asset's cost basis.

A plant asset diminishes in value during use of the passage of time. An allocated portion of the asset's original cost, called a depreciation expense, must be removed at periodic intervals. The amount of the depreciation expense is determined in part by the method of depreciation that is selected for the asset and in part by the estimated economic life of the asset. These depreciation amounts are included in the record of the individual plant asset as well as in adjustments to general ledger accounts.

When their economic lives have come to an end, plant assets are either sold, retired, or exchanged for replacement assets. These disposals, like acquisitions, require the approval of management. They also lead to the removal of asset amounts from the general ledger accounts.

## 20.6 ORGANIZATIONAL CONTEXT OF THE CYCLE

The facilities management system functions are under the direction of the finance/accounting organizational unit of the firm. The key departments involved are budgeting, accounts payable, cash disbursements, property accounting, and general ledger. The budgeting department develops capital expenditures budgets and coordinates these budgets with the short-range and cash budgets. The accounts payable department approves the suppliers' invoices pertaining to plant assets for payment. The cash disbursements department, an arm of the cashier, prepares checks for disbursement to suppliers of plant assets. The property accounting department establishes and maintains the records concerning plant assets. The general ledger department maintains control over all asset, equity, expense, and income accounts.

Other units of the organization are involved to various degrees. Higher-level managers from various organizational functions (e.g., production), in addition to the finance and accounting functions, must approve the acquisition and disposal of plant assets. As might be expected, the process is subject to considerable organizational politics. As in the case of the expenditure cycle, the purchasing and receiving departments are responsible for ordering and receiving the plant assets.

## 20.7 DOCUMENTATION NEEDED FOR THE CYCLE

The system's documentation is mainly based on inputs from the managers in departments needing new plant assets. Other sources are the plant assets records maintained by accounting departments. Source documents typically used in the management of facilities include the following:

*Capital investment proposal.*   The initiating form is the capital investment proposal, also called the property expenditure request. As the form indicates, it is accompanied by a capital investment analysis form. This latter form lists all future cost and benefit flows that are expected to accrue from the asset investment, with the net cash flows being discounted to present values. The proposal package, including both forms, is forwarded to higher-level managers—such as the controller, vice-president, and president—for approval. On approval, a copy of the proposal is sent to the purchasing department, where it serves as a requisition.

*Plant asset change form.*   A plant asset change form is used as the basis for transferring plant assets from one department to another, or for retiring, selling, or trading in plant assets. It lists net book values of the assets and the amounts to be received (if disposed of). It also provides spaces for justifying the disposal and for the approval signatures of higher-level managers.

*Other source documents.*   Since expenditures are involved, additional documents include the request for quotation, purchase order, bill of lading, receiving report, supplier's invoice, disbursement voucher, check voucher, and journal voucher. These documents were discussed in Chapter 17.

## 20.8 ACCOUNTING FOR THE FACILITIES MANAGEMENT CYCLE

The acquisition of plant assets, for instance a drill press with a useful life of three years, is recorded by an entry such as the following, to record the acquisition of machinery having an estimated life of three years:

Dr. Capital Asset        XXX
Cr. Accounts Payable   XXX

The depreciation of plant assets is recorded periodically by an entry such as the following (using machinery as the example) to record the depreciation expense for the year:

Dr. Depreciation Expense                        XXX
Cr. Accumulated Depreciation—Machinery   XXX

The disposition of plant assets is recorded by an entry like the following, assuming that machinery is sold for its book value, to record the disposal of machines acquired on (date):

Dr. Cash                                          XXX
Dr. Accumulated Depreciation–Machinery   XXX
Cr. Machinery                                    XXX

## 20.9 DATABASE FOR THE FACILITIES MANAGEMENT CYCLE

The distinctive files needed in managing facilities are the plant assets master file and the plant assets transaction file. Other files are those used in all expenditure transactions and described in Chapter 17, such as the supplier master, open purchase order, open voucher, and check disbursement files. An important concern in this cycle, as in the others, is keeping the records and their permanent data up-to-date. When an asset is relocated, for instance, the location code should be promptly changed. When a new plant asset is approved and acquired, a new record should appear in the file.

**(a) Plant assets master file.** A key file needed for the cycle is the plant assets master file, a subsidiary ledger that supports the plant assets control accounts in the general ledger. The contents of a typical record for a single plant asset will include the plant asset number, a unique identifier that generally serves as the primary and sorting key. The asset type code identifies the major classification of plant assets (e.g., land, buildings, equipment) to which the individual asset belongs. The location code refers to the department or physical site to which the asset is assigned.

**(b) Plant assets transaction file.** A plant assets transaction file contains transactions pertaining to new acquisitions, sales of currently held plant assets, retirements, major additions to asset costs or to economic lives, and transfers between locations. It is needed if plant asset transactions are accumulated for a period of time (e.g., a week) and then processed in a batch. If the transactions are posted to the records as they arise, the file will likely not exist in a physical sense. Transactions that allocate depreciation expense for each plant asset are not included in this file. Instead, they are included in the adjusting journal entries at the end of each accounting period.

## 20.10 DATA FLOWS AND PROCESSING

To help illustrate the data flows and processing for the cycle, two processing procedures are described. The first description emphasizes the logical sequence and the key source documents. The second focuses on the processing of the master file within a computer-based system.

Exhibit 20–1 presents a data flow diagram of the plant assets activity. In essence, it follows the functions discussed earlier, although it shows the installation as a separate process.

1. Acquisition of plant assets. To begin the acquisition, a manager in a user department prepares a request. Together with a capital investment analysis, this form is forwarded to senior-level management. After the request is reviewed and approved, it is distributed to the purchasing and accounts payable departments. Then the regular purchasing, payables, and cash disbursements steps are followed. These

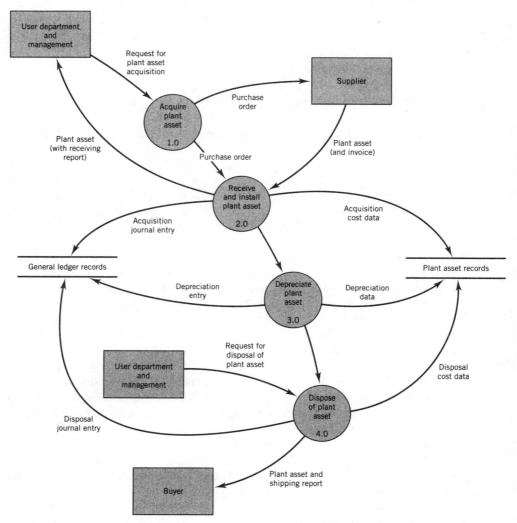

**Exhibit 20–1. A data flow diagram pertaining to plant assets activity.**
Adapted from J. Wilkinson, *Accounting Information System: Essential Concepts & Applications*, 2nd. ed. (NY: John Wiley & Sons).

steps are not included in the dataflow diagram, so that we may focus on the distinctive steps involving plant assets.

2. Receipt and installation of plant assets. On receiving the ordered plant asset from the supplier, the plant asset is sent to the requesting department and installed. The receiving report, which accompanies the plant asset, is signed in the using department to acknowledge receipt. Then, a suitable department within the accounting function, such as the property accounting department, assigns a number to the new plant asset and prepares a record that shows the relevant details. A journal voucher is prepared and forwarded for posting to the general ledger accounts.

3. Depreciation of plant assets. At the end of each accounting period, an amount representing the depreciation expense, determined in accordance with the specified depreciation method and economic life, is computed. An adjusting journal entry is prepared and posted to the general ledger accounts.

4. Disposal of plant assets. The disposal procedure likewise begins with a request. After the request is approved, the plant asset is shipped to the person or firm who has agreed to accept the plant asset. Based on a copy of the approved request form, a property accounting clerk posts and removes the appropriate record from the plant assets file. Then the clerk prepares a journal voucher that reflects the final depreciation expense, actual salvage value (if any), and the gain or loss on the disposal.

Exhibit 20–2 shows a system flowchart of a procedure involving plant assets transactions. The on-line method has been selected for discussion because the number of plant asset transactions is relatively small in many firms and the records can easily be kept up-to-date. The flowchart begins at the point when the property accounting department has been notified of an acquisition or disposal. A clerk uses the department terminal to enter data from each transaction document when received. The entered data are first validated by programmed checks. Then the data are immediately posted by an updating program to the appropriate record in the plant assets master file. If the transaction affects general ledger accounts, the program also prepares a journal voucher and stores it on the general ledger transaction file. At the end of the accounting period (e.g., month) a print program generates useful reports. It also prepares journal vouchers that reflect depreciation entries.

## 20.11 CONTROLS NEEDED FOR FACILITIES MANAGEMENT

Exhibit 20–3 lists representative risks to which the facilities management system is exposed and the consequent exposures due to these risks.

(a) **General Controls.** These general controls help counteract risks due to exposures:

- Within the organization, the managers who approve requests relating to plant assets should be separated from the users of the plant assets and from all units involved in the processing of expenditures and disposal. Otherwise, the organizational segregation described for the expenditure cycle pertains.
- Complete and up-to-date documentation should be available concerning plant assets transactions.

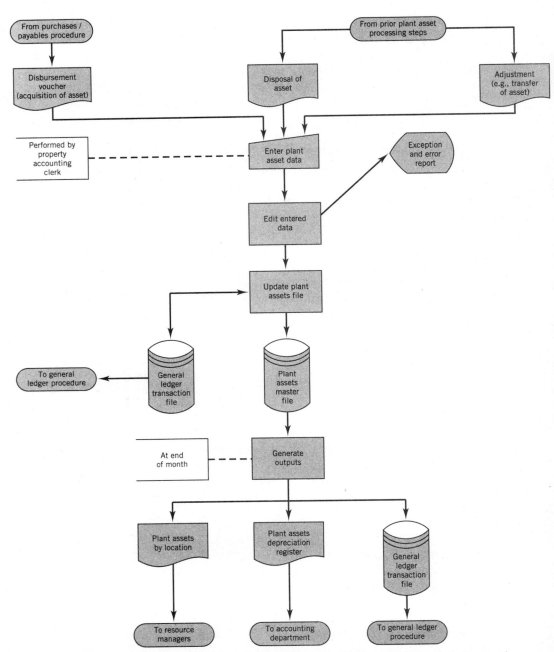

**Exhibit 20–2.   A system flowchart of an on-line computer-based plant assets transaction processing procedure. Adapted from J. Wilkinson,** *Accounting Information System: Essential Concepts & Applications,* **2nd. ed. (NY: John Wiley & Sons).**

| Risk | Exposure(s) |
|------|-------------|
| 1. Improper acquisition of plant assets | 1. Excessive costs for plant assets |
| 2. Improper disposal of plant assets | 2. Loss of productive capability; loss of disposal values |
| 3. Theft or loss of plant assets | 3. Loss in plant-asset values |
| 4. Errors in billing for ordered plant assets | 4. Possibility of excessive costs for plant assets |
| 5. Errors in balances in plant-asset accounts | 5. Over- or undervaluation of total assets |

**Exhibit 20–3.   A table of risk exposure within a facilities managed system.**
Adapted from J. Wilkinson, *Accounting Information System: Essential Concepts & Applications*, 2nd. ed. (NY: John Wiley & Sons).

- Security should be maintained (in the case of on-line systems) by techniques such as (a) requiring that clerks enter assigned passwords before accessing plant assets files, (b) employing terminals with restricted functions for the entry of plant assets transactions, (c) generating audit reports (access logs) that monitor accesses of system files, and (d) dumping the plant assets files onto magnetic tape backups.
- Operating practices relating to processing schedules, reports, changes in programs, and other matters should be clearly established.

**(b) Transaction (Application) Controls.**   The following controls and control procedures are applicable to transactions involving plant assets:

- Documents relating to requests for acquisitions and disposal are prenumbered and well-designed. Also, they are approved by responsible higher-level managers before being issued.
- Acquisitions of plant assets are required to follow the same purchasing, receiving, payables, and cash disbursements procedures employed for merchandise, raw materials, and supplies.
- A unique identification number is assigned to each plant asset, and a tag bearing this number is affixed to the asset.
- Detailed and up-to-date records of plant assets are maintained.
- Balances in the plant assets subsidiary ledger (master file) are reconciled at least monthly with the balances of the plant assets control accounts in the general ledger.

## 20.12 REPORTS AND OTHER OUTPUTS

Managers track the cycle with the help of the following reports.

- The plant assets register is a listing of all plant assets, arranged by plant asset numbers, and showing book and/or tax values of the assets.
- The plant assets acquisition listing shows all assets acquired during an accounting period, including capitalized amounts and estimated salvage values.

- The assets retirement register shows all assets disposed of during the accounting period.
- A plant assets change register for an accounting period is very useful. It tallows accountants to review all transactions that affect plant assets and can serve as an audit trail.
- The plant assets depreciation expense report lists depreciation expenses for every plant asset for the current accounting period, plus related costs and accumulated depreciation amounts.
- Certain reports are also needed to fulfill information requirements of the Securities and Exchange Commission, the Internal Revenue Service, and local property tax authorities. An example is a summary of all acquisitions, transfers, and retirements during a year.

In addition to the above reports, the system should be able to provide additional reports of interest to managers, like those showing plant assets reported by location or by department. Other reports can show maintenance schedules and costs and projected depreciation expenses.

### 20.13  FUNDS MANAGEMENT SYSTEM

Within a firm, the funds resource consists of the working capital and long-term investment assets. The funds resource, therefore, extends beyond cash to include receivables, prepaid expenses, short-term marketable securities, and investments in the long-term bonds and stocks of corporations. Even payables and current accrued liabilities are encompassed, since they enter the computation of working capital. Moreover, the term "funds" can be broadly viewed as including the net capital of the owners of a firm; in the case of a corporation, the capital consists of bonds, common and preferred stock, contributed capital, and retained earnings. Although the net capital is an equity of the firm, rather than a resource, it does give rise to transactions and does need to be managed.

### 20.14  OBJECTIVES OF THE SYSTEM

The major purpose of the funds management system, also known as the treasury or the financial or investment cycle, is to facilitate the inflows and outflows of the funds needed to maintain a business. Key objectives within this system are to (1) assure that an adequate quantity of funds is available to meet all legitimate needs, (2) reflect reasonable values for funds at all times, and (3) safeguard funds from theft and loss. Among the functions related to funds are acquisition of needed funds, uses of funds, and maintenance of control over funds.

(a) **Acquisition of Funds.**   Sources of funds include sales to customers, loans from banks, issues of bonds, sales of stock, sales of plant assets, dividends and interest, sales of investments, factoring of accounts receivable, and delays in payments on accounts payable.

Two key procedural concerns when acquiring funds are to (1) require clear authorization for the acquisition by a responsible manager, and (2) ascertain that the intermediaries are qualified and responsive to the best interests of the acquiring firm. Sound funds management depends on careful planning. A time-tested tool is the funds-oriented budget, such as a projected sources and uses of funds statement. In addition, projected cash flow statements can be quite useful. By knowing a firm's needs in advance, the financial managers are more likely to obtain funds at the lowest feasible rates. Funds management should also consider long-term needs and opportunities. For instance, by means of foresight a number of firms have accrued the funds to reacquire their own stock at advantageous prices when the financial markets have become depressed.

**(b) Uses of Funds.**   Funds may be disbursed for purchase of goods and services, conversion into plant assets or investments, repurchase of the firm's own stock (as noted above), retirement of loans and bond issues, payment of dividends and interest, and redemption of stock rights or options. Procedures involved in the use of funds were discussed in Chapter 17, as well as in Chapters 18 and 19. As in the case of acquiring funds, planning is important to the sound uses of funds.

**(c) Maintenance of Controls over Funds.**   Accounting controls are needed to assure the reliability of financial statement balances and to safeguard the funds. Although most of the controls have been listed earlier, several of the most desirable controls involve:

- Close supervision of all activities involving funds
- Prompt endorsement and recording of all cash receipts, with the deposit of all received cash intact
- Tight physical security over liquid funds, including the use of locked cash registers, safes, lockboxes, and safe deposit boxes
- Separate imprest bank accounts for payroll expenditures
- Issuance of prenumbered checks for all expenditures (except petty cash items)
- Reconciliation of each bank account monthly by a person not otherwise involved in cash transactions
- Separate ledgers for long-term investments, notes receivable, bonds, stock (or stockholders), and other sizable funds
- Audits, on a surprise basis, of all cash and securities

## 20.15 CONCLUSION

This chapter brings to completion the subject of resources management by discussing facilities management and funds management cycle. A firm's resources include not only its employees and its inventories, the topics of the previous chapters, but also comprise its property, plant, and equipment, as well as its working capital.

The facilities management system includes the functions of acquiring, maintaining, and disposing of plant assets. This system involves such organizational units as budgeting, accounts payable, cash disbursements, property accounting, and general ledger. Key inputs are the capital investment proposal and plant assets change form. The database includes the plant assets master and transaction files. Outputs include the plant assets register, retirement and depreciation expense registers, reports of plant assets by location, and inquiry screens concerning individual plant assets.

The funds management system includes the functions of acquiring funds, using funds, and maintaining control over funds. This system involves such organizational units as finance, credit, budgeting, and cash receipts and disbursements. Many of the system components are those pertaining to cash receipts and disbursements, although other funds such as long-term investments require their own master files.

# CHAPTER 21

# USER-FRIENDLY ACCOUNTING MANUALS

## 21.1 INTRODUCTION

Historically, accounting manuals fall into the category of imposing professional references that sit on shelves, rarely used, their bulky binders never opened. This "HANDS-OFF" policy is in large measure responsible for the lack of attention accorded the accounting manuals. If accounting documentation is to be illuminative, it must be usable and useful. The big binders that contain accounting manuals are among the impediments that make the manual less usable.

Fortunately, the advent of desk-top publishing, has helped remedy the formidable look of accounting documentation by giving manuals user-friendly appeal. Even now, it is entirely possible to think about the accounting manual as a part of the information bank on client server systems, linked to far-flung users on the information highway. Today's information technology links data, jobs, and employees in ways that was not possible until recently. Now there is software that interprets and extracts information from many sources; there is also software for relational databases, structured query language, hypertext, and GUI. Given such technological tools, the black and white, yellowing, small prints of accounting manuals are a thing of the past.

This concluding chapter is about the accounting manual itself. It describes the content, the organization, the logistics, and the mechanics involved in documenting an accounting manual.

## 21.2 THE MESSAGE AND THE MEDIUM

A well-developed policy and procedures system facilitates the performance of accounting functions. The accountant's role as a scorekeeper, attention director, and problem solver can be accomplished more effectively in an environment where policy and procedures are well-defined, documented, and accessible. The accounting documentation, an integral part of an effective accounting department, is extremely useful in the preparation of reports in which relevance, uniformity, and continuity are important. Management may and should use the manual to establish uniform standards for evaluating the results of different divisions, plants, and products. If the same system of accounting is used in all locations, it is much easier to identify and analyze variances. In fact, if a company operates on the management by exception principle, common bases are absolutely essential. Without a common frame of reference, it would be impossible to classify results as standard or nonstandard.

The value of effective documentation is not limited to the accounting department. It benefits the entire organization. Effectively communicated policies and procedures enable all segments of operations to start from a commonly understood base. This ensures that comparisons are accurate between plants, departments, people, and products, enhancing performance evaluations. Time requirements and errors are reduced when proper procedures can be confirmed by consulting the manual.

However, the policies and procedures outlined in the manual must be followed in order to realize the benefits of a streamlined, efficient process. This requires that process be constantly reviewed. Revisions and improvements should be made periodically so that the policies and procedures do not become obsolete.

## 21.3 PROGRAM FOR STARTING AN ACCOUNTING MANUAL

The scope of a policy and procedures manual is determined by the complexity of the organization. The greater the number of departments or plants, the diversity of products or procedures, and the echelons of management, the more thorough the manual must be. The planning process is outlined in the Exhibit 21–1 and the critical factors that must be taken into account are given in Exhibit 21–2. The decision to prepare or to revise an accounting manual must have a commitment from management, or the manual prepara-

---

Outline of The Manual Planning Process

Planning the actual writing of the manual
- Defining the subject matter
- Deciding who will collect the data
- Deciding who will write the manual
- Establishing a timetable for data collection and writing

Planning the mechanics of production
- Developing an efficient clearance and review process
- Deciding on layout format and binders
- Determining distribution
- Determining method of reproduction
- Establishing a control point
- Developing and distributing a work schedule
- Anticipating and preventing problems

Collecting and organizing the data
- Collecting the data
- Who should collect the data
- Methods of data collection
- Anticipating and preventing potential data collection problems

Organizing the data in a working outline
- The importance of outlining
- Guidelines for outlining

Explore the use of information technology

---

**Exhibit 21–1.   Planning an accounting manual.**

IDENTIFYING YOUR MANUAL NEEDS

Benefits that manuals provide to the organization
Different types of manuals in use
Determining your manuals specific objectives

DESIGNING AN ATTRACTIVE, READABLE MANUAL

Choosing the appropriate format
Using layout to increase readability
Selecting the appropriate binder, index tabs, and paper

PREPARING THE MANUALS INTRODUCTORY SECTION

Preparing a table of contents
Developing an index
Writing effective introductory material

WRITING FOR MANUALS

Reviewing the principles of effective writing
Using step-by-step listing to simplify your writing
Using formats that clarify procedures

USING VISUAL SUPPORT TECHNIQUES IN MANUALS

Using flowcharts in manuals
Using forms effectively
Determining when to use special exhibits

THE REVIEW PROCESS

The reviewer's role and responsibilities
Scheduling reviews
Techniques for ensuring that reviewers meet deadlines

THE PRODUCTION, STORAGE AND DISTRIBUTION OF MANUALS

The role of word processing in manual production
The printing of manuals
Determining the appropriate distribution
Handling the storage of extra copies

**Exhibit 21–2.  The critical factors impacting documentation of the manuals.**

tion process will lack the resources it needs to become fully established and operative. A person or committee should be designated to develop and manage the process.

The next step is to obtain the necessary personnel and resources to develop the manual. This requires selection of an editor; designation of the persons who will review the final drafts to see that errors have not slipped through; designation of persons who are to receive and maintain the manual; and selection of persons responsible for the filing and maintenance of copies in all locations. The program should also specify those who may issue, revise, or rescind bulletins.

After the project has been properly staffed, the next step is to select and organize the topics to be covered in the manual. A company-wide manual may have the following general sections:

- General matters
- Operations

- Sales
- Accounting
- Credit and collections

Accounting manuals in theory do not devote much space to topics that do not concern them directly. However, there is very little that occurs within an organization that does not eventually flow into the accounting system. Consequently, it is wise to include peripheral procedures that may affect the accounting department.

## 21.4  ORGANIZATION OF THE ACCOUNTING MANUAL

There are a number of ways in which an accounting procedures manual can be organized. It may be split into two volumes: one volume describes the accounting system in terms of the chart of accounts, the coding scheme, account descriptions and the general accounting principles and policies being followed; the second volume deals with the specific procedures that are in place.

Other manuals may follow a scheme where the material is divided into several volumes. One way to divide the material is as follows:

- Overview of accounting system. Gives information about the organization of the system and the chart and description of accounts; discusses the organizational structure and its relationship with accounting information system; provides a directory of the employees.
- Policy and procedure statements. Gives the actual policies and procedures that are in place in the system; describes how the various accounting subsystems work and the procedures they employ to accomplish their objectives.
- Budget manual. Gives information about the budget process as well as identifies the components of the budget.
- Year-end procedures. Gives the procedures to be used for the year-end work that must be done in order to generate the financial statements.
- Data processing manual. Gives information about the data processing needs of the accounting system; facilitates the interactions between accounting and systems.
- Forms manual. Collects the forms and work sheets that are used by the accounting system to process its transaction and facilitate the data collection.
- Information release. Collects directives about topics of interest to employees.

The topics that may be included in an accounting manual are listed in Exhibit 21–3.

## 21.5  SYSTEM MAINTENANCE

It is not enough merely to know how to initiate a manual system for policies and procedures; it is equally important to establish a system that will maintain the manual. To accomplish this there must be a person in charge of the documentation who will have the master copy and will see to it that it is being kept up-to-date. This responsibility is best delegated to a person who has the knowledge and ability to write new bulletins or revise existing ones. In a large company this is a full-time job; in smaller companies the job is often handled by the controller or the treasurer.

---

ACCOUNTING MANUAL
  Overview of accounting system
  Need for and source of accounting policies
    Scope of an organization's accounting policies
  Writing and updating accounting policies
    Drafting, gaining approval, and updating policies
  Internal control policies
  Code of business and personal conduct, incidental gifts, managerial authorizations,
      entertainment, and cost transfers
  Accounting and financial management policies
  Signature authorities, cash management, investments, revenue recognition, inventory
      valuation, inventory adjustments, incurred costs, multi-year budgets, audits, and
      record retention
  Cost accounting policies
    Labor, material, direct/indirect costs, other direct costs, compensated absences,
    leases, capitalization, depreciation, insurance, pensions, and credits
  Employee compensation and disbursement policies
    Timecard preparation, overtime, bonuses, 401 (K) plans, relocation, exempt/non-
    exempt employees, employees vs. consultants, travel reimbursement, and frequent
    flyer awards
  Property management policies
    Policies regarding organization-owned, leased, and small-dollar items, computer
    hardware, software, and physical counts
  Purchasing policies
    Ethical conduct, full and open competition, adequate number of bidders, source
    selection, cost and price analysis and purchase order processing

---

**Exhibit 21–3.   The contents of an accounting manual.**

    The person in charge of documentation may write the proposed material and submit it to management for review and approval. Following approval, the material is issued under the name of the individual who has authority over the particular policy or procedure. This practice is important, since it establishes the source of authority over documentation. In some instances, management prepares its own material, and after approval it is given to the controller for proper indexing, classification, and inclusion in the manual.

    All of the documentation maintained in respective departments of an individual plant or location need not be complete, since each individual needs only the section of the manual that pertains to his job. However, one complete and intact copy should be in each location. One person at each location should be responsible for the maintenance of manuals in all departments. This method is more effective than permitting each department to maintain its own manual, which results in the task of filing changes and adding new material being performed unevenly throughout the organization.

## 21.6 REVIEW AND MAINTAINENANCE OF THE MANUAL

As a matter of general policy, each bulletin should be periodically reviewed for revision. The person in charge of the manual should schedule and record such reviews. The date of any revision should be shown in the body of the particular bulletin, and the date of review should also be noted in order to facilitate later reviews. The department that suggests the revision should review the bulletin at the same time it is reviewed by the manual manager.

There are always changes being made to procedures and the organizational structure also keeps changing. All such changes require adaptations in the manual. Such maintenance will ensure greater respect for the manual as well as greater adherence to the procedures they document.

## 21.7 POLICING THE MANUAL

There may be times when employees seek to do things their own way. Such customization of work may not be the optimum way to achieve the organization's objectives. Given this tendency, it is important to police the compliance with standard operating procedures.

Independent auditors can use company manuals to check the performances of various units against prescribed policies and procedures. A company's internal audit staff also checks on compliance with standard procedures as a part of their audit program. A third method of policing is to set up a task force whose sole function is to check on compliance throughout the company. If a program is well maintained, cooperation and communication already exists between the plants and the general office. There will be little danger of noncompliance because all will be aware of the benefits to be derived from the system. If the system is not well maintained at its source, then trouble may be expected from the field.

## 21.8 THE PHYSICAL ATTRIBUTES

Careful attention to the physical characteristics of the accounting manual can go a long way in making the manual user-friendly. Following are some factors to be considered when planning the manual:

- Type of binder—should be loose-leaf post or ring-type to allow for insertions
- Type of page presentation—stationery should be identifiable as pertaining to the manual
- A system of indexing
- Page format
- Numbering of bulletins, pages, and paragraphs
- Method of reproduction

The manual must allow for expansion, as revisions will certainly come as the system progresses.

## 21.9 THE CORE OF THE ACCOUNTING MANUAL

The heart of the accounting documentation is the portion dealing with policies and procedures. It is quite important to understand the distinction between policies and procedures. The topics covered in a manual can be either a matter of policy or a matter of procedure, or they may be both. All policy matters are approved by the appropriate level of management, but procedures may be approved by lower management as long as they do not contradict the policy they are intended to serve. The distinction between them is important because the person who sets policy generally does not perform the related procedures. This separation is used to argue that policy and its related procedures should exist as separate documents. Even if policies and procedures are to be combined in a manual, a clear distinction should always exist between a statement of policy and a statement describing the procedural details.

## 21.10 DIRECTIVES

Directives are an important tool for the accounting department in communicating with both internal and external customers. They are meant to inform about pending policy changes, staff changes, organization changes, holiday schedule, and to announce minor procedural changes. Directives may also be issued as reinforcement to recent policy changes. They should be communicated on official stationery that will identify it as a directive. They may be numbered so that a record can be maintained.

## 21.11 POLICY STATEMENTS

Policy statements, or bulletins, are those documents that concern matters of policy, both external and internal. They have official standing and they are formal documents. They have direct impact on the organizational culture of an institution, and at the same time are the outgrowth of the organizational culture. The environment may also be influenced with policy statements. In an environment dominated by litigation, there may be more of them.

Common subject matter for these bulletins may include:

Absence from work due to: accident, sickness, approved leaves, personal business, jury duty, military service, death of a relative, National Guard duty, tardiness, pregnancy

Advancement and promotion

Car pooling

Coffee or rest breaks

Company stationery and logo

Disabled employees

Discipline and discharge

Educational assistance to employees

Employee purchases

Ethical codes

Expense allowances

Garnishments

Holidays

Insurance

Intoxicants

Inventions and patents

Keys

Mail

Moving expenses

New equipment acquisitions

Office or plant hours

Outside business or employment

Overtime

Parking

Pensions and retirements

Personal telephone calls

Personal work

Picnics and parties

Rehiring

Relatives (Nepotism)

Repairs and maintenance

Safety

Seniority

Service recognition

Smoking

Tools and tool boxes

Travel time

Unclaimed wages

Use of personal car on business

Vacations

Visiting and visitors

## 21.12 PROCEDURES

Procedure bulletins cover the procedures required to implement policy but, as stated previously, they should be distinguished from policy. Procedures change more often than policies, and exact details are spelled out in procedures, whereas in policies, broad outlines are drawn.

The documentation of procedures can take a number of forms. It could be designed to follow the document flow that occurs in the performance of a given procedure. Or it may describe individual transactions in a narrative style, set off by the person responsible for the task. Following is an example of the narrative layout—the step being described is from the procedure for cash collection during registration at a university:

1. Assistant Controller   Will prepare a cash change fund to give back change to students registering. This will be handed over to Head Cashier for use during registration.

2. Head Cashier   Will sign receipt for cash received. Prepares change drawer for each station as the shift begins.

In another arrangement, called a playscript format, the above material is written thus:

1. Assistant Controller   Will prepare a cash change fund to give back change to students registering. This will be handed over to Head Cashier for use during registration.

2. Head Cashier   Will sign receipt for cash received. Prepares change drawer for each station as the shift begins.

The range of procedure bulletins is unlimited but, to ensure company-wide comparability of results and uniformity in the execution of policies, should include the following:

| | |
|---|---|
| Absence from work | Job evaluation and rating |
| Accidents and disability benefits and claims | Key control |
| Address changes | Memberships and fees |
| Admittance to plant and office | Moving expenses |
| Advances on pay | New products |
| Applications and employment | Overtime |
| Badges and numbers | Part-time employees |
| Bathing and clothes change | Payroll deductions |
| Company tools | Payroll rates, ranges, classifications, changes |
| Contributions and fund drives | Personnel records |
| Discharges and dismissals | Repair and maintenance |
| Employee purchases | Requisitions and returns to stock |
| Employment benefit plans | Safety and safety equipment |
| Empty bottles | Scrap |
| Endorsing, cashing, and handling paychecks | Stationery and supplies |
| Error reports | Storeroom |
| Errors in pay | Suggestions or idea awards |
| Expense reimbursement | Supper or meal allowance |
| Filing | Toolroom |
| First aid and dispensary | Unclaimed wages |
| Insurance coverage for personnel | Workmen's compensation |

## 21.13 POLICY AND PROCEDURE STATEMENTS

The policy/procedure statements (P/PS) combine the description of the policy and a summary of the procedures being used. The description of the procedure is aimed at the users, and generally leaves out the actual processing details. Even the simplest form of procedures, with just one form, may require fifteen to twenty identifiable steps by several different clerical employees to be performed. Omitted from P/PS are details concerning tasks performed, such as initiating, checking, approving, distributing, calculating, balancing, cross-checking, reviewing, and filing. Such omissions make the document easy to read, however, these omissions make it difficult to analyze those tasks from a value-adding or cost-benefit perspective.

An example provided in the appendix for Chapter 10 serves to illustrate the form and content of a policy/procedure statement. Readers are referred to it. The range that can be covered by policy/procedure statements is illustrated in Exhibit 21–3.

## 21.14 ILLUSTRATING ACCOUNTING MANUALS PREPARATION

To help illustrate the process involved in the preparation of an accounting manual, two articles that have described the experiences of two major corporations with regard to their respective accounting manuals are summarized here. Between the two, almost all the issues concerning the logistics of accounting manuals are illustrated.

**(a) The Coca-Cola Company.** The use of a comprehensive, easy-to-use accounting manual became an important tool at the Coca-Cola Company for maintaining strong financial controls in a globally active organization, whose overseas operation is staffed with individuals whose native language is not English and who have no knowledge of the U.S. GAAP. Management's demands for timely, reliable financial information required that the world-wide accounting reports be uniform.

Such reporting required a universal chart of accounts to facilitate consolidation, auditing, and communication of accounting reports. It also mandated standard practices and procedures throughout the organization and a uniform data collection system.

In response to such needs, the company developed an all-purpose accounting manual for use in foreign offices. The manual is divided into two parts: the first part contains a chart of accounts along with account definitions, and the second part contains standard practices and procedures.

The initial step in the preparation of the manual was the appointment of a project team charged with writing the manual. The project team initially sought to rehash the old manual, but they found that simply editing the old manual was not helpful. The company's needs had changed since the previous manual was issued, and functions that were deemed important only a few years ago were no longer relevant in an automated environment with high-speed bottling equipment, pre-selling, automated warehousing, and satellite communications.

The project team sought to determine what to include in the manual. One essential was a universal chart of accounts. It became a project in itself because of the new emphasis on automation. After deciding on a logical numbering scheme, the chart of accounts was created. It was detailed enough to identify activities important to the company so that collection of data would not require extensive account analysis. Information that was not reported on a regular basis was maintained at the subaccount level, whereas main accounts were set up to meet management and tax reporting needs. The accounts were described, as was the expense account distribution. The chart of accounts and the explanation became a volume called *Accounting Manual Part I.*

The writing of the policies and procedures became the next concern. These procedures became *Accounting Manual Part II.* The procedures' arrangement followed the structure of financial statements. Therefore, cash was the first section, followed by accounts receivable, inventories and cost accounting, properties, and sections on internal control and the translation of foreign currency statements into U.S. dollars. A "general" section contained procedures on record retention, insurance, travel expenses, a glossary of terms, and a few other areas of concern to international businesses.

The company used tabs to separate each section and wrote the procedures in outline format so that inquiries from field operations or auditors were easily referenced as to section and paragraph.

The next step was to determine the content of each section.

- The section on cash included procedures dealing with the opening and closing of bank accounts and the handling of cash receipts along with remittances to headquarters.
- The accounts receivable section dealt with the company's procedures for aging, collection, extension of credit, order entry, and invoicing.
- Inventory procedures dealt in general terms, leaving specifics of inventory management to the discretion of the operating unit. Areas that were detailed were physical security of the inventories, physical counts, and reconciliation. A uniform costing procedure was also included here. In an extensive manufacturing environment, cost accounting procedures would be in a separate manual.
- The properties section included property control, asset depreciation rates, and property disposal procedures.
- The procedure for translating local currency statements into U.S. dollars took up another section.
- A separate section dealing with considerations such as the Foreign Corrupt Practices Act and security over automated systems, financial records, data files, and user documentation, was included.

The writing was assigned equally among the team members, as was editing for content and form. Once drafts were completed, they were reviewed by the audit, legal, and tax managers. Then the drafts were distributed to a select number of field accounting managers for comments, since they were the actual users of the document. The writing style was determined to a great extent by the company's personality, but the goal was to keep the prose simple, especially since, for the majority of users in a multinational company, English is a second language.

The distribution of the volume also was important. Because the manual contained certain sensitive information, it had to be restricted to a select number of recipients. The list needed to be periodically reviewed and updated.

It took twelve work-months from the initial planning stage to design, write, edit, and distribute the finished product.

Adapted from "Coca-Cola Writes an Accounting Procedures Manual" by Andrew L. Nodar in *Management Accounting*, October 1986, pp. 52–53.

**(b) The Dow Chemical Company.** The philosophy that guides Dow Chemical's documentation is driven by the following perceptions:

- It provides improved, consistent decision making
- It helps delegate accounting functions
- It enhances communication
- It provides coordination across and within sub-units

Another factor that affects Dow's documentation is its belief that the procedure documentation should be detailed enough to meet both local and corporate reporting requirements.

Dow's first accounting manual consisted of a chart of accounts and an accumulation of policy and procedure letters written over a period of time. It provided no formal mechanism for updating the documentation, nor was it possible to determine if all concerned parties, functions, and locations were operating in accordance with the most recent version of a policy or procedure. During the late 1960s, Dow was becoming a global company and needed a uniform system of accounting policies and procedures with worldwide application. To address these needs, Dow's *Accounting Policy and Procedures Manual* was formalized and became effective as of January 1, 1971. It consisted of two volumes: *Policies and Procedures* (AP Volume) and *Chart of Accounts* (AC Volume). It was anticipated that the revised coding arrangement of the accounts would aid in the accumulation of information and facilitate the preparation of Dow's financial reports. In 1990, after a twenty-years trial, it was reported to have been effective in meeting these objectives.

Dow's *Accounting Policy and Procedures Manual* (AP) contains eighteen sections. The sections are family groupings within which related subjects appear and each family or major category has been assigned a section number. The first section, numbered 000–099, provides an overview of Dow as an organization by discussing locations, divisions, subsidiaries, and related companies. This discussion of the firm is followed by a corporate policy section, numbered 100–199, highlighting such areas as internal control, code of conduct, transfer pricing, and product policies. The third section discusses personnel practices, such as employee transfers and Dow's stock purchase plan. The remaining sections present Dow's policies and procedures relating to various financial statement items in addition to their cost accounting policies.

The *Chart of Accounts* (AC Volume) outlines the intended use of each account. The general ledger accounts are assigned by the corporate controller's office to provide for all eventualities for which a valid need is disclosed. The account number also serves as a document control number and is the principal reference number as indicated on each page. Since the manual undergoes continual changes, the pages are not numbered consecutively. Rather, each subject has its own document number with the pages numbered sequentially. With few exceptions, the assignment of subaccounts to the *Chart of Accounts Manual* is at the discretion of the local units of Dow.

All pages of the manual include a format section for the date of the current revision and a second formatted section for the date of the superseded section. Thus, the manual is always current, as superseded sections are removed and the current revisions inserted. In addition, the manual's index lists the most recent issue of a policy, procedure, or account revision against which any page may be compared. The dating feature facilitates worldwide communication by quickly enabling corresponding parties to determine if they are discussing a subject from the point of view of the most recent revision.

In some cases there is need for interim communication to help with rapid dissemination of procedural information, including clarification of new policies or the opening of new accounts. Such interim communication is followed with formal documents at the next official manual update. This interim method of communication was used recently to disseminate the new procedures required to implement SFAS No. 95, Statement Cash Flows. Several new general ledger accounts were opened to collect dollar amounts on investment and financial activities (new items from receipts or repayments) and also to segregate opening balances from current year activities. Dow has issued forty-four transmittal letters since January 1971, which translates into slightly more than two revisions per year. Using interim communications allowed for data exchange between corporate and area representatives, which resulted in the modification of certain procedures before they were included in the manual as policy.

As an aid to the reader, revision symbols are used to identify and assist the manual user in locating the changes in the updated material. They include:

>>>>   Denotes a new or revised paragraph or that something has been added to the index.

<<<<   Points to the approximate location where something was removed.

<<>>   Points to a location where one thing has been added or revised and another has been removed.

In addition to these indicators, a brief explanation for the change is provided with the revision. Such explanations are dispatched to various locations via a numbered letter of transmittal.

The documentation is stored on floppy disks: about eight hundred pages are stored on twenty-six disks in five hundred files. As of 1990, Dow was searching for a package to help move the documentation on-line with worldwide access. Editing the documentation to ensure current usage of terms or proper names is done with the help of a software package that can examine all the files and report on the number of times a term is used. When the firm went to quarterly closing from monthly closings, it used a file search to help change to quarterly the word monthly when used in reference to closings. When the employee relations department changed its name to human resources department, a computer search helped with the revision. The search revealed that the words employee relations were used twelve times and those references were changed to human resources.

Dow has distributed 507 copies of its manuals worldwide. They are issued to the locations rather than individuals, so that when managers transfer, the manuals can remain at the location. A periodic audit is done to verify the existence of the documentation.

Adapted from "An Inside Look at Dow Chemical's Policy and Procedures Manual," by T. R. Weirich and D. F. Leneschmidt in *Corporate Controller*, July/August 1990, pp. 5–11.

## 21.15 THE BUDGET MANUAL

The preparation of the budget is a major task in the lives of the controllers and their staff. It is important, but is only an annual event for most of those involved in the preparation of the budget. Not everyone remembers all the bits and pieces that are a part of the "how to" of budgeting. Because of its infrequency, and given its importance, budget preparation requires that a major portion of the accounting manual be devoted to it. We recommend that there be a distinct volume entirely devoted to the budget process.

Such a manual will inform and instruct those involved in the process about the organizational policies pertaining to its budget preparation. The manual will also describe the procedures and the tasks that are necessary in putting together the budget. Such documentation will ensure that data flowing in will be processed uniformly. Since preparing a budget involves organizing, analyzing, and reporting large amounts of information, it is easy to overlook important details and tasks. If those tasks are prerequisite for subsequent processes, then the entire process may face roadblocks. Preventing such lapses due to the frailty of human memory is a function of the budget manual. It should list the various phases of the process, as well as the tasks specific to each phase. It should also indicate who was assigned various tasks in previous years.

## 21.16 CONTENTS OF A BUDGET MANUAL

Organizations approach budget manuals in different ways. The approaches range from having no manual at all to having a very detailed, task-by-task set of instructions. Moreover, the contents of the budgets themselves will vary from one organization to another, and the process will vary from industry to industry. It will also be affected by the size of the firm and its organizational culture. Notwithstanding such diversity, a budget manual should contain:

- A statement of the budget policy for the CEO.
- A guide describing how to communicate the organizational goals for the budget period. The section could also contain the organizational goals used for previous budget periods. The presence of such historic awareness will benefit managers in dealing with the present and the future.
- Definition of the responsibilities involved in the process.
- The procedures devoted to budget preparation and its review.
- The tasks involved in data collection and its consolidation.
- Principle forms and the worksheets used, as well as a description of the software packages used. Preparing a budget manually for even a mid-sized firm can be a challenging experience.
- Calendar and timelines used for the budget.
- Procedures for changing budgets
- Distribution of the budget document.
- Explanation of how of the budget is to be used by managers and supervisors.

## 21.17 CONCLUSION

Business firms must realize that operating manuals can be useful, productive tools. It will take an investment of resources to develop better manuals, but this investment will more than pay for itself by making operations smoother and more efficient. To live up to their potential usefulness, accounting manuals must be current, available, readable, and easy to understand.

In the spirit of continuous improvement and of keeping up with technological changes, it may be advisable to rely on emerging information technology to make documentation less archival and more interactive. In keeping with the new emphasis on the process on-line documentation is wise investment.

# APPENDIX: REFERENCES FOR FURTHER READING

The following list provides sources for further study of topics in the book. Besides the items listed here, readers are recommended to browse through *Management Accounting*, published by the Institute of Management Accountants located in Montvale, New Jersey for additional material on the accounting topics covered here. *Harvard Business Review, Sloan Management Review*, and *Fortune* are also good sources for further reading.

Abrams, M. H. *The Mirror and the Lamp: Romantic Theory and the Critical Tradition*. New York: Norton, 1958.

Argyris, C. *Knowledge for Action*. San Francisco: Jossey-Bass, 1993.

Atkinson, H., et. al. *Linking Quality to Profits: Quality-Based Cost Management*. Montvale, NJ: Institute of Management Accountants, 1994.

Bhote, K. "Improving White Collar Productivity Can Enhance Profitability." *Corporate Controller*, (May/June 1991): 39–46.

Blau. P. and M. W. Meyer. *Bureaucracy in Modern Society*. New York: Random House, 1971.

Bothwell, C. "How to Improve Financial Planning with a Budget Manual." *Management Accounting*, (December 1984): 34–39.

Brown, H. *Design and Maintenance of Accounting Manuals*. New York: John Wiley and Sons, 1988.

Cushing, B. E. and M. B. Romney. *Accounting Information Systems*, 5th ed. Reading, Massachusetts: Addison Wesley, 1990.

Davenport, T. Book review of *Reengineering the Corporation. Sloan Management Review*, (Fall 1993): 103–104.

———, T. *Process Innovation: Reengineering Work Through Information Technology*. Boston: Harvard Business School Press, 1993.

——— and J. Short. "The New Industrial Engineering: Information Technology and Business Process Redesign." *Sloan Management Review*, (Summer 1990): 11–27.

de Geus, A. "Planning as Learning." *Harvard Business Review*, (March/April 1988): 70–74.

Dobyns, L. "Because Better Costs Less." *Smithsonian*. (August 1990): 74–83.

Drucker, P. "The New Society of Organizations." *Harvard Business Review*, (September/October 1992): 95–105.

Dumaine, B. "The Bureaucracy Busters." *Fortune*, (June 17, 1991): 35–50.

———. "Mr. Learning Organization." *Fortune*, (October 17, 1994): 147–160.

Ehrbar, A. "The Price of Progress." The *Wall Street Journal*. (March 16, 1993): 1.

Felker, D. (ed.). *Document Design: A Review of Relevant Research*. Washington, D.C.: American Institute of Research, 1989.

Felker, D., et. al. *Guidelines for Document Designers*. Washington, D.C.: American Institute of Research, 1989.

Fiol, C. M. and M. Lyles. "Organizational Learning." *Academy of Management Review*, (1985): V 10(4), 803–813.

Freedman, D. "Is Management Still a Science?" *Harvard Business Review*, (November/December, 1992): 26–38.

Galloway, D. *Mapping Work Process*. Milwaukee, WI: ASQC Quality Press, 1994.

Garvin, D. "Building a Learning Organization." *Harvard Business Review*, (July/August 1993): 78–92.

Gildersleeve, T. R. *Successful Data Processing System Analysis*. Englewood, New Jersey: Prentice-Hall, 1978.

Graham, M. "The Power of Information." *Sloan Management Review*, (Winter 1991): 120–121.

Gray, M. and K. R. London. *Documentation Standards*. Princeton, New Jersey: Auerbach, 1969.

Hammer, M. "Reengineering Work: Don't Automate, Obliterate." *Harvard Business Review*, (July/August 1990): 104–112.

———— and J. Champy. *Reengineering the Corporation*. New York: HarperCollins, 1993.

Henkoff, R. "Make Your Office More Productive." *Fortune*, (February 25, 1991): 72–84.

Huey, J. "The New Post-Heroic Leadership." *Fortune*, (February 21, 1994): 42–53.

Imai, M. *Kaizen*. New York: McGraw-Hill, 1986.

Johnson, H. T. and R. S. Kaplan. *Relevance Lost: The Rise and Fall of Management Accounting*. Boston, Massachusetts: Harvard Business School Press, 1987.

Kiechel III, W. "The Organization That Learns." *Fortune*, (March 12, 1990): 133–136.

Kim, D. "The Link Between Individual and Organizational Learning." *Sloan Management Review*, (Fall, 1993): 37–50.

Knowlton, C. "Shell Gets Rich by Beating Risk." *Fortune*, (August 26, 1991): 79–82.

Krass, P. "Building a Better Mousetrap." *Information Week*, (March 25, 1991): 24–30.

Leonard-Barton, D. "The Factory as a Learning Laboratory." *Sloan Management Review*, (Fall, 1992): 23–38.

Main, J. "How to Battle Your Own Bureaucracy." *Fortune*, (June 29, 1981): 52–58.

Miles, R. *Coffin, Nails and Corporate Strategies*. Englewood Cliffs, New Jersey: Prentice-Hall, 1982.

Murtuza, A. "Operating Manuals as Productive Tools." *Arizona Business and Economic Review*, (1984): 11(1), 4–5.

————. "Procedures Documentation Ought to Be Illuminative, Not Just Archival." *Management Communication Quarterly*, (November, 1994): 225–243.

Nelson, B. H., D. W. Luse, and D. D. DuFrene. "The Structure and Content of the Introductory Business Communication Course." *The Bulletin of the Association for Business Communication*, (1992): 55(3), 7–14.

Nodar, A. "Coca-Cola Writes an Accounting Procedures Manual." *Management Accounting*, (October 1986): 52–53.

Nonaka, I. "The Knowledge Creating Company." *Harvard Business Review*, (November/December 1991): 96–104.

Normann, R. and R. Ramirez. "From Value Chain to Value Constellation: Designing Interactive Strategy. *Harvard Business Review*, (July/August 1993): 65–77.

Prentice-Hall Editorial Staff. *Handbook of Successful Operating Systems and Procedures with Forms*. Englewood Cliffs, NJ: Prentice-Hall, 1964.

Quinn, J. B. *Intelligent Enterprise*. New York: Free Press, 1992.

Randall, C. B. and S. W. Burgly. *Systems and Procedures for Business and Data Processing*. Cincinnati, Ohio: South-Western, 1968.

Redish, J. *How to Write Regulations and Other Legal Documents in Clear English*. Washington, D.C.: American Institutes for Research, 1991.

Senge, P. M. *The Fifth Discipline: The Art and Practice of the Learning Organization*. New York: Doubleday, 1990.

Stalk, G. and A. Webber. "Japan's Dark Side of Time." *Harvard Business Review*, (July/August 1993): 93–102.

Stata, R. "Organizational Learning—The Key to Management Innovation." *Sloan Management Review*. (Spring 1989): 63–74.

Stewart, T. "Reengineering: The Hot New Managing Tool." *Fortune*, (August 23, 1994): 41–48.

Taylor, A. "Why Toyota Keeps Getting Better and Better and Better." *Fortune*, (November 19, 1990).

Turk, W. "Management Accounting Revitalized: The Harley Davidson Experience." *Journal of Cost Management*, (Winter 1990): 28–39.

Turney, P. B. B., ed. *Performance Excellence in Manufacturing and Service Organizations*. Sarasota, Florida: American Accounting Association, 1990.

Tyre, M. and W. Orlikowski. "Exploiting Opportunities for Technological Improvement in Organizations." *Sloan Management Review*, (Fall 1993): 13–26.

Walton, R. *Up and Running: Integrating Information Technology and the Organization*. Boston, Massachusetts: Harvard Business School Press, 1989.

Waterman Jr., R. H. *The Renewal Factor*. New York: Bantam Books, 1987.

Weirich, T. R. and D. W. Leneschmidt. "An Inside Look at Dow Chemical's Policy and Procedures Manual." *Corporate Controller*, (July/August 1990): 5–11.

Yates, J. "The Emergence of the Memo as a Managerial Genre." *Management Communication Quarterly*, (1989): 2(4), 485–510.

———, J. *Control Through Communication: The Rise of System in American Management*. Baltimore, Maryland: Johns Hopkins University Press, 1989.

——— and W. J. Orlikowski. "Genres of Organizational Communication: A Structurational Approach to Studying Communication and Media." *Academy of Management Review*, (1992): 299–326.

Zuboff, S. *The Age of Smart Machine*. New York: Basic Books, 1987.

# Index

# Date Due

| MCK DUE | NOV 2 2 1995 MCK RTD | DEC 1 5 1999 |
|---|---|---|
| MCK RTD OCT 2 2 1995 N | | |
| | MCK DUE OCT 2 2 1996 | |
| MCK RTD | SEP 2 7 1996 | |
| | MCK DUE MAY 0 1 1998 | |
| MCK RTD | SEP 2 1 1997 | |
| MCK DUE MAR 19 1998 MCK RTD | MAR 0 2 1998 | |
| MCK DUE JUN 1 6 1999 MCK RTD | APR 07 1999 | |
| MCK DUE DEC 1 9 1999 | | |
| JAN 3 1 2000 | | |